Post-genomic Approaches in Drug and Vaccine Development

RIVER PUBLISHERS SERIES IN RESEARCH AND BUSINESS CHRONICLES: BIOTECHNOLOGY AND MEDICINE

Volume 5

Combining a deep and focused exploration of areas of basic and applied science with their fundamental business issues, the series highlights societal benefits, technical and business hurdles, and economic potentials of emerging and new technologies. In combination, the volumes relevant to a particular focus topic cluster analyses of key aspects of each of the elements of the corresponding value chain.

Aiming primarily at providing detailed snapshots of critical issues in biotechnology and medicine that are reaching a tipping point in financial investment or industrial deployment, the scope of the series encompasses various specialty areas including pharmaceutical sciences and healthcare, industrial biotechnology, and biomaterials. Areas of primary interest comprise immunology, virology, microbiology, molecular biology, stem cells, hematopoiesis, oncology, regenerative medicine, biologics, polymer science, formulation and drug delivery, renewable chemicals, manufacturing, and biorefineries.

Each volume presents comprehensive review and opinion articles covering all fundamental aspect of the focus topic. The editors/authors of each volume are experts in their respective fields and publications are peer-reviewed.

For a list of other books in this series, visit www.riverpublishers.com
http://riverpublishers.com/series.php?msg=Research and Business Chronicles: Biotechnology and Medicine

Post-genomic Approaches in Drug and Vaccine Development

Kishore R Sakharkar

OmicsVista, Singapore

Meena K Sakharkar

Department of Pharmacy and Nutrition,
University of Saskatchewan, SK, Canada

Ramesh Chandra

Department of Chemistry, Delhi University, India
B. R. Ambedkar Center for Biomedical Research,
University of Delhi, India

Published, sold and distributed by:
River Publishers
Niels Jernes Vej 10
9220 Aalborg Ø
Denmark

ISBN: 978-87-93102-84-2 (Hardback)
978-87-93102-85-9 (Ebook)

Contents

Series Note

The deciphering in 2003 of the nucleotide sequence of the human genome, which followed the determination of the chromosomal sequences of an array of model microorganisms including *Escherichia coli*, *Bacillus subtilis*, and *Saccharomyces cerevisiae*, is a landmark in biology that has paved the way for a revolution in research and development in biotechnology and pharmaceutical sciences. Moreover, several novel discovery tools have since emerged, including dramatically enhanced computers, sequencing instruments with dramatically higher throughout and decreased costs, softwares conferring the ability to generate and manage very large amounts of data, and systems biology tools which have enabled *in silico* experiments and the creation of virtual patients or virtual microbes to both accelerate and increase the scope of pharmaceutical and biotechnological research. Whereas more than 10 years have already elapsed since this major scientific milestone, the translation into novel products, perhaps best exemplified by the current focus of pharmaceutical companies on personalised medicine, has only reached mainstream at the beginning of the present decade.

The impact of post-genomic approaches in drug and vaccine development is the focal point of the present monograph. A scientific journey is proposed here from the basic principles of drug discovery, to data mining and systems biology, using in watermark the development of conventional therapeutics and vaccines. The ultimate purpose of the journey is to accelerate the development of disease-modifying pharmaceuticals, and answer unmet medical needs to enable patients worldwide to live healthier, longer, and better lives.

Alain Vertès, Basel, Switzerland
Pranela Rameshwar Rutgers, USA
Paolo di Nardo, Roma, Italy

Preface

The genome sequencing projects and the associated efforts over the past decade have enabled an amazing increase in our understanding of the human genome and pathogenesis of human diseases and infections. These efforts have facilitated the discovery of several novel disease targets and the approval of several innovative drugs. To gain a more complete understanding of human genome and diseases, and help discover more effective treatments and biomarkers, extensive research networks are conducting large-scale integrative and data-mining studies for annotation of human and bacterial genomes. This main objective of this book is dissemination of knowledge related to the development of new drugs and vaccines with focus on recent advances in the field. The goal is to bridge these disciplines, look beyond boundaries and to explore the impact genomics has had on the discovery and choice of drug targets, selection of antigenic determinants for vaccine development, diagnostics, and our understanding of pathogenesis and the pathogen. It is evident that the fundamental nature of target identification/validation and drug discovery has radically changed. The wealth of information and tools available have added significant insights to our knowledge of protein function and pathogen physiology. The contribution of these findings for the discovery of novel drugs is continuous.

Chapter 1, Chapter 2, Chapter 3 and Chapter 4 discuss diseases, targets and their disease associations. Deciding precisely which targets or drugs to direct efforts against is difficult and complex. Experimental consideration for successful application of chemical cross-linking structural data on drugs and targets in biological studies is elaborated in Chapter 5, Chapter 6 and Chapter 7. Vaccines are the most effective tools to prevent diseases. Selecting the optimal antigen represents the cornerstone in vaccine design. Towards this end, computational prediction methods and genomics can strongly reduce time and costs for vaccine development by utilising high throughput technologies and systems biology approaches in biotechnology research and discovery of vaccines. Chapter 8, Chapter 9, Chapter 10, Chapter 11 and Chapter 12 illustrate these aspects.

Natural product-based medicines have been in effect owing to the diverse biological activities and medicinal potentials of natural products. The importance of phytochemicals present in fruits and vegetables in deceasing infections and disease risk are elaborated in Chapter 13. The spatiotemporal expression of genes in proper amounts is a requirement for ensuring appropriate functional outcome. Gene regulatory networks (GRNs) are used to visualize the regulatory relationships between genes and their regulators. Chapter 14 describes the use of *in silico* approaches for dealing with gene regulatory networks to understand molecular mechanisms of immunity. Such studies provide insight into the transcriptional mechanisms and their relationships to genotypic and phenotypic robustness and variability. There are a number of critical issues that must be considered as strategies are developed to elucidate the inherited determinants of drug effects. Chapter 15 illustrates a statistical method for prediction of drug side effects. The authors of the different chapters represent a group of professionals in the field who generously joined this project, giving their effort, knowledge, experience, and valuable time. Drug discovery encompasses several diverse disciplines united by a common goal of development of novel therapeutic agents. In putting together this book, we have tried to bring to table the contributions of various experts towards some key aspects in drug discovery. As editors of this book, we are grateful to all the contributors who have made this book possible.

Acknowledgements

On behalf of all the authors, we would like to thank all our mentors, colleagues and friends who instilled in us the culture of science. Without support from them, we could not have written this book. The unconditional love and support from our families is gratefully acknowledged.

Finally, we would like to take this opportunity to acknowledge the services of the team of River Publishers and everyone who collaborated in producing this book.

Kishore R. Sakharkar

Meena K. Sakharkar

Ramesh Chandra

List of Figures

List of Tables

1

Drug Discovery: Diseases, Drugs and Targets

Kishore R Sakharkar[1], Meena K Sakharkar[2] and Ramesh Chandra[3,4]

[1]OmicsVista, Singapore
[2]Department of Pharmacy and Nutrition, University of Saskatchewan, SK, Canada
[3]Department of Chemistry, Delhi University, India
[4]B. R. Ambedkar Center for Biomedical Research, University of Delhi India

1.1 Introduction

The elucidation of genetic causation of human diseases at the molecular level provides crucial information for developing target specific therapeutic approaches with tremendous implications for disease prevention and treatment. Mutations in many essential genes have been linked to hereditary disorders. Concomitantly, a large body of literature has documented human disease associated mutations, normal sequence variations, and alterations that acquire pathological significance when combined with other deleterious alleles or second-site mutations (1–4). Such studies acquire additional discriminatory power with the availability of multiple genome sequences of model organisms that have contributed substantially to our understanding of the etiology of human diseases and facilitated the development of new treatment modalities. To that end, high-throughput genome sequencing and systematic experimental approaches are spawning strategic programs designed to investigate gene functions at biochemical, cellular and the whole organism levels. With all this information compiled into organized databases (5–7) it is now possible to conduct large scale comprehensive analyses of human disease genes. In light of the above, a comprehensive list of functionally annotated

Post-genomic Approaches in Drug and Vaccine Development, 1–22.

human disease genes and their analyses could lead to greater integration of medicine and biology. Conversely, human genome annotation and analyses are limited for various reasons, including the inability to infer the existence of splice variants or interactions between the encoded proteins from gene sequences alone, and the fact that the function of most of the DNA in the genome being unclear (8–9). Also, many of the computationally derived annotations in the databases are either minimal or incorrect (apart from a carefully manually-curated database such as Swiss-Prot). Additionally, as annotation of genes is provided by multiple public resources using different methodologies, the resultant information may be similar but not always identical (10). Hence, the potential targets provided by the genome projects are not endowed with elaborate background knowledge. However, accurate information on gene architecture and gene annotation is essential for drug discovery as this allows insight into splice variants. This increase in the number of targets and the lack of functional knowledge about them is generating a bottleneck in the target validation process (11).

1.2 Drug Discovery

Drug discovery is a time consuming, increasingly risky and costly process and one of the prime reasons for this is a lack of reliable drug target prediction methods as reflected by the low clinical target validation success rate (12–13). The general timeline to develop one new medicine from the time its therapeutic properties are identified to when it is available for treating patients is about 10–15 years. The average cost to research and develop each successful drug is estimated to be about $1 billion (14). Here, it must be emphasised that only 1 chemical entity gets FDA approval for every 5,000–10,000 compounds that enter the research and development (R&D) pipeline (15). Knowledge on understanding the disease to be treated in as much detail as possible, and to unravel the underlying cause of the condition is the fundamental basis for treatment of the disease condition (16). Information on how the genes are altered, how that affects the proteins they encode and how those proteins interact with each other in living cells, how those affected cells change the specific tissue they are in and finally how the disease affects the entire patient is essential and key first step in drug development process. Towards this end, advances in genomics research have boosted progress in the discovery of susceptibility genes and fuelled expectations about opportunities of genetic profiling for personalizing medicine. The key steps in the drug discovery process are outlined below:

Target Identification: A single, or sometimes multiple molecule(s), e.g. gene or protein, involved in a particular disease and can be targeted by a drug/potential new drug molecule.

Target Validation: Confirmation of target and validation for its role indisease pathogenesis and modulation by the drug by complicated experiments in both computational studies, *in vitro* experiments such as in cell lines and animal models.

Drug Discovery: After determining the key targetfor disease interception and course alteration, the next step is to find a lead molecule that could become a drug compound. There are several sources of lead compounds:

- *Nature*: Several phytochemicals e.g. Digoxin and digitoxin from foxglove, or *Digitalis lanata*, bacteria found in soil and fungi e.g. penicillin from penicillium fungi
- *De novo*: synthetic molecules can be created using sophisticated computer modeling to predict potentially effective molecules.
- *High-throughput Screening*: Advances in robotics and computationalpower screen a large number of compounds againstthe target to identify lead molecules. Promising molecules are selected for further studies.
- *Biotechnology*: Living systems can be genetically engineered to produce disease-fighting biological molecules.

Lead screening: Lead compounds undergo safety assessment.Their pharmacokinetics must be tested. Successful drugs must be:

- Absorbed into circulation at proper bio availability,
- Distributed and redistribution to the proper site or tissue of action in the body taking into consideration plasma protein binding,
- Metabolized efficiently and effectively in order to make nonpolar compounds polar and prevent their excretion through the renal tubules,
- Successfully excreted from the systemic circulation, and
- Demonstrated to be non-toxic at therapeutic dosage.

These ADME/Tox studieshelp in prioritization of lead compounds early in thepreclinical phase of the drug discovery process.

Lead Optimization: Screened compounds are altered by structure modification to change their properties and increase efficacy and safety. The resulting compound is the candidate drug.

The pharmacy aspect of the drug i.e. formulation (including excipients), drug delivery (enteral, parenteral, special drug delivery systems like nanocapsules) and large-scale manufacturing is also given importance.

Preclinical Testing: Lab and animal testing determines whether the drug can proceed to human testing taking into consideration most importantly its safety.

1.2.1 Phase 1 Clinical Trial: Perform initial human testing in a small group of healthy volunteers

Phase 1 trials constitute the first human administration of the candidate drug. Qualified clinical pharmacologists or trained physicians carry out these studies in a setting where all vitals are monitored and emergency resuscitation is available. These studies are usually conducted on 20 to 100 healthy volunteers in turn, but in special circumstances patients for which the drug is indicated are used. Lowest estimated dose is administered and dosage is slowly and stepwise increased to achieve minimum therapeutic dosage. The emphasis is on safety and tolerability, while the purpose is to observe absorption and the pharmacodynamic effects in man.The study is open label.

1.2.2 Phase 2 Clinical Trial:Test in a small group of patients

In Phase 2 trials physicians trained as clinical researchers evaluate the candidate drug's effectiveness in about 100 to 500 patients selected according to specific inclusion and exclusion criteria, and establishment of therapeutic efficacy, dose range and ceiling effect in a controlled setting. Researchers also analyze optimal dose strength and schedules for using the drug. The expected mechanism of action is confirmed or rejected and effectiveness in treating the condition in question is determined. The study may be extended to pharmacokinetics and side effects, tolerability and risks associated with the drug. The study may be blinded or open label and is conducted at 2-4 centres.

1.2.3 Phase 3 Clinical Trial:Test in a large group of patients to show safety and efficacy

In Phase 3 trials multicentric randomized double blind comparative trials studying the candidate drug candidate in about 1,000–5,000 patients are conducted by several physicians to generate statistically significant data about the overall benefit-risk ratio and value of the drug as compared to the existing drugs in the market. This phase of research is key in determining safety and tolerability, and possible drug interactions of drug in a comparatively wider, more variable population, as well as obtaining pharmacokinetic data. Indications and instructions for proper drug usage, for example its use with other drugs, are finalized on this basis, and guidelines for therapeutic use are formulated. During the Phase 3 trial researchers conduct plans for full scale production and preparation of the complex application required for FDA approval, who if convinced give marketing permission. (Figure 1.1 for summary).(12, 17).

- Phase 4 Clinical Trials:

Post marketing surveillance/studies

After the drug enters the market, practicing physicians are identified and data are collected through them on a structured proforma about the efficacy, acceptability and adverse effects of the drug (similar to prescription event monitoring). The sample size is the largest – patients treated in the normal course of the disease. Bizarre and delayed adverse drug effects, as well as unsuspected drug interactions are detected at this stage. Patterns of drug utilization and additional indications (off label usage) may emerge from the surveillance data.

Further therapeutic trials involving special groups like children and elderly (physiological differences in different age groups), pregnant/lactating women (transplacental and transcolostral transport of drug and their effects on the foetus or infant), patients with renal/hepatic disease (pathological states may alter bio transformation and excretion), etc. (which are generally excluded during clinical trials) must be undertaken at this stage. Modified or sustained release dosage forms, additional routes of administration, fixed dose drug combinations, etc. can be explored. As such, many drugs continue their development or are replaced with more effective compounds even after marketing or after several years in the market in order to continually increase efficacy and minimize adverse drug effects as some effects can only be seen after long term use of the drug. Some may be with drawn after many years in the market due to teratogenic effects or permanent disability occurring to the patient.

1.3 Drug and Targets

A critical point in the development of a novel drug is the elucidation of its molecular targets, as this is crucial for understanding the cellular processes taking place in the response to the drug. Failure in correct identification of targets is a significant factor in the low drug approval rate stemming from safety issues (17–18). Specifically, the association of a drug with additional

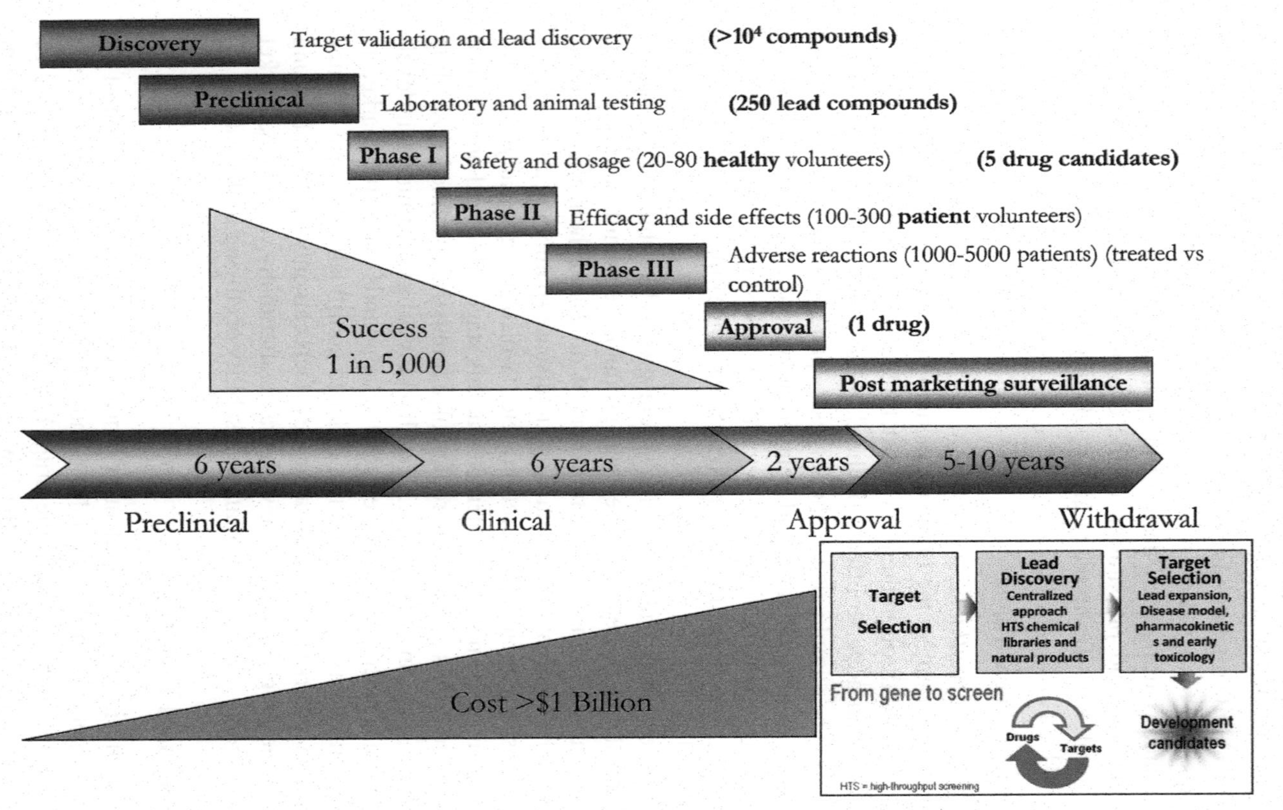

Figure 1.1 Summary of the drug discovery process.

Table 1.1 Drug statistics based on drugbank

Total # Drugs	**6825**	100.00
Drugs with at least 1 target	6180	90.55
Drugs with only 1 target	4056	59.43
Drugs with > 1 target	2124	31.12
Drugs with no target defined	**645**	9.45
Approved drugs	1691	24.78
Experimental drugs	5082	74.46

targets, beyond its direct ones may give rise to hazardous side effects, precluding further drug development and usage (19).

Almost all drug discovery now begins with activity of molecules on a molecular target. However, at least half of drugs date from the premolecular era, have no targets defined as their action was explored against whole tissues, rarely on isolated proteins, and target identities were only inferred from tissue-based responses. Gregori-Puigjané *et al.* reported that 7% of approved drugs are purported to have no known primary target, and up to 18% lack a well-defined mechanism of action (20). Even today, 37% of first-in-class drugs derive from phenotypic screens (21), and the targets for some new drugs remain unknown. A quick analyses of Drugbank shows that out of 6825 drugs, 645 of the drugs have no targets defined (Out of these 645, 553 were experimental drugs and 92 were FDA approved). 3414 drugs target genes associated with diseases (data based on genecards) (Out of these were 1218 FDA approved drugs and 2196 were experimental).

1.3.1 Drugs/target and Targets/drug

A distribution of drug per target shows that 4056 out of 6825 drugs ($\sim$ 60%) bind to only one target. However, 2124 drugs ($\sim$ 31%) have multiple targets (2 or more targets). The number of targets decreases as the number of drugs increase (Figure 1.2).

35 drugs target more than 25 genes. Targets with > 25 drugs are listed in Table 1.2. It must be noted that though each of these drugs has the same target, they might have different tissue distribution because of difference in physical properties or may require have a specific pro-drug status and may require a specific metabolism for activation (22). Also, interactions with other targets, which are not currently well defined, may create differences in the pharmacological effect of the drug.

So, when these drugs are administered there will be simultaneous changes in several biochemical signals, and there will be feedback reactions of the

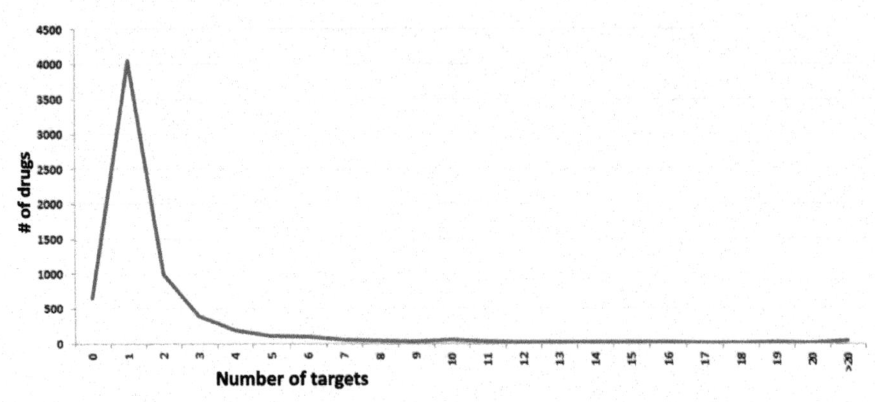

Figure 1.2 Drugs per targets. 500+ drugs have no target defined.

pathways disturbed for these set of drugs. The drug Nicotinamide-Adenine-Dinucleotide and NADH target maximum number of genes (143 and 144 genes). It must be noted that the interaction parameters with the target protein may be different (e. g. Ki – drug binding affinity). Earlier, it has been proposed that drugs with multiple targets might have a better chance of affecting the complex equilibrium of whole cellular networks than drugs that act on a single target (23).

A distribution of targets per drug shows that most of the targets are targeted by one drug (1530 genes) (Figure 1.3). 1889 genes are targeted by more than 20 drug. 110 genes targeted by more than 20 drugs are listed in Table 1.3.

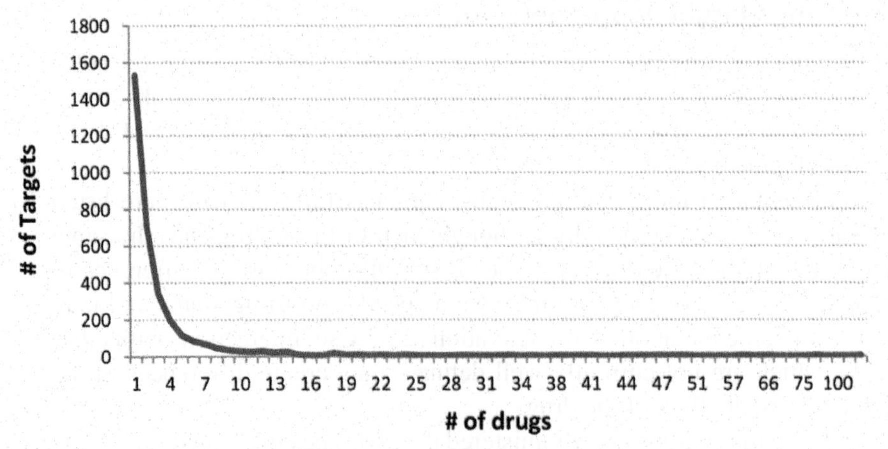

Figure 1.3 Targets per drug.

Table 1.2 Drugs with more than 25 targets

Name	#Targets
NADH	144
Nicotinamide-Adenine-Dinucleotide	143
Adenosine-5'-Diphosphate	121
Heme	120
Beta-D-Glucose	92
Flavin-Adenine Dinucleotide	85
Alpha-D-Mannose	76
Pyridoxal Phosphate	74
L-Glutamic Acid	70
Citric Acid	69
2'-Monophosphoadenosine 5'-Diphosphoribose	60
Riboflavin Monophosphate	55
Formic Acid	51
Guanosine-5'-Diphosphate	51
Phosphoaminophosphonic Acid-Adenylate Ester	51
Lauryl Dimethylamine-N-Oxide	50
Coenzyme A	45
MYRISTIC ACID	45
N-Formylmethionine	43
Glutathione	40
2-(N-Morpholino)-Ethanesulfonic Acid	40
S-Adenosyl-L-Homocysteine	36
Glycine	34
Adenosine triphosphate	33
Zonisamide	31
2-METHYLTHIO-N6-ISOPENTENYL-ADENOSINE-5'-MONOPHOSPHATE	27
Vitamin C	26
Succinic acid	26
Clozapine	26
Quetiapine	26
Ziprasidone	25
Olanzapine	25
Aripiprazole	25
(Hydroxyethyloxy)Tri(Ethyloxy)Octane	25
UBIQUINONE-2	25

Gag-pol has maximum number of drugs available for it – 46 drugs. A distribution of disease genes is shown in Table 1.4. A mapping of the 3414 targets identified above onto the disease genes in genecards shows that 970 of these targets are etiological targets and 2445 (3416–970) of the targets are palliative targets. It is interesting to see 877 approved drugs target 187 disease genes and 239

Table 1.3 Targets with more than 20 drugs

Gene		Gene		Gene		Gene	
gag-pol	136	GABRA2	46	OPRK1	32	HTR1D	24
CDK2	136	HSP90AA1	45	GABRA6	32	GABRG2	24
PRSS1	101	GABRA5	45	CHEK1	32	DRD4	24
F2	100	GABRA3	45	OPRD1	31	def	24
CA2	96	SRC	44	MAOB	31	CSNK2A1	24
HRH1	79	PRKACA	44	CHRM5	31	cobT	24
ADRA1A	79	ADRB1	44	PKIA	30	CALM1	24
CHRM1	75	ADRA1B	43	ESR2	30	CA1	24
DRD2	68	pbpA	42	SCN5A	29	NCOA2	23
GABRA1	67	NOS3	41	map	29	GABRR1	23
PTGS1	66	AR	41	E	29	GABRG1	23
HTR2A	66	OPRM1	40	BCHE	29	GABRE	23
CHRM2	61	NR3C1	40	AKR1B1	29	GABRD	23
ESR1	60	NOS2	39	RNASE1	28	TK	22
ADRA2A	60	F10	39	DRD3	28	tgt	22
PTGS2	57	PPARG	38	DHFR	28	pbp3	22
CHRM3	56	HTR2C	38	HTR1B	27	MMP3	22
PTPN1	55	GRIA2	38	GABRB3	27	GABRR3	22
DPP4	51	bla	36	PDE4B	26	GABRR2	22
SLC6A2	50	thyA	35	GABRB1	26	GABRP	22
PYGM	50	CHRM4	35	BACE1	26	GABRG3	22
MAPK14	50	ADRA2B	35	ADRA1D	26	CELA1	22
ADRB2	48	SLC6A3	34	NOS1	25	PIM1	21
ACHE	48	PLAU	33	NA	25	PDE4D	21
DRD1	47	GABRA4	33	IGHG1	25	NCOA1	21
ampC	47	CCNA2	33	GABRB2	25	KCNH2	21
SLC6A4	46	ADRA2C	33	DHODH	25	GRIK2	21
HTR1A	46			TOP2A	24		

non-disease genes. 62 experimental drugs target 172 disease targets and 481 non-disease genes. It is worth mentioning that bio-drugs (FDA-approved biotech (protein/peptide) drugs) target 31 disease genes and 54 non-disease genes.

Additionally, drug target identification can broaden the therapeutic areas of extant drugs. Concurrently, identification of the target(s) of a bioactive compound with an unclear or unknown mode of action is a common challenge in drug discovery (24).

Our dataset set has 20 targets for diabetes and 112 drugs available against them (Table 1.4). Multiple drugs against same target are available. Some of the targets for diabetes e. g. peroxisome-proliferator–activated receptor γ (PPARγ) are targeted by 38 drugs, Since, these 112 drugs act on 20

Table 1.4 Drugs available against diabetic genes

Gene	Drugs name (total # of targets for the drug)
ABCC8	Adenosine triphosphate (32)
ABCC8	Glimepiride (3)
ABCC8	Chlorpropamide (1)
ABCC8	Nateglinide (2)
ABCC8	Repaglinide (2)
ABCC8	Glyburide (8)
ABCC8	Glipizide (2)
ABCC8	Gliclazide (2)
ABCC8	Tolbutamide (2)
ABCC8	Gliquidone (2)
ABCC8	Mitiglinide (2)
ABCC8	Glycodiazine (2)
AKT2	N-[(1S)-2-amino-1-phenylethyl]-5-(1H-pyrrolo[2,3-b]pyridin-4-yl)thiophene-2-carboxamide (2)
AKT2	4-(4-CHLOROPHENYL)-4-[4-(1H-PYRAZOL-4-YL)PHENYL]PIPERIDINE (4)
AKT2	ISOQUINOLINE-5-SULFONIC ACID (2-(2-(4-CHLOROBENZYLOXY)ETHYLAMINO)ETHYL)AMIDE (4)
AKT2	(2S)-1-(1H-INDOL-3-YL)-3-{[5-(3-METHYL-1H-INDAZOL-5-YL)PYRIDIN-3-YL]OXY}PROPAN-2-AMINE (4)
AVPR2	Desmopressin (3)
AVPR2	Vasopressin (3)
AVPR2	Conivaptan (2)
AVPR2	Terlipressin (3)
AVPR2	Tolvaptan (2)
CCR5	Maraviroc (1)
CTLA4	Alpha-D-Mannose (73)
ENPP1	Ribavirin (6)
ENPP1	Amifostine (2)
GCGR	Glucagon recombinant (3)
GCK	Beta-D-Glucose (81)
GCK	2-amino-N-(4-methyl-1,3-thiazol-2-yl)-5-[(4-methyl-4H-1,2,4-triazol-3-yl)sulfanyl]benzamide
GCK	3-[(4-fluorophenyl)sulfanyl]-N-(4-methyl-1,3-thiazol-2-yl)-6-[(4-methyl-4H-1,2,4-triazol-3-yl)sulfanyl]pyridine-2-carboxamide
HNF1A	Norleucine (3)
GCK	2-(methylamino)-N-(4-methyl-1,3-thiazol-2-yl)-5-[(4-methyl-4H-1,2,4-triazol-3-yl)sulfanyl]benzamide
GCK	2-AMINO-4-FLUORO-5-[(1-METHYL-1H-IMIDAZOL-2-YL)SULFANYL]-N-(1,3-THIAZOL-2-YL)BENZAMIDE
HNF4A	Myristic acid (31)
IL2RA	Denileukin diftitox (3)

(Continued)

Table 1.4 Continued

Gene	Drugs name (total # of targets for the drug)
IL2RA	Aldesleukin (3)
IL2RA	Basiliximab (13)
IL2RA	Daclizumab (12)
IL6	Ginseng (3)
INS	M-Cresol (1)
INS	MYRISTIC ACID (31)
INSR	Insulin Regular (12)
INSR	Insulin Lispro (2)
INSR	Insulin Glargine (2)
INSR	Insulin, porcine (14)
INSR	Mecasermin (4)
INSR	Insulin Aspart (1)
INSR	Insulin Detemir (1)
INSR	Insulin Glulisine (1)
INSR	Adenosine-5'-[Beta, Gamma-Methylene]Triphosphate (14)
INSR	[4-({5-(AMINOCARBONYL)-4-[(3-METHYLPHENYL)AMINO]PYRIMIDIN-2-YL}AMINO)PHENYL]ACETIC ACID (3)
INSR	Insulin, isophane
IRS1	[4-({5-(AMINOCARBONYL)-4-[(3-METHYLPHENYL)AMINO]PYRIMIDIN-2-YL}AMINO)PHENYL]ACETIC ACID (3)
KCNJ11	Glimepiride (3)
KCNJ11	Ibutilide (10)
KCNJ11	Verapamil (16)
KCNJ11	Levosimendan (4)
KCNJ11	Glyburide (8)
KCNJ11	Diazoxide (6)
KCNJ11	Thiamylal (3)
MAPK8IP1	2,6-Dihydroanthra/1,9-Cd/Pyrazol-6-One (4)
MAPK8IP1	6-CHLORO-9-HYDROXY-1,3-DIMETHYL-1,9-DIHYDRO-4H-PYRAZOLO[3,4-B]QUINOLIN-4-ONE (2)
MAPK8IP1	N-(4-AMINO-5-CYANO-6-ETHOXYPYRIDIN-2-YL)-2-(4-BROMO-2,5-DIMETHOXYPHENYL)ACETAMIDE (2)
MAPK8IP1	5-CYANO-N-(2,5-DIMETHOXYBENZYL)-6-ETHOXYPYRIDINE-2-CARBOXAMIDE (2)
MBL2	O3-Sulfonylgalactose (2)
MBL2	Methyl alpha-D-mannoside (1)
MBL2	O4-Sulfonylgalactose (1)
MBL2	Alpha-D-Mannose (73)
MBL2	Beta-L-Methyl-Fucose (1)
MBL2	O-Sialic Acid (15)
MBL2	Alpha-L-Methyl-Fucose (2)
MBL2	Alpha-L-1-Methyl-Fucose (1)

(Continued)

Table 1.4 Continued

Gene	Drugs name (total # of targets for the drug)
MBL2	Alpha-Methyl-N-Acetyl-D-Glucosamine (2)
OAS1	Cysteine-S-Acetamide (2)
PPARG	Icosapent (10)
PPARG	Troglitazone (6)
PPARG	Mesalazine (8)
PPARG	Indomethacin (7)
PPARG	Rosiglitazone (2)
PPARG	Nateglinide (2)
PPARG	Sulfasalazine (8)
PPARG	Repaglinide (2)
PPARG	Telmisartan (2)
PPARG	Balsalazide (4)
PPARG	Ibuprofen (8)
PPARG	Glipizide (2)
PPARG	Pioglitazone (1)
PPARG	Mitiglinide (2)
PPARG	Bezafibrate (3)
PPARG	(S)-3-(4-(2-Carbazol-9-Yl-Ethoxy)-Phenyl)-2-Ethoxy-Propionic Acid (1)
PPARG	2-{5-[3-(6-BENZOYL-1-PROPYLNAPHTHALEN-2-YLOXY)PROPOXY]INDOL-1-YL}ETHANOIC ACID (1)
PPARG	(2S)-3-(1-{[2-(2-CHLOROPHENYL)-5-METHYL-1,3-OXAZOL-4-YL]METHYL}-1H-INDOL-5-YL)-2-ETHOXYPROPANOIC ACID (2)
PPARG	(9Z,11E,13S)-13-hydroxyoctadeca-9,11-dienoic acid (1)
PPARG	2-{5-[3-(7-PROPYL-3-TRIFLUOROMETHYLBENZO[D]ISOXAZOL-6-YLOXY)PROPOXY]INDOL-1-YL}ETHANOIC ACID
PPARG	(4S,5E,7Z,10Z,13Z,16Z,19Z)-4-hydroxydocosa-5,7,10,13,16,19-hexaenoic acid
PPARG	(5R,6E,8Z,11Z,14Z,17Z)-5-hydroxyicosa-6,8,11,14,17-pentaenoic acid
PPARG	(8E,10S,12Z)-10-hydroxy-6-oxooctadeca-8,12-dienoic acid
PPARG	(8R,9Z,12Z)-8-hydroxy-6-oxooctadeca-9,12-dienoic acid
PPARG	(9S,10E,12Z)-9-hydroxyoctadeca-10,12-dienoic acid
PPARG	difluoro(5-{2-[(5-octyl-1H-pyrrol-2-yl-kappaN)methylidene]-2H-pyrrol-5-yl-kappaN}pentanoato)boron
PPARG	(2S)-2-ETHOXY-3-{4-[2-(10H-PHENOXAZIN-10-YL)ETHOXY]PHENYL}PROPANOIC ACID
PPARG	3-(5-methoxy-1H-indol-3-yl)propanoic acid
PPARG	3-{5-methoxy-1-[(4-methoxyphenyl)sulfonyl]-1H-indol-3-yl}propanoic acid (4)
PPARG	(2S)-2-(4-ethylphenoxy)-3-phenylpropanoic acid
PPARG	2-chloro-5-nitro-N-phenylbenzamide (3)
PPARG	(2S)-2-(biphenyl-4-yloxy)-3-phenylpropanoic acid

(Continued)

Table 1.4 Continued

Gene	Drugs name (total # of targets for the drug)
PPARG	3-[5-(2-nitropent-1-en-1-yl)furan-2-yl]benzoic acid
PPARG	2-[(2,4-DICHLOROBENZOYL)AMINO]-5-(PYRIMIDIN-2-YLOXY)BENZOIC ACID (3)
PPARG	(5E,14E)-11-oxoprosta-5,9,12,14-tetraen-1-oic acid
PPARG	3-FLUORO-N-[1-(4-FLUOROPHENYL)-3-(2-THIENYL)-1H-PYRAZOL-5-YL]BENZENESULFONAMIDE
PPARG	(2S)-2-(4-chlorophenoxy)-3-phenylpropanoic acid
PPARG	Aleglitazar (3)
SLC2A2	Streptozocin (2)

different targets, possibly in different ways to lower blood glucose levels, they may be used together. Though taking more than one drug can be more costly and can increase the risk of side effects, combining oral medications can improve blood glucose control when taking only a single pill does not have the desired effects (23). Conversely, with the administration of multiple drugs simultaneously, possible drug–drug interactions multiply, leading to unexpected adverse effects that may be difficult to trace (25). The combination of drugs may enhance the efficiency of drug treatment, but selection of the optimal combination and the optimal doses remains a matter of conjecture and complexity that needs to be resolved by trial and error. Thus, further understanding of the mechanisms involved in many different diseases, and the development of new therapeutic strategies, can be gained by i) studying genes and their products in the context of the molecular networks in which they function, and ii) investigating how such networks are altered in disease cells compared to their unaffected counterparts (26). In light of the above, the networks presented here may be able to predict adverse or integral effects of drugs working on different targets. These networks are also important keeping in mind the occurrence of affinity enhancing and affinity diminishing mutations, which may occur in the genome.

1.4 Disease Genes – Targets and Drugs

In the development of new chemotherapeutic agents, several issues need to be addressed, including improved and durable therapeutic efficacy, reduction of toxicities, which can prevent effective dosing of potentially efficacious drugs, and prevention of drug resistance caused by the inherent genomic instability of several diseases and associated pathogens (27). Nothing provides more compelling validation for a target than knowledge of the human genetics of a specific disease. Solving a compound's mechanism of action (MOA) is one of the most challenging task researchers face in drug discovery. As bioactive compounds have multiple, often pleiotropic effects on tissues and organs, a truly systemic experimental approach, comprehensive knowledge resources and analytical tools are needed.

Though, the conventional 'one drug, one target, one disease' approach continues to dominate pharmaceutical thinking, and has led to the discovery of many successful drugs and drug target, it has been challenged (28–32). Several diseases like cancer, cardiovascular disease, and depression tend to result from multiple molecular abnormalities i. e. not from a single target/gene defect. Earlier, Goh *et al.*, elaborated on the human disease gene networks (30) and demonstrated that in many, the clinical effect is caused by patterns of target interactions. Recently, Bauer-Mehren et al reported that a core set of biological pathways is found to be associated with most human diseases. They also got similar results when studying clusters of diseases, suggesting that related diseases might arise due to dysfunction of common biological processes in the cell. In line with this, network models have suggested that partial inhibition of a small number of targets can be more efficient than the complete inhibition of a single target (33). Altogether, pinpointing a single target is unlikely to help effectively because cells can often find ways to compensate for a protein whose activity is affected by a drug (redundancy). Therefore, the new drug discovery paradigm is shifting from 'one drug, one-target, one disease' to 'one drug, one disease, multiple-targets' and 'multiple drugs, one disease, multiple targets'. Sometimes drugs targeted to particular diseases end up being marketed for unexpected and unrelated effects that are identified during development or evaluation of the drug. Thus, there is a 'one drug, multiple target, multiple diseases', paradigm as well.

It is becoming increasingly clear that genes and their products interact in complex biological networks with local and global properties and perturbations of these networks contribute to the disease state. The disease gene network demonstrates how the aberrant network gene/protein nodes contribute to disease(s) and defines the multiple protein nodes that can be targeted to rewire the network. This is an example of integrating molecular and physiological information to capture diverse characteristics of complex disorders.

In line with previous reports the disease gene network shows a genetic origin for the fact that obesity is a risk factor for diabetes (Figure 1.4). A quick look at the list of genes associated with these two diseases indicates that two genes, including ectoenzyme nucleotide pyrophosphate phosphodiesterase (ENPP1),

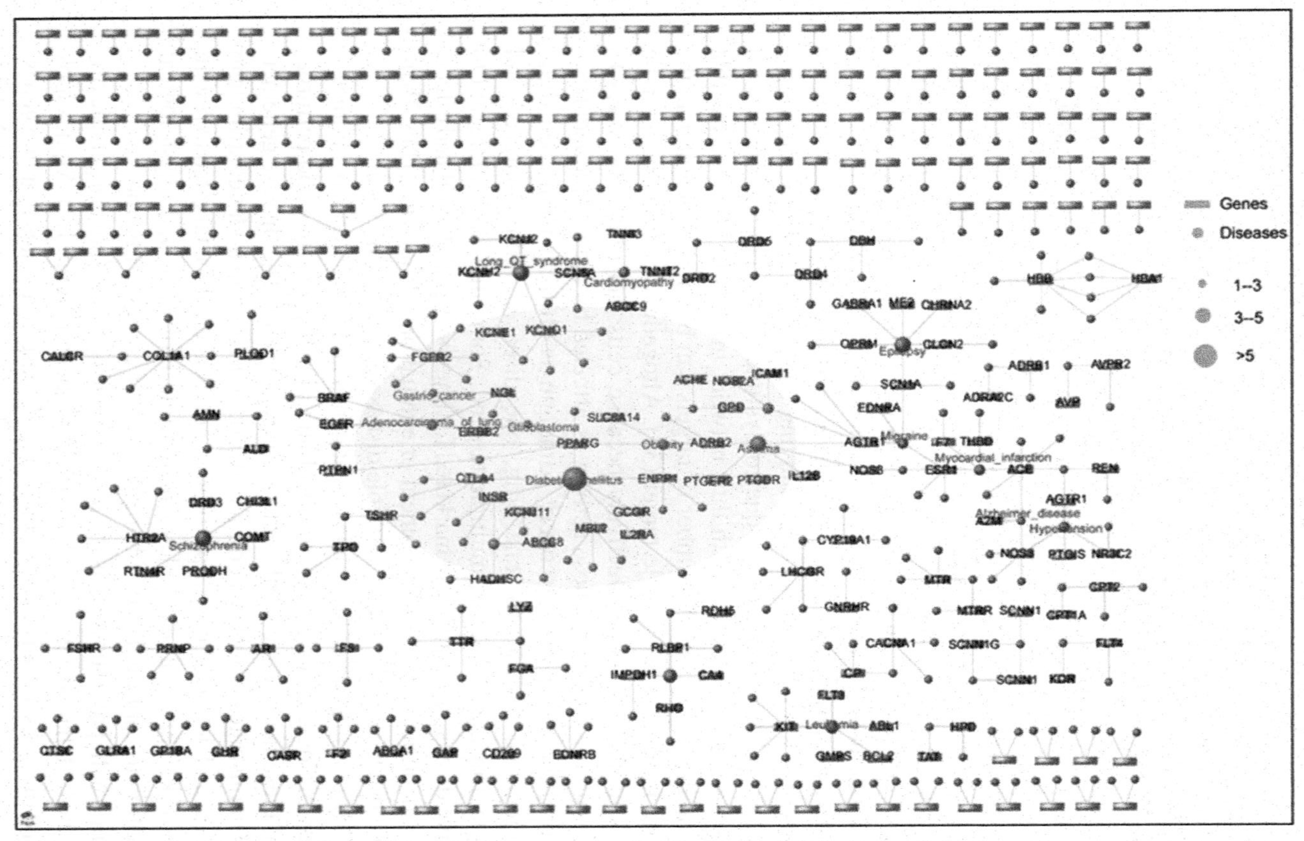

Figure 1.4 Disease gene network. Obesity is a risk factor for diabetes.

peroxisome-proliferator–activated receptor-γ (PPARγ) may be implicated in both diseases. In addition to the well-known link between diabetes and obesity, the large number of genes shared by often quite distinct disorders indicates that these diseases may have common genetic origins, for example, ADRB2 is implicated both in Asthma and Obesity. Thus, the gene disease network may help in understanding how the human disease network is configured based on integrating data available from literature and databases.

1.5 Conclusions

With biotech steadily uncovering more drug targets and mechanisms, and finding new ways to exploit them, there has been great progress in understanding the molecular basis of drug action and in elucidating genetic determinants of disease pathogenesis and drug response (34). The knowledge about relationships between diseases and genes, and drugs and targets is essential for the understanding disease phenotypes and drug response (35–36). These relationships are targeted towards advancing human pharmacology.

However, information on disease gene relationships and drug targets is spatially and temporally heterogeneous and complex. Hence, although it is possible to create an abstract representation of these associations, the heterogeneity of the interactions necessitates appropriate sampling and analysis complemented by systematic biological, physiological, chemical and clinical data integration. Establishing connectivity between diseases and genes and drugs and their biological targets and mapping of the two networks is the major challenge that remains. Concurrent mapping of protein interaction networks will also be crucial for understanding of the biological system and for predicting the drug response. Having said that, it will be interesting to see how far these kinds of analyses take us in the real world. How much will 'off-target' effects affect the analysis and to what extent is the mapping between network topologies and response surfaces a one-to-one function. (37–38)

Botstein *et al.* suggest that the disease genes discovered to date likely represent the easy ones, and that discovering the genetic basis of the remaining Mendelian and complex disorders like cancer will be more challenging, perhaps due to the rarity of the phenotypes, due to genetic heterogeneity, or because of complex genetics, i. e., multiple genes and modifiers contributing to a phenotype (39). On the other hand a huge reservoir of putative drug targets and their possible interactions (several thousands), hitherto remains to be explored. This is also because of the fact that human gene repertoire remains phenotypically uncharacterized (40). It is also uncertain if the network analysis can help us predict the side-effects and can provide us information on patterns of multiple drugs acting on one target.

Finally, connecting phenotypes induced by different drugs and the understanding on their molecular mechanisms will drive the identification of novel multidrug treatments, side effect predictions, adverse reaction predictions and the discovery of novel drug targets. It should also be kept in mind that a chemical with certain reactivity or binding property is used as a drug because of its clinical effects, but it should be stressed that it can be challenging to prove that a certain molecular interaction is indeed the one triggering the effect(s) (41). With the current state of the data, it is important to be very cautious about inferring biological significance from large-scale topological parameters. Only in this way can we ensure that network analysis and biological understanding are connected.

References

[1] Boby T, Patch AM, Aves SJ. TRbase: a database relating tandem repeats to disease genes for the human genome. Bioinformatics. 2005 Mar; 21(6): 811–6. Epub 2004 Oct 12.

[2] Zhidong Tu, Li Wang, Min Xu, Xianghong Zhou, Ting Chen and Fengzhu Sun Further understanding human disease genes by comparing with housekeeping genes and other genes. BMC Genomics 2006, 7: 31

[3] Dhillon VS, Fenech M. Mutations that affect mitochondrial functions and their association with neurodegenerative diseases. Mutat Res. 2013 Sep 17.

[4] Cooper DN, Kehrer-Sawatzki H. Exploring the potential relevance of human-specific genes to complex disease. Hum Genomics. 2011 Jan; 5(2): 99–107.

[5] Safran, M., Solomon, I., Shmueli, O., Lapidot, M., Shen-Orr, S., Adato, A., et al. (2002). GeneCards 2002: towards a complete, object-oriented, human gene compendium. Bioinformatics, 18, 1542–1543.

[6] Stenson, P. D., Ball, E. V., Mort, M., Phillips, A. D., Shiel, J. A., Thomas, N. S., et al. (2003). Human Gene Mutation Database (HGMD): 2003 update. Hum. Mutat., 21, 577–581.

[7] Treacy, E., Childs, B.,&Scriver, C. R. (1995). Response to treatment in hereditary metabolic disease: 1993 survey and 10-year comparison. Am. J. Hum. Genet., 56, 359–367.

[8] Brodniewicz T, Grynkiewicz G. Preclinical drug development. Acta Pol Pharm. 2010 Nov-Dec; 67(6): 578–85.

[9] Imming P, Sinning C, Meyer A. Drugs, their targets and the nature and number of drug targets. Nat Rev Drug Discov. 2006 Oct; 5(10): 821–34.
[10] Sakharkar MK, Sakharkar KR, Pervaiz S. Druggability of human disease genes. Int J Biochem Cell Biol. 2007; 39(6): 1156–64.
[11] Ofran, Y., Punta, M., Schneider, R., & Rost, B. (2005). Beyond annotation transfer by homology:novel protein-function prediction methods to assist drug discovery. Drug Discov. Today, 21, 1475–1482.
[12] Issa NT, Byers SW, Dakshanamurthy S. Drug Repurposing: Translational Pharmacology, Chemistry, Computers and the Clinic. Curr Top Med Chem. 2013 Sep 20.
[13] Chen YA, Tripathi LP, Mizuguchi K. TargetMine, an integrated data warehouse for candidate gene prioritisation and target discovery PLoS One. 2011 Mar 8; 6(3): e17844.
[14] DiMasi JA, Hansen RW, Grabowski HG. The price of innovation: new estimates of drug develop ment costs. J Health Econ. 2003 Mar; 22(2): 151–85.
[15] Martell RE, Brooks DG, Wang Y, Wilcoxen K. Discovery of novel drugs for promising targets. Clin Ther. 2013 Sep; 35(9): 1271–81.
[16] Janssens AC, van Duijn CM. Genome based prediction of common diseases: advances and prospects. Hum Mol Genet. 2008 Oct 15; 17(R2): R166–73.
[17] Kraljevic S, Stambrook PJ, Pavelic K. Accelerating drug discovery. EMBO Rep. 2004 Sep; 5(9): 837–42.
[18] M. A. Lindsay, Target discovery, Nature Reviews Drug Discovery, 2 (2003), pp. 831–838.
[19] M. Campillos, M. Kuhn, A. C. Gavin, L. J. Jensen and P. Bork, Drug target identification using side-effect similarity, Science, 321 (2008), pp. 263–6.
[20] Elisabet Gregori-Puigjané, Vincent Setola, Jérôme Hert, Brenda A. Crews, John J. Irwin, Eugen Lounkine, Lawrence Marnett, Bryan L. Roth, Brian K. Shoichet Identifying mechanism-of-action targets for drugs and probes Proc Natl Acad Sci U S A. 2012 July 10; 109(28): 11178–11183.
[21] Swinney DC, Anthony J. How were new medicines discovered? Nat Rev Drug Discov.2011; 10: 507–519. [PubMed]
[22] Ma'ayan A, Jenkins SL, et al. (2007) Network analysis of FDA approved drugs and their targets. Mt. Sinai J. Med. 74(1): 27–32
[23] Csermely P, Agoston V et al. (2005) The efficiency of multi-target drugs: the network approach might help drug design. Trends Pharmacol Sci. 26(4): 178–182.

[24] Feng Cong, twood K. Cheung, and Shih-Min A. Huang Chemical Genetics–Based Target Identification in Drug Discovery Annual Review of Pharmacology and Toxicology Vol. 52: 57–78
[25] Sadée W, Dai Z (2005) Pharmacogenetics/genomics and personalized medicine. Hum. Mol. Genet. Oct 15; 14 Spec No. 2: R207–14.
[26] Barabási AL (2007) Network medicine–from obesity to the "diseasome". N. Engl. J. Med. 357(4): 404–7.
[27] Gibbs JB. Mechanism-based target identification and drug discovery in cancer research. Science. 2000 Mar 17; 287(5460): 1969–73.
[28] Law MR, Wald J, et al. (2003) Value of low dose combination treatment with blood pressure lowering drugs: analysis of 354 randomised trials. Br. Med. J. 26(7404): 1427–1431.
[29] Keith CT, Borisy AA et al. (2005) Multicomponent therapeutics for networked systems. Nature Rev. Drug Discov. 4(1): 71–78.
[30] Goh K, Cusick ME et al. (2007) The human disease network. Proc. Natl. Acad. Sci. USA. 104(21) 8685–8690
[31] Lamb J (2007) The Connectivity Map: a new tool for biomedical research. Nat. Rev. Cancer 7(1): 54–60.
[32] Sakharkar MK, Sakharkar KR. Targetability of human disease genes. Curr Drug Discov Technol. 2007 Jun; 4(1): 48–58. Review.
[33] Bauer-Mehren A, Bundschus M, Rautschka M, Mayer MA, Sanz F, Furlong LI. Gene-disease network analysis reveals functional modules in mendelian, complex and environmental diseases. PLoS One. 2011; 6(6): e20284.
[34] McLeod HL and Evans WE (2001) PHARMACOGENOMICS: Unlocking the Human Genome for Better Drug Therapy. Annu. Rev. Pharmacol. Toxicol. 41: 101–21
[35] Tzouvelekis A, Patlakas G, et al. (2004) Application of microarray technology in pulmonary diseases. Respir Res. 5: 26.
[36] Hyman SE (2008) A glimmer of light for neuropsychiatric disorders. Nature 455, 890–893
[37] Yildirim MA, Goh KI, et al. (2007) Drug-target network. Nat. Biotechnol. 10: 1119–26.
[38] Pamela Yeh and Roy Kishony Networks from drug–drug surfaces. Mol Syst Biol. 2007; 3: 85.
[39] Botstein D, Risch N (2003) Discovering genotypes underlying human phenotypes: past successes for mendelian disease, future approaches for complex disease. Nat. Genet. 33 (Suppl.): 228–37

[40] Brinkman RR, Dubé MP, et al. (2006) Human monogenic disorders- a source of novel drug targets. Nat Rev Genet. 7(4): 249–60

[41] Imming P, Sinning C, et al. (2006) Drugs, their targets and the nature and number of drug targets. Nat. Rev. Drug. Discov. 5(10): 821–34.

[42] RRR

[43] Kamerman, A., "Spread Spectrum Schemes for Microwave-Frequency WLANs," Microwave Journal, Vol. 40, No. 2, February 1997, pp. 80–90.

[44] Pahlavan, K., and A. H. Levesque, Wireless Information Networks, New York, NY: Wiley, 1995.

[45] Kamerman, A., and L. Monteban, "WaveLAN-II: A High-Performance Wireless LAN for the Unlicensed Band," Bell Labs Technical Journal, Vol. 2, No. 3, 1997, pp. 118–133.

[46] van Nee, R., G. Awater, M. Morikura, H. Takanashi, M. Webster and K. Halford, "New High Rate Wireless LAN Standards," IEEE Communications Magazine, Dec. 1999, pp. 82–88.

[47] ISO/IEC 8802–11, ANSI/IEEE Std 802.11, First Edition 1999–00–00, Information Technology – Telecommunications and information exchange between systems – Local and metropolitan area networks – Specific requirements - Part 11: Wireless LAN Medium Access Control (MAC) and Physical Layer (PHY) specifications.

[48] IEEE Std 802.11b-1999, Supplement to Standard for Information Technology – Telecommunications and information exchange between systems – local and metropolitan area networks – Specific requirements-, Part 11: Wireless LAN Medium Access Control (MAC) and Physical Layer (PHY) specifications: Higher speed Physical Layer (PHY) extension in the 2.4 GHz band.

[49] Chen, K. C., "Medium Access Control of Wireless LANs for Mobile Computing," IEEE Network, September/October 1994, pp. 50–63.

[50] Hall, M. P. M., and L. W. Barclay, Radiowave Propagation, IEE Electromagnetic Waves Series 30, London, 1991.

[51] Rappaport, T. S., Wireless Communications: Principles & Practice, NJ: Prentice Hall, 1996.

[52] Prasad, R., Universal Wireless Personal Communications, Norwood, MA: Artech House, 1998.

[53] Mehrotra, A., Cellular Radio Performance Engineering, Artech House, 1994.

[54] Haartsen, J. C., "The Bluetooth Radio System," IEEE Personal Communications, Vol. 7, No. 1, February 2000, pp. 28–36.

2

Target Identification and Validation in Microbial Genomes

Kishore R Sakharkar[1], Meena K Sakharkar[2], Deepak Perumal[3], Ramesh Chandra[4,5] and Chow T. K Vincent[6]

[1]OmicsVista, Singapore
[2]College of Pharmacy and Nutrition, University of Saskatchewan, SK, Canada
[3]H. Lee Moffitt Cancer Center and Research Institute, Tampa, FL
[4]Department of Chemistry, Delhi University, India
[5]B. R. Ambedkar Center for Biomedical Research, University of Delhi, India
[6]Associate Professor, Department of Microbiology, National University of Singapore, Singapore

2.1 Introduction

Antibiotics are crucial in the fight against infectious diseases caused by bacteria and other microbes. Currently, about 70% of the bacteria that cause infections in hospitals are reported as resistant to atleast one of the drugs most commonly used for treatment (1). These disease-causing microbes that are resistant to antibiotic therapy pose increasing public health challenges globally. Alongside, the pace of development of novel antibiotics is dawdling (2). The absolute number of new licensed antibiotics has declined, and there is a shortage of new agents specifically against the Gram-negative bacteria. The prime reasons for drying up of the antibiotic pipeline, and for big pharmaceutical companies moving away from antibiotic research and development include: (a) aggressive price control of antibiotics for their being "life-saving" drugs, and short-course therapy; (b) greater chance of liability claims for adverse

Post-genomic Approaches in Drug and Vaccine Development, 23–52.

reactions; and (c) less lucrative than drugs for life-style and chronic conditions, e.g. drugs for hypertension, diabetes or Alzheimer's disease. Hence, there is an urgent need for novel therapies against microbial pathogens. Hence, the discovery of new antimicrobial targets and consequently new antimicrobial agents is of immediate concern. Towards this end, the availability of several complete microbial genome sequences is facilitating computational approaches to understand bacterial genomes and DNA structure/function relationships. One of the most interesting applications of genomic data of pathogenic bacteria is in the *in silico* identification of novel targets by genome exploration and the design of novel antimicrobial molecules against these targets (3).

A target is a defined molecule or structure within the organism, which is linked to a particular disease. For disease intervention, the drug target may either be blocked, inhibited/inactivated or activated by a drug (small organic molecules, antibodies, therapeutic proteins). Drug molecules can physically attach to a drug target, triggering a cascade of intracellular biochemical reactions, followed by a cellular response. Drug target generally identification involves acquiring a molecular level understanding of a specific disease state or organism and includes analysis of gene sequences, protein structures and structure, protein interactions and metabolic pathways (4). Towards this end, techniques to analyze the genomic sequences to identify target genes, their functions, and their possible relationships to health and disease processes are gaining momentum. Once key genes relevant to disease pathogenesis are shortlisted, the focus moves on to the discovery and development of small molecules, antibodies, proteins, or a combination of these, in search of drugs that will target the consequences of the virulence genes or dysregulated pathways to overcome the therapeutic challenge of infectious diseases. In addition, the genomics approaches facilitate a better understanding of the molecular mechanisms of infectious diseases, with the possibility of exploiting genomic information for the discovery of vaccine candidates leading to a new paradigm in vaccine development (discussed elsewhere in the book). Roemer et al. earlier proposed that coupling the dual processes of antimicrobial small-molecule screening and target identification in a whole-cell context is essential to empirically annotate 'druggable' targets and advance early stage antimicrobial discovery (5). Here, we highlight some target prioritization strategies for the discovery of novel antimicrobial targets that may potentially be explored for antimicrobial drug discovery using *Pseudomonas aeruginosa* as a case study.

2.2 Strategies for Target Identification in Silico

Global efforts on sequencing the genomes of microbes have focused primarily on pathogens that encompass the majority of all genome projects, and have generated an enormous amount of raw material for *in silico* analysis (6). These data pose a major challenge in the post-genomic era, i.e. to fully exploit this treasure trove for the identification of novel putative targets for therapeutic intervention. Genomics can be applied to evaluate the suitability of potential targets using two main criteria, i.e. "essentiality" and "selectivity" (7–8). Essential genes are defined as being vital for growth in nutrient medium in the laboratory but could also include other factors, such as survival in a host. Here, it must be mentioned that, Gene essentiality is increasingly being viewed as contextual, with decreased nutrient levels, changes in carbon sources, or environmental stress (e.g., change of temperature) altering the set of genes required for growth (9–11). Conservation of the drug target among different species of bacteria (antibiotics), fungi (antifungal) or parasites (antiparasitics), and little or no conservation in humans (selectivity) are required for broad spectrum antimicrobial agents (12). With the growth of microbial sequence databases, it is possible to use *in silico* comparison among genomes to identify potential targets in the pre-clinical stages of the drug discovery process. Some strategies for target identification in silico are outlined below.

2.2.1 Essential genes

Identification of essential genes has been one of the primary contributions of genomics to antibiotic target discovery. The target must be essential for the growth, replication, viability or survival of the microorganism of interest, i.e. encoded by genes critical for pathogenic life-stages. A gene is deemed to be essential if the cell cannot tolerate its inactivation by mutation, and its status is confirmed using conditional lethal mutants (6). Several methods are employed for the prediction, identification and prioritization of essential genes as discussed below. These may work in combinations or alone.

2.2.2 Conserved genes

Identification and characterization of essential genes for the establishment and/or maintenance of infection may be the basis to elucidate novel and effective antimicrobials against bacteria, especially if these genes are conserved in various bacterial pathogens. The functions encoded by essential genes are considered to constitute the foundation for life of the organism, and are therefore likely to be common to all cells. Being indispensable, essential genes are more evolutionarily conserved than non-essential genes (13). This is because negative selection acting on essential genes is expected to be more stringent than on non-essential genes. Genes that are conserved in different genomes often turn out to be essential.

2.2.3 Disordered proteins

Targets have been identified among intrinsically disordered proteins proteins (IDPs) that have disordered regions. IDPs frequently function by molecular recognition, and usually undergo a binding-induced folding transition upon binding to a suitable partner. They are reported to be interaction hubs in complex protein networks (14–15). It is proposed that proteins that are most essential for bacterial cell functioning and viability have larger numbers of interactions, i.e. they are well-connected proteins or hubs in protein interaction networks (PINs). The advantage of using such hub proteins as drug targets lies in their essentiality, non-replaceable position in the PINs, and lower rate of mutation, which can help to counter bacterial resistance.

2.2.4 Load point & Choke points, and FBA

Network models are crucial for understanding complex networks, and help to explain the origin of observed network characteristics. It is reported that highly interconnected proteins play important roles in the central metabolism of the bacterial cell, and are several times more essential than proteins that interact with only a few other neighbors.

The "load-point" of a metabolite in a metabolic network is defined as the ratio of the number of k-shortest paths passing through the metabolite and its nearest neighbor links (16). These load-point values provide a global view of the metabolic network, and aid in the analysis of the metabolic pathway reactions. Pathways that are highly connected in cellular metabolism tend to have high load values. Moreover, the lethality of an enzyme depends on the number of connections it has in the whole metabolic network (17). Enzymes with large numbers of connections are observed to be highly essential "hubs", and hence targeting them would result in disruption of the entire metabolic network. Highly interconnected proteins would play an important part in the central metabolism of the bacterial cell and are found to be three times more essential than proteins that interact with only a few other neighbors (17).

On the other hand, "choke-point" enzymes are those that participate in a reaction that consumes a unique specific metabolite (substrate) or uniquely produces a specific metabolite (product) in the metabolic network. These choke-point enzymes are crucial points in the metabolic pathway and inactivation of these important enzymes may lead to the disruption of the metabolic network of the bacterium. Thus, choke point enzymes are proposed as potential drug targets (18). Blocking the action of chokepoint enzymes, which catalyze producing or consuming chokepoint reactions, can definitely result in paucity of that specific product or accumulation of that specific substrate (18). Concomitantly, it is crucial to define metabolic pathways in the context of network framework to determine the importance of enzymes and metabolites of the metabolic network as connections between biochemical reactions via substrate and product metabolites develop complex metabolic networks that may be analyzed to gather information on protein structure/function and metabolite properties.

The advent of flux balance analysis (FBA) and related *in silico* approaches contribute novel techniques to predict cellular phenotypes using genomescale metabolic reconstructions under different environmental conditions. Constraint-based flux analysis is utilized to deduce the metabolic phenotype from the genotype, and this plays a critical role in drug targeting. The widely-used FBA method assumes that the metabolic network will reach a steady state that satisfies certain constraints (e.g. mass balance and flux limitations), and maximizes biomass production. FBA is shown to accurately predict essential genes in yeast and Escherichia coli. Thus, it can serve as a useful tool for the rational identification of drug targets in pathogens. A list of essential proteins in some bacteria including *Pseudomonas aeruginosa* metabolism predicted by FBA of the genome-scale metabolic model iMO1056 is available (19).

2.2.5 Virulence proteins

The adaptive evolution of benign bacteria into pathogens usually involves the acquisition of foreign genes encoding for specific virulence factors (20). An alternate or additional evolutionary mechanism employed by pathogenic bacteria is the subtle modification of an existing gene or genes through point mutations that confer a selective advantage in the virulence niche. Virulent factors provide the bacterium the ability to invade the host niche; preferentially colonize a specific host organ or tissue (i.e. strategies for tissue tropism); effectively consume available nutrients; evade host antibacterial defences, and inflict damage to the host (21). Virulent factors may thus be essential for survival of the bacterium, and contribute towards the elucidation of pathogenic mechanisms in infectious diseases and to the development of novel approaches for disease treatment and prevention.

2.2.6 *In vitro* experiments

A list of all currently available essential genes is compiled into the database of essential genes (DEG), that includes the essential genes identified in the genomes of *Mycoplasma genitalium, Haemophilus influenzae, Vibrio cholerae, Staphylococcus aureus, Escherichia coli, and Saccharomyces cerevisiae.*(22). A blast search of bacterial protein sequences against the DEG database can help identify putative targets.

2.2.7 Overlapping genes/proteins

Overlapping genes are a common occurrence in prokaryotic genomes (23). In overlapping genes, the same DNA sequence encodes two proteins using different reading frames. Overlapping gene pairs can assume one of three structures, namely, 'convergent' (→←), 'unidirectional' (→→ or ←←), or 'divergent' (←→) (24). Clark et al. (2001) suggested that overlapping genes occur as a result of mutational bias towards deletion. Overlapping genes are more conserved between species than non-overlapping genes (25), mostly because a mutation in the overlapping region causes changes in both genes, and selection against such mutations should therefore be stronger. Miyata and Yasunaga (26) reported that rates of evolution are slower in overlapping genes.

Overlapping genes may evolve as a result of the extension of an ORF caused by a switch to an upstream initiation codon, substitutions in initiation or termination codons, and deletions and frameshifts that eliminate initiation or termination codons (24). Comparative analysis of the genomes of *Mycoplasma genitalium* and *Mycoplasma pneumoniae* showed that most overlapping genes are generated by mutations at the end of coding regions (27). The analysis revealed that the loss of a stop codon causes the gene to elongate to the next stop codon. The role of overlapping genes in obligatory intracellular parasites of humans, and their evolution and dynamics showed several overlapping genes to be involved in essential cellular processes (7). Analysis of overlapping genes is hampered by sequencing and annotation errors present in genomes, and by the limitations of gene-finding algorithms to handle multiple reading frames (28). The authors thus advised caution in drawing inferences from these data.

2.2.8 Fusion proteins

The event of bringing together two separate genes into a single gene (gene fusion) has long been identified as a potentially important evolutionary phenomenon (29–30). Gene fusion events have been proposed to represent a valuable "Rosetta stone" information for the identification of potential protein–protein interactions and metabolic or regulatory networks (31–32). Fusion genes gain added advantage by coupling biochemical reactions through tight regulation of fusion partners, compared to individual partners (33). Yanai *et al.* (34) used gene fusion to establish links between fusion genes and functional networks with their involvement. Gene fusion has also been used to illustrate novel gene function (35), enhanced substrate specificity (36) and multi-functional enzyme specificity (37). Fusion genes have been revealed to be essential and hence the potential of the products of fusion genes as putative microbial drug targets is suggested.

2.2.9 Hub proteins and networks

Essentiality is almost perfectly predicted by the lack of an alternative pathway. Most crucial points in the metabolic networks correspond to the enzymes at the periphery of the biological system (41). Lethality corresponds to the lack of alternate paths in the perturbed network linking the nodes affected by the removal of the enzyme. Cellular metabolism is described and interpreted in terms of the biochemical reactions that make up the metabolic network. These network frameworks take into consideration different aspects of metabolic chemistry and demonstrate the importance of metabolic biochemistry on the local and global properties of cellular metabolism. The integration of these metabolic pathway approaches along with systematic classification of the chemical structure of metabolites will not only enhance understanding of metabolic pathways, but will also improve ability to predict enzyme function and novel potential drug targets. Graph theory based pathway analyses methods are useful in analyzing metabolic networks consisting of reactions, metabolites and enzymes (38–40). Here, metabolic networks are represented as a metabolite graph consisting of nodes (metabolites) and edges (reactions) with large number of connecting links. Such representation of network allows the characterization of the metabolic pathways with respect to degree of metabolite (nodes) connectivity defined as possible number of reactions by a metabolite; and the degree of interconnectivity or average network diameter defined as the average shortest path length (39).

2.2.10 Membrane proteins, uncharcterised essential genes, species specific genes as targets

Galperin and Koonin (42) alluded to the demonstration of essentiality of a particular gene as the first step towards using it is a possible drug target, and suggested the selection of membrane proteins, uncharacterized essential genes and species-specific genes as drug targets.

2.2.11 Structural genomics

Protein with a high degree of sequence identity (>30%) depending on length may have a similar fold. Thus, a powerful approach for assigning previously unknown molecular function to a protein is to determine the 3D structure of the protein and compare this, rather than the amino acid sequence, with those of the protein structure database (PDB). If there are significant structural homologs, the protein under investigation is predicted to have similar molecular properties. These predictions can then be tested experimentally (43). However, the power of the structural genomics approach will increase when a larger number of protein structures become available. Furthermore, some computational modeling strategies indicate that this approach provides improved functional predictions, compared to alignments of amino acid sequences (44).

Three-dimensional protein structures help in detailed understanding of the molecular basis of protein function. Sequence information along with 3D structure gives significant insights for the development of rational methods for experiments such as site directed mutagenesis, mutation studies or structure based design of potential inhibitors. It is therefore likely that structural genomics will carry on present efforts and eventually substantial knowledge will be gathered in the near future about the increasing number of potential drug targets.

Availability of high resolution structure of a target can provide substantial advantages in several phases of the drug discovery process. However, certain aspects of the discovery process are not currently aided by possession of a high resolution structure of the target. Nonetheless, efficiencies can be realized at many points in the process when using the structure of a protein target to guide efforts. This is still an active field of research, in which the tools for deriving advantage from the structures are still emerging, and in which there is still substantial room for improvement of the techniques that are being used (45).

Recent developments in X-ray crystallography and Nuclear Magnetic Resonance techniques have given rise to the expectation that high resolution 3D structure or reliable homology modeling of target proteins can be achieved in a reasonably short time. Additional methods are likely to emerge from the increasing numbers of protein structures and co-structures available, and the wealth of high throughput screening information that has been gathered over the past decade. The information databases linking structure and ligand binding information will also aid this effort. Computational models will allow focusing efforts on proteins that are more likely to yield high affinity ligands for drug discovery. Furthermore, the large numbers of essential bacterial proteins with high resolution structures provide an excellent testing ground for such methods. The availability of a high resolution protein structure at the beginning of a drug discovery project allows *in silico* screening methods to supplement experimental high throughput screening. *In silico* screening uses high resolution protein structure information to computationally test and identify small molecules more likely to bind the protein. In the process, more number of compounds can be screened *in silico* and subsequent experimental testing of compounds identified from a virtual screen can be undertaken using high throughput screening methods (HTS) (46).

2.2.12 Selectivity

The microbial target for treatment should not have any well-conserved homolog in the host, in order to address cytotoxicity issues. This can help to avoid expensive dead-ends when a lead target is identified and investigated in great detail only to discover at a later stage that all its inhibitors are invariably toxic to the host (7). It is also possible to identify targets that are conserved indifferent strains of one specific pathogen. On the other hand, drugs designed for specific pathogens may be desired when long-term therapy for a known pathogen is required, to minimize effects on the body's normal flora and the development of drug resistance. Since the human gut and oral flora consist of microbes that are considered to influence the physiology, nutrition, immunity, and development of the host, interfering with their metabolism may have adverse effects. In line with this concept, an interesting approach designated "differential genome display" is proposed for the prediction of potential drug targets. This approach relies on the fact that genomes of parasitic microorganisms are generally much smaller, and encode fewer proteins than the genomes of free-living organisms. The genes that are present in the genome of a parasitic bacterium, but absent in the genome of a closely related free-living bacterium, are therefore likely to be important for pathogenicity and may be considered candidate drug targets. A complementary approach to target identification by bioinformatics is described in a concordance analysis of microbial genomes. A simple and efficient computational tool has been developed that can determine concordances of putative gene products, unraveling sets of proteins conserved across one set of user-specified genomes, but that are not present in another set of user-specified genomes. However, there are no guarantees that if there is no homology, toxicity is not observed. An example is the large ribosomal subunit which is targeted by chloramphenicol. Despite the difference in the structure of the prokaryotic and eukaryotic ribosomes, a functionally conserved area is found in the ribosomes of mitochondria and bacteria resulting in adverse effects in certain patients.

2.2.13 Target prioritization in completely sequenced *Pseudomonas aeruginosa* genomes

Pseudomonas aeruginosa species are important pathogenic bacteria. *Pseudomonas aeruginosa* is an opportunistic respiratory tract infection causing ubiquitous bacterium. The situation for patients with *Pseudomonas aeruginosa* infections is particularly problematic since this organism is inherently resistant to many drug classes and is able to acquire resistance to all effective antimicrobial drugs (47). *Pseudomonas* has the ability to adapt and thrive in many ecological niches, including humans. Once infections are established, *Pseudomonas aeruginosa* produces a number of toxic proteins which cause not only extensive tissue damage, but also interfere with the human immune system's defense mechanisms. In addition, some *Pseudomonas* strains can inactivate the drugs that threaten them by using enzymes to modify the drug. Most *Pseudomonas* strains have diverse metabolic capabilities that provide environmental variability. Besides, *Pseudomonas aeruginosa* continues to be a major pathogen among patients with immuno-suppression, cystic fibrosis, malignancy, and trauma (48). A challenge for the biotechnological industry is the development of effective prevention methods based on novel fundamental research in biotechnology and information technology. The development of wide range of bio-therapeutic antibiotics and vaccines using genomic data are under progress. In this chapter, we demonstrate the use of insilico approaches for enabling identification of potential drug targets and validation of two of the selected drug targets as novel potential targets (Figure 2.1).

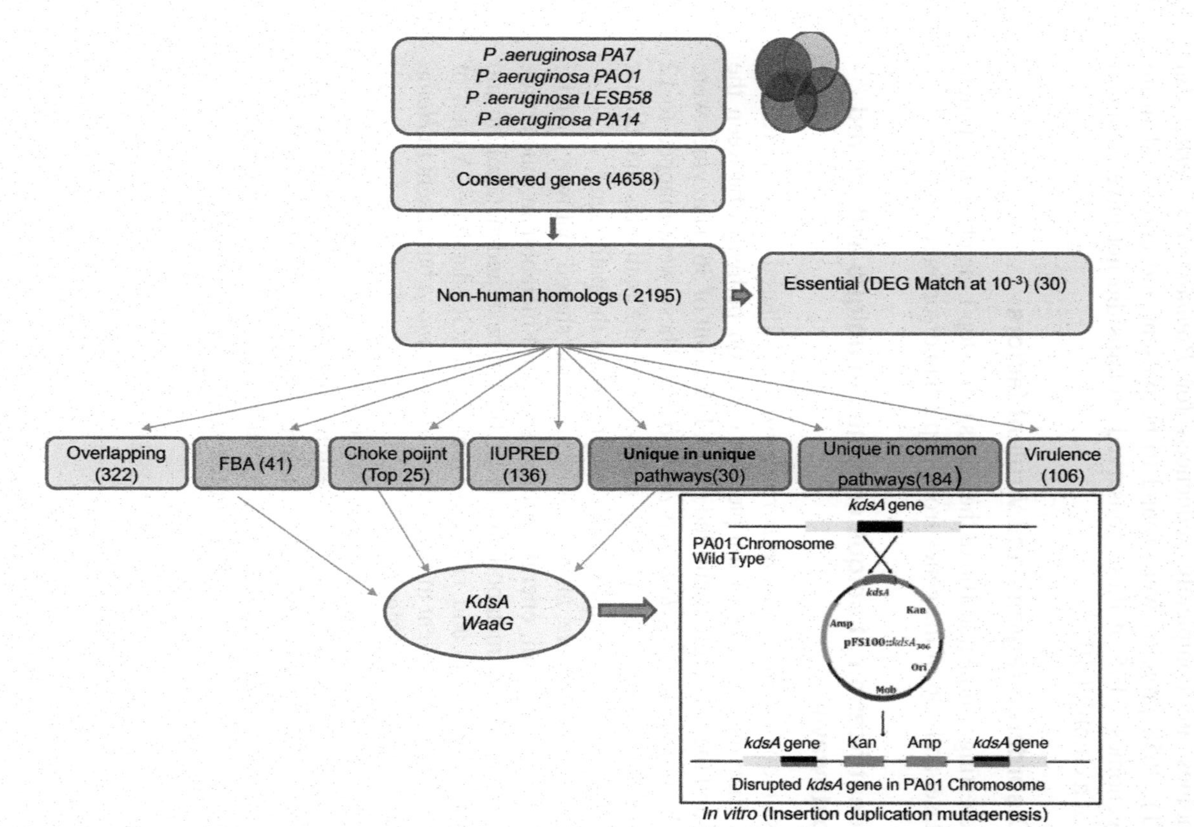

Figure 2.1 A schematic representation of the various genome mining methods for the identification of drug targets in *Pseudomonas aeruginosa.*

2.2.14 Genomes/Conserved proteins/Non-human homologs

Four genomes are completely sequenced for *Pseudomonas aeruginosa* (Table 2.1). 4658 proteins were found to be present and conserved in all the four genomes. 2195 of these conserved proteins did not have a human homolog at a cutoff of 10^{-2}.

2.2.15 Disordered regions and Virulent factors

It was interesting to note that 106 of the genes from step 1 matched the virulent genes dataset and 136 of the genes from step 1 had disordered regions.

2.2.16 Identification of unique enzymes in unique and shared pathways

Pathways unique to *Pseudomonas aeruginosa* were filtered out using KEGG database. Enzymes from both unique and common pathways between the pathogen and the host were identified (361). Fifty (out of 361) enzymes were found to be unique to *Pseudomonas aeruginosa*. These were mapped to 12 unique pathways (Table 2.2). Unique pathways are the pathways that do not appear in the host (*Homo sapiens*) but are present in the pathogen.

Unique enzymes were also identified among shared pathways under carbohydrate metabolism, energy metabolism, lipid metabolism, nucleotide metabolism, amino acid metabolism, glycan biosynthesis and metabolism and metabolism of cofactors and vitamins from the KEGG database. A total of 311 enzymes that are present in *Pseudomonas aeruginosa* but absent in *Homo sapiens* were obtained (Table 2.3).

The protein sequences for these 361 (311+50) unique enzymes were retrieved and were subject to BLAST search against human protein sequence database at an E-value cutoff of 10^{-2}. Removing enzymes from the pathogen that share a similarity with the host protein ensures that the targets have nothing in common with the host proteins, thereby, eliminating undesired

Table 2.1 Completely sequenced genomes of *P. aeruginosa*

Name	Size	GC%	Genes	Proteins
P. aeruginosa PA7	6.59	66.4	6369	6286
P. aeruginosa PAO1	6.26	66.6	5682	5571
P. aeruginosa LESB58	6.6	66.3	6061	5925
P. aeruginosa	6.54	66.3	5977	5892

Table 2.2 Pathways unique to *Pseudomonas aeruginosa*

S. No	Pathways and their enzymes	Gene	EC #
1	**Polyketide sugar unit biosynthesis**		
	Glucose 1-phosphate thymidylyltransfease	rmlA	2.7.7.24
	dTDP-D-Glucose 4,6 dehydratase	rmlB	4.2.1.46
	dTDP-4-dehydrorhamnose 3,5 epimerease	rmlC	5.1.3.13
	dTDP-4-dehydrorhamnose reductase	rmlD	1.1.1.133
2	**Biosynthesis of siderophore group**		
	nonribosomal peptides		
	Isochorismate synthase	pchA	5.4.4.2
	Isochorismate pyruvate lyase	pchB	4.1.99.-
3	**Toluene and xylene degradation**		
	Catechol 1,2-dioxygenase	catA	1.13.11.1
4	**1,2 Dichloroethane degradation**		
	Quinoprotein alcohol dehydrogenase	exaA	1.1.99.8
	Probable aldehyde dehydrogenase		1.2.1.3
5	**Type II secretion system**		
	Two-component sensor PilS	pilS	2.7.3.-
	Leader peptidase (prepilin peptidase) / N-methyltransferase	pilD	3.4.23.43
	Methyltransferase PilK	pilK	2.1.1.80
6	**Type III secretion system**		
	Flagellum-specific ATP synthase FliI	fliI	3.6.3.14
7	**Phosphotransfease system (PTS)**		
	Phosphotransferase system, fructose-specific IIBC component	fruA	2.7.1.69
	Probable phosphotransferase system enzyme I		2.7.3.9
8	**Bacterial Chemotaxis**		
	Methyltransferase PilK	pilK	2.1.1.80
	Two-component sensor PilS	pilS	2.7.3.-
	Probable methylesterase		**3.1.1.61
9	**Flagellar Assembly**		
	ATP synthase in type III secretion system		3.6.3.14
10	**D-Alanine metabolism**		
	D-alanine-D-alanine ligase A	ddlA	6.3.2.4
	Biosynthetic alanine racemase	alr	5.1.1.1
11	**Lipopolysaccharide Biosynthesis**		
	Probable glucosyltransferases		2.4.-
	3-deoxy-manno-octulosonate cytidylyltransferase	kdsB	**2.7.7.38
	Putative 3-deoxy-D-manno-octulosonate 8-phosphate phosphatase		3.1.3.45
	Tetraacyldisaccharide 4'-kinase	lpxK	**2.7.1.130
	Lipid A-disaccharide synthase	lpxB	**2.4.1.182
	Lipopolysaccharide core biosynthesis protein WaaP	waaP	**2.7.-.-

(Continued)

Table 2.2 Continued

S. No	Pathways and their enzymes	Gene	EC #
	Poly(3-hydroxyalkanoic acid) synthase 1	phaC1	2.3.1.-
	UDP-glucose:(heptosyl) LPS alpha 1,3-glucosyltransferase WaaG	waaG	**2.4.1.-
	UDP-2,3-diacylglucosamine hydrolase		**3.6.1.-
	UDP-3-O-acyl-N-acetylglucosamine deacetylase	lpxC	3.5.1.-
	UDP-N-acetylglucosamine acyltransferase	lpxA	2.3.1.129
	ADP-L-glycero-D-mannoheptose 6-epimerase	rfaD	5.1.3.20
	2-dehydro-3-deoxyphosphooctonate aldolase (KDO 8-P synthase)	kdsA	**2.5.1.55
12	**Two component system**		
	Two-component sensor PilS	pilS	2.7.3.-
	Probable 2-(5”-triphosphoribosyl)-3’-dephosphocoenzyme-A synthase		2.7.8.25
	Serine protease MucD precursor	mucD	3.4.21.-
	Probable acyl-CoA thiolase		2.3.1.9
	Glutamine synthetase	glnA	6.3.1.2
	Citrate lyase beta chain		4.1.3.6
	Protein-PII uridylyltransferase	glnD	2.7.7.59
	Beta-lactamase precursor	ampC	3.5.2.6
	Anthranilate synthase component II	trpG	4.1.3.27
	Anthranilate phosphoribosyltransferase	trpD	2.4.2.18
	Indole-3-glycerol-phosphate synthase	trpC	4.1.1.48
	Tryptophan synthase alpha chain	trpA	4.2.1.20
	Potassium-transporting ATPase	kdpA	3.6.3.12
	Probable methylesterase		3.1.1.61
	Alkaline phosphatase	phoA	3.1.3.1
	Respiratory nitrate reductase alpha chain	narG	1.7.99.4

** Enzymes that matched with list of experimentally validated essential genes

host protein-drug interactions. A total of 214 enzymes resulted that had “no hits” in BLAST search against human. Thirty of these 214 “no hits” belong to the unique pathways set and the remaining 184 belong to unique enzymes in shared pathways. The distribution of these into different pathways is presented in Figure 2.2.

2.2.17 Comparison of unique enzymes to essential gene data

Further comparison of the 214 (30 + 184) unique enzymes to the list of candidate essential genes of *Pseudomonas aeruginosa* obtained from transposon

Table 2.3 Unique enzymes from common pathways for *Homo sapiens* and *P. aeruginosa*

EC No	Gene	EC No	Gene	EC No	Gene	EC No	Gene
1.2.1.12		6.2.1.17		4.6.1.1	*cyaA*	1.4.1.9	*ldh*
4.1.2.13	fda	*1.1.1.157		*2.7.7.6	rpoA	1.3.99.2	
5.4.2.8	algC	1.2.1.16	gabD	*2.7.7.7	dnaN	1.1.1.86	ilvC
*1.3.99.1	sdhC	2.8.3.12		3.1.3.5	surE	4.2.1.33	leuD
4.1.1.49	pckA	3.1.3.25		2.4.2.7	apt	1.1.1.85	leuB
*4.1.2.14		3.1.4.3	plcH	3.2.2.4	amn	4.2.1.9	ilvD
2.7.1.45		2.7.4.1	ppk	3.2.2.1		6.3.2.10	murF
4.2.1.12	edd	*1.10.3.-	cioB	6.3.5.2	guaA	2.7.2.4	lysC
2.7.1.12		2.7.2.2	arcC	3.5.1.5	ureA	4.2.1.52	
1.1.5.2	gcd	4.2.1.104	cynS	3.5.3.19		1.3.1.26	dapB
*1.1.99.3		*1.7.1.4	nirD	3.1.7.2	spoT	2.3.1.117	dapD
2.7.1.17	mtlY	*1.7.99.4	napA	2.7.6.5	relA	2.6.1.17	
2.7.1.4	mtlZ	1.7.2.1	nirS	3.6.1.41		3.5.1.18	dapE
4.2.2.3	algL	4.3.1.1		3.1.5.1	dgt	5.1.1.7	dapF
2.7.1.56	fruK	3.5.1.1	ansA	1.17.4.2		4.1.1.20	lysA
1.1.1.132	algD	1.4.1.4	gdhA	3.5.3.4	alc	6.3.2.13	murE
4.1.2.17		*1.4.1.13	gltB	2.1.3.2	pyrB	1.2.1.-	aruD
2.7.11.5	aceK	1.13.12.-	pvdA	3.5.2.3	pyrC	*2.3.1.109	aruF
*1.14.-.-		*6.3.5.4		4.1.3.30	prpB	4.3.1.12	
1.1.1.158	murB	3.5.1.38	ansB	3.5.4.13		3.5.1.4	
2.5.1.7	murA	*1.14.12.1	antA	*2.4.2.9	pyrR	3.5.3.6	arcA
2.3.1.157	glmU	1.7.99.6	nosZ	3.5.4.1	codA	3.4.11.5	
2.6.1.16	glmS	2.3.1.31	metX	4.1.1.23	pyrF	2.6.1.11	aruC
4.1.1.31	ppc	1.8.4.8	cysH	5.1.1.3	murI	3.5.3.8	hutG
*1.1.99.16	mqoA	*2.7.1.25	cysC	2.4.2.14	purF	3.5.2.7	hutI
2.7.9.2	ppsA	*2.5.1.47	cysM	1.2.1.16	gabD	2.4.2.17	hisG
2.3.3.9	glcB	2.3.1.30	cysE	4.3.1.1		1.1.1.23	hisD
*1.1.2.3	lldA	*1.8.1.2	cysI	4.1.1.11		2.6.1.9	hisC1
1.2.2.2	poxB	1.3.99.-		2.6.1.76		4.2.1.19	hisB
2.3.1.8	pta	1.18.1.1		*1.2.1.8	betB	5.3.1.16	hisA
*6.2.1.1	acsA	1.3.99.7	gcdH	4.1.1.65	psd	3.5.4.19	hisI
2.3.3.13	leuA	1.17.1.2	lytB	*4.1.2.5	ltaA	3.6.1.31	hisE
*6.4.1.2	accD	1.17.4.3		*6.1.1.14	glyS	*4.1.3.-	hisF2
2.7.2.1		4.6.1.12		2.7.2.4	lysC	*2.4.2.-	hisH1
3.6.1.7		2.7.7.60		2.7.8.8	pssA	*1.14.13.3	pvcC
4.1.3.1		2.7.1.148	ipk	4.3.1.19	ilvA1	4.1.2.-	
*3.5.1.10	purU1	1.1.1.267	dxr	4.2.3.1	thrC	*5.3.3.10	
*1.2.1.2	fdnH	2.2.1.7	dxs	6.3.2.6	purC	2.6.1.57	phhC
1.1.1.81		2.5.1.10		1.2.1.11	asd	4.2.1.-	hpcG
1.1.1.60		*1.6.5.2		2.7.1.39	thrB	4.1.1.68	
4.1.1.47	gcl	2.3.1.51		1.1.1.3	hom	1.13.12.-	pvdA
*3.1.3.18		2.1.2.2	purN	1.1.1.81		1.2.1.60	hpcC
2.7.1.31		6.3.3.1	purM	4.3.1.18	dsdA	1.13.11.27	hpd
1.1.1.29	hprA	2.1.2.3	purH	4.2.1.22		*1.4.99.1	dadA

(Continued)

Table 2.3 Continued

EC No	Gene	EC No	Gene	EC No	Gene	EC No	Gene
EC No	Gene	EC No	Gene	EC No	Gene	EC No	Gene
2.3.3.5	prpC	*4.1.1.21	purK	2.5.1.47	cysM	5.3.1.24	trpF
*2.6.1.18		2.7.7.8	pnp	2.3.1.30	cysE	2.4.2.18	trpD
*4.1.3.27	trpE	*2.4.2.11	pncB2	1.13.11.1	catA	4.2.1.46	rmlB
*5.4.99.5	pheA	3.5.1.19		3.8.1.2		2.7.7.24	rmlA
*4.2.1.51	pheA	3.2.2.1		1.14.13.2	pobA	4.1.99.-	pchB
1.3.1.12		1.6.1.1	sth	*1.14.13.82	vanA	2.7.4.16	thiL
2.7.1.71	aroK	2.7.1.24		*1.13.11.3	pcaH	2.5.1.3	thiE
1.1.1.25	aroE	4.1.1.36	dfp	2.8.3.12		2.7.1.49	thiD
*4.2.1.10	aroQ2	1.1.1.86	ilvC	1.14.12.10	xylX	*3.5.4.25	ribA
4.2.3.4	aroB	4.2.1.9	ilvD	2.3.1.174	pcaF	2.5.1.9	ribC
*2.5.1.54		*2.1.2.11		1.1.99.-		3.5.4.26	ribD
2.7.2.8	argB	*1.1.1.169		5.5.1.2	pcaB	2.4.2.21	cobU
1.2.1.38	argC	3.5.1.6		*3.1.1.24	pcaD	*1.1.1.262	pdxA
1.2.1.41	proA	6.3.4.15	birA	*4.1.1.44	pcaC	1.1.1.-	pdxB
*3.5.1.5	ureA	2.8.1.6	bioB	5.3.3.4	catC	2.7.7.18	
*3.5.1.16		6.3.3.3	bioD	4.1.1.7	mdlC	6.3.5.10	cobQ
*2.3.1.35	argJ	2.6.1.62	bioA	2.8.3.-		5.4.1.2	cobH
1.1.1.35		2.3.1.47	bioF	3.4.23.36		6.3.5.9	cobB
6.3.2.1		6.3.5.8		*2.7.7.7	dinP	*6.6.1.2	cobN
3.5.2.2		4.1.3.38		3.6.3.41	ccmA	*5.4.3.8	hemL
4.1.1.11		2.5.1.15	folP	3.6.3.29	modC	1.2.1.70	hemA
1.1.99.-		2.7.6.3	folK	*3.6.3.34		2.1.1.133	cobM
2.6.1.37	phnW	4.1.2.25	folB	3.6.3.30		2.1.1.132	cobL
*1.8.1.2	cysI	1.3.99.22	hemN	3.6.3.36		1.3.1.54	
2.9.1.1	selA	6.3.1.10	cobD	3.6.3.25	cysA	2.1.1.64	ubiG
*2.5.1.47	cysM	2.4.2.21	cobU	3.6.3.27	pstB	*4.1.1.-	
6.3.2.9	murD	2.7.8.26	cobV	3.6.3.32		5.4.4.2	pchA
6.3.2.8	murC	2.7.1.156	cobP	4.2.3.5	aroC	2.5.1.31	uppS
*6.3.2.4	ddlA	*2.1.1.107	cysG	*3.6.3.31		*3.5.1.28	amiB
3.6.1.27	bacA	2.1.1.131	cobJ	*3.6.3.21	aotP	*2.4.1.129	mrcB
2.4.1.227	murG	1.14.13.83		*3.6.3.17		1.16.8.1	
2.7.8.13	mraY	*6.6.1.1		4.2.1.51	pheC	2.1.1.130	cobI
2.7.8.25	prpD	3.4.21.-	mucD	2.1.1.14	metE	*1.14.12.-	
4.2.1.79		*4.3.2.2	purB	2.5.1.49	metY		

Unique enzymes among shared pathways under carbohydrate metabolism, energy metabolism,lipid metabolism, nucleotide metabolism, amino acid metabolism, glycan biosynthesis & metabolism and metabolism of cofactors and vitamins from the KEGG database. A total of 311 enzymes that are present in Pseudomonas aeruginosa but absent in Homo sapiens are tabulated. *Enzymes present in more than one pathway. Genes hpcG (EC 4.2.1. -) and pnc B2 (EC 2.4.2.11) are under study by the New York Structural Genomics Research Consortiu m (NYSGRC)

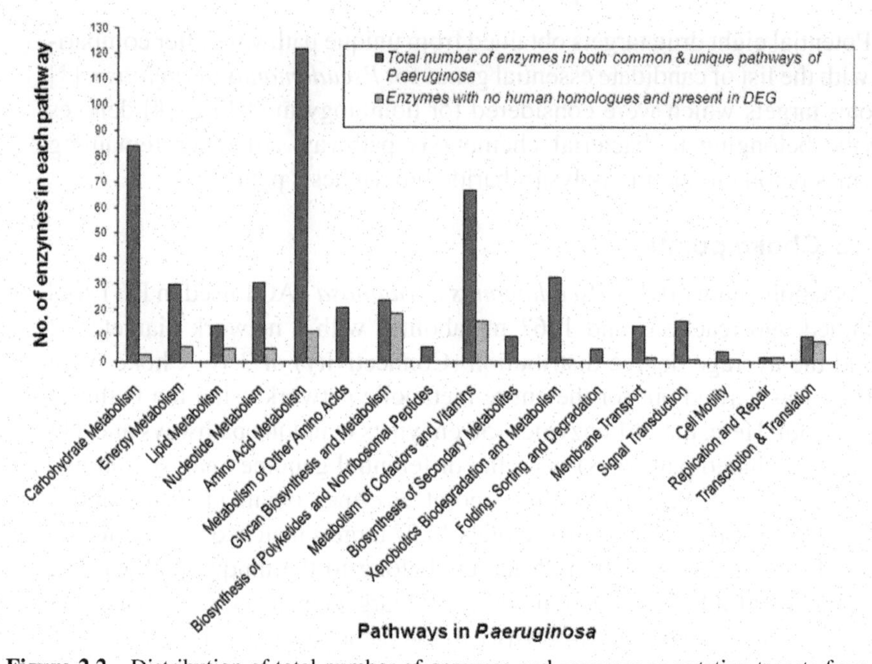

Figure 2.2 Distribution of total number of enzymes and enzymes as putative targets from different metabolic pathways.

mutagenesis studies was done. Eighty three of the 214 enzymes (8 enzymes from unique pathways and 75 enzymes from shared pathways) mapped to the essential gene dataset. It is noteworthy that seven of the eight enzymes from the unique pathways map to a single pathway that of lipopolysaccharide biosynthesis (Table 2.4).

Table 2.4 List of potential eight targets

EC No	Protein name	Gene
3.6.1.-	Conserved hypothetical protein	ybbf
2.7.7.38*	3-deoxy-manno-octulosonate cytidylyltransferase	kdsB
2.7.1.130	Tetraacyldisaccharide 4-kinase	lpxK
2.5.1.55*	2-dehydro-3-deoxyphosphooctonate aldolase	kdsA
2.4.1.182	Lipid A-disaccharide synthase	lpxB
3.1.1.61#	Probable methylesterase	-
2.7.-.-	LPS biosynthesis protein RfaE	rfaE
2.4.1.-*	UDP-glucose: (heptosyl) LPS alpha 1,3-glucosyltransferase WaaG	waaG

Potential eight drug targets obtained from unique pathways after comparison with the list of candidate essential genes for *Pseudomonas aeruginosa.* [*] denotes targets which were considered for homology modeling. [#] denotes enzyme belonging to bacterial chemotaxis pathway. All other remaining enzymes belong to the lipopolysaccharide biosynthesis pathway.

2.2.18 Choke points

The metabolic network for *Pseudomonas aeruginosa* PAO1 used in PHT tool contained 996 reactions and 1063 metabolites with a network diameter of 34 and the average degree distribution (Connectivity) of 3.16. Choke point analysis was carried out for the entire metabolic network using the Pathway Hunter Tool. Here, the 361 enzymes (50 enzymes in unique pathways and 311 enzymes in shared pathways) from the differential genome analysis (above) was compared to that of the choke point enzymes obtained in the above metabolic network tool analysis. A total of 227 targets matched the previously reported 361 targets of which 25 targets belong to the unique pathways of the pathogen (Table 2.5). The remaining 202 targets matched with that of the enzymes in the shared pathways between the pathogen and human.

2.2.19 FBA

The FBA approach was used to determine the set of essential enzymes for growth on rich medium within the 1030 proteins accounted in the *Pseudomonas aeruginosa* genome-scale metabolic model iMO1056. Growth was simulated on rich medium by allowing uptake of all external metabolites in the iMO1056 model. Further determined the set of essential enzymes for growth on rich medium in the *Pseudomonas aeruginosa* metabolic network using two slightly different FBA formulations. Using the first method, 116 of the 1030 proteins were found to be essential for growth on rich medium. Using the second method, found 113 of the 1030 proteins to be essential for growth on rich medium. The set of 113 essential proteins obtained using the second method is a subset of 116 essential proteins obtained using the first method. 41 essential enzymes of PAO1 were found to have no homology at cutoff value of 10^{-2} with the UP-UC enzymes in human (49).

The availability of a target's 3D structure would aid in rational drug design in turn providing a strong practical advantage in high throughput docking and other structural studies. The 41 putative target enzymes and their sequences when searched against DrugBank show that 9 enzymes have approved drugs in the DrugBank database. All 41 targets (inclusive of these 9 targets with

Table 2.5 Choke point enzymes in pathways unique to *Pseudomonas aeruginosa*

EC #	Pathways and their enzymes	Gene Id	Load value (in)	Load value (out)	*k*-Shortest paths (in)	*k*-shortest paths (out)
	Polyketide sugar unit biosynthesis					
5.1.3.13	dTDP-4-dehydrorhamnose 3,5 epimerease	PA5164	1.11	1.11	4463	4463
1.1.1.133	dTDP-4-dehydrorhamnose reductase	PA5162	0.063	0.75	2086	2086
	Biosynthesis of siderophore group nonribosomal peptides					
5.4.4.2	Isochorismate synthase	PA423	–0.399	0.005	984	984
	Toluene and xylene degradation					
1.13.11.1	Catechol 1,2-dioxygenase	PA2579	–0.75	–0.75	229	229
	1,2 Dichloroethane degradation					
1.1.99.8	Quinoprotein alcohol dehydrogenase	PA1982	–0.52	–0.29	1450	1450
1.2.1.3	Probable aldehyde dehydrogenase	PA0219	–0.25	–0.02	11410	11410
	Phosphotransfease system (PTS)					
2.7.1.69	Phosphotransferase system, fructose-specific IIBC component	PA3560	0.53	0.53	2509	2509
	D-Alanine metabolism					
6.3.2.4	D-alanine-D-alanine ligase A	PA4201	–0.27	1.32	3699	3699
5.1.1.1	Biosynthetic alanine racemase	PA4930	1.39	0.29	3934	3934
	Lipopolysaccharide Biosynthesis					
2.7.7.38	3-deoxy-manno-octulosonate cytidylyltransferase	PA2979	1.14	2.24	4611	4611

(Continued)

Table 2.5 Continued

EC #	Pathways and their enzymes	Gene Id	Load value (in)	Load value (out)	*k*-Shortest paths (in)	*k*-shortest paths (out)
3.1.3.45	Putative 3-deoxy-D-manno-octulosonate 8-phosphate phosphatase	PA4458	1.78	1.78	2909	2909
2.7.1.130	Tetraacyldisaccharide 4'-kinase	PA2981	–0.86	–0.86	414	414
2.4.1.182	Lipid A-disaccharide synthase	PA3643	–1.02	–1.01	883	883
2.7.-.-	Lipopolysaccharide core biosynthesis protein WaaP	PA5009	–1.58	–2.51	4496	4496
2.3.1.-	Poly(3-hydroxyalkanoic acid) synthase 1	PA0011	0.01	0.52	2478	2478
2.4.1.-	UDP-glucose:(heptosyl) LPS alpha 1,3-glucosyltransferase *WaaG*	PA5010	1.09	0.4	7310	7310
3.6.1.-	UDP-2,3-diacylglucosamine hydrolase	PA1792	–1.72	–0.54	2268	2268
3.5.1.-	UDP-3-O-acyl-N-acetylglucosamine deacetylase	PA4406	1.67	–1.15	5238	5238
2.3.1.129	UDP-N-acetylglucosamine acyltransferase	PA3644	0.014	1.4	1986	1986
5.1.3.20	ADP-L-glycero-D-mannoheptose 6-epimerase	PA3337	0.424	0.424	748	748
2.5.1.55 6.3.1.2 4.2.1.20 3.1.3.1 1.7.99.4	2-dehydro-3-deoxyphosphooctonate aldolase (KDO 8-P synthase) Two component system Glutamine synthetase Tryptophan synthase alpha chain Alkaline phosphatase Respiratory nitrate reductase alpha chain	PA3636 PA0296 PA0035 PA3296 PA1174	0.19 –1.22 1.16 1.35 –4.8	1.8 –1.53 0.82 1.35 –5.49	2978 4328 18904 18944	2978 4328 18904 18944

Table 2.6 List of 41 targets compared with DrugBank and PDB databases

S.NO	Protein or Gene Product	EC numbers	Drug Bank ID	Approved Drug Targets based on DrugBank database	PDB Structure Hits (>70% sequence identity)	PDB ID
1	AccA	EC-6.4.1.2	DB00121; DB00161	Biotin, L-Valine	6	1VRG,1XOU,1XNV, 2BZR,2F9I,2 F9Y
2	AccB	EC-6.4.1.2	DB00121	Biotin	6	1 BDO,1 DD2,2ENB,2JKU, 2QF7,3 BG5
3	AccC	EC-6.4.1.2	DB00121; DB00173; DB00128; DB00130	Biotin,Adenine, L- Aspartic Acid, L- Glutamine	14	1A9X,1 GSO, 1 UC8,1 ULZ, 1W96,2 CQY,2DZD,2IP4, 2QF7,2UPQ,2W 7O,2YW2,3BG5, 3GLK
4	AccD	EC-6.4.1.2	DB00121; DB00161	Biotin,L-Valine	7	1ON3,1VRG,1XOU, 1XNV,2BZR, 2F9I,2F9Y
5	AceK	EC-2.7.11.5	NIL	NIL	NIL	NIL
6	AroA	EC-1.3.1.12	NIL	NIL	NIL	NIL
7	AroB	EC-4.2.3.4	NIL	NIL	8	1 DQS,1 UJN,1XAH,2GRU, 3CLH
8	AroC	EC-4.2.3.5	NIL	NIL	6	1QIL,1QXO,1R53, 1SQ1,1UMO,2 QHF

(Continued)

Table 2.6 Continued

S.NO	Protein or Gene Product	EC numbers	Drug Bank ID	Approved Drug Targets based on DrugBank database	PDB Structure Hits (>70% sequence identity)	PDB ID
9	AroE	EC-1.1.1.25	NIL	NIL	11	INUT,INYT,1P77, 1VI2,2D5C,2EG G,2HK9,2NLO,2O7S, 3DON
10	BacA	EC-3.6.1.27	NIL	NIL	NIL	NIL
11	DapB	EC-1.3.1.26	NIL	NIL	4	1DRW,1VM6,1YL7, 3IJP
12	DapD	EC-2.3.1.117	NIL	NIL	2	2RIJ,3FSY
13	DapF	EC-5.1.1.7	NIL	NIL	5	2GKE,2OTN,3EDN, 3EJX,3FVE
14	FolB	EC-4.1.2.25	NIL	NIL	1	2O9O
15	FolP	EC-2.5.1.15	DB00263; DB00576;DB01298; DB00250;DB00891; DB00664	Sulfisoxazole, Sulfamethizole, Sulfacytine, Dapsone, Sulfapyridine, Sulfametop yrazine	8	1AD1,1AJ2,1EYE, 1TX2,2BMB,2 DQW,2VEF,2VP8
16	HemA	EC-1.2.1.70	NIL	NIL	1	1GPJ
17	KdsA	EC-2.5.1.55	NIL	NIL	5	1VR6,1ZCO,3EOI, 3E9A,3FS2
18	KdsB	EC-2.7.7.38	NIL	NIL	3	1H7E,1VH1,1VIC

19	LpxA	EC-2.3.1.129	NIL	NIL	5	1J2Z,1XHD, 2IU8,2JF2,3CJ8,SFS 8 NIL
20	LpxB	EC-2.4.1.182	NIL	NIL	NIL	
21	LpxC	EC-3.5.1.-	NIL	NIL	2	2GO3,2VES
22	LpxK	EC-2.7.1.130	NIL	NIL	NIL	NIL
23	LysC	EC-2.7.2.4	NIL	NIL	10	1YBD,2AIF,2BRX, 2CDQ,2DT9,2 DTJ,2JOW,2REI, 3CIM,3EK6
24	MurA	EC-2.5.1.7	DB00828	Fosfomycin	3	1EJD,1G6S,2YVW
25	MurB	EC-1.1.1.158	NIL	NIL	3	1HSK,1UXY,2GQT
26	MurC	EC-6.3.2.8	NIL	NIL	8	1GG4,1J6U,1P3D,2AMI,2FOO,2 JFG,3EAG,3HN7
27	MurD	EC-6.3.2.9	NIL	NIL	1	2JFG
28	MurE	EC-6.3.2.13	NIL	NIL	1	1E8C
29	MurG	EC-2.4.1.227	NIL	NIL	1	1FOK
30	PanB	EC-2.1.2.11	NIL	NIL	4	1M3U,1O66,1OYO,3EZ4
31	PanC	EC-6.3.2.1	NIL	NIL	4	1IHO,1V8F,2EJC,3COV

(Continued)

Table 2.6 Continued

S.NO	Protein or Gene Product	EC numbers	Drug Bank ID	Approved Drug Targets based on DrugBank database	PDB Structure Hits (>70% sequence identity)	PDB ID
32	PanE	EC-1.1.1.169	NIL	NIL	4	1KS9,2EW2,3EGO,3HWR
33	PhoA	EC-3.1.3.1	DB00848; DB01143	Levamisole, Amifostine	5	1K7H,1ZED,2IUC,3BDG,3E2D
34	PssA	EC-2.7.8.8	NIL	NIL	NIL	NIL
35	RibB	EC-3.5.4.25	NIL	NIL	5	1G57,1K4J,1SNN,ITKS,2BZ1
36	RibC	EC-2.5.1.9	DB00140	Riboflavin	4	1I8D,1KZL,1PKU,3DDY
37	RibD	EC-3.5.4.26	NIL	NIL	6	1Z3A,2AZN,2B3Z,2G6V,2HXV,2NX8
38	RibH	EC-2.5.1.9	DB00140	Riboflavin	10	1C2Y,1C41,1DIO,1EJB,1KZ1,1NQU,1RVV,2C92,2FS9,2OBX
39	RmlC	EC-5.1.3.13	NIL	NIL	10	1DZR,1EPO,1NXM,1OI6,1UPI,1WLT,2B9U,2COZ,2IXC,2IXK
40	WaaG	EC-2.4.1.-	NIL	NIL	4	2GEK,2IW1,2JJM,3C48
41	WaaP	EC-2.7.-	NIL	NIL	NIL	NIL

approved drugs) are non-homologous to human protein sequences and can be considered as novel potential drug targets. One of the drugs fosfomycin observed in this method has been used as a broad spectrum antibiotic. If a broad spectrum antibiotic is sought, targets can be weighted heavily for having close homolog conserved across a range of pathogens. This analysis would provide significant progress in identifying many such broad spectrum novel antibacterial targets.

The 41 target enzymes obtained from the FBA and homology analysis were subjected to search against the DrugBank database to identify any approved drugs. Sequence search was carried out and the drugs were identified. Out of the 41 target enzymes 9 enzymes had approved drugs from the DrugBank database. The number of PDB structure hits for the targets are also shown in the table (Table 2.6).

2.2.20 Mutagenesis

Two genes *kdsA* and *waaG* encoding 2-dehydro-3-deoxyphosphooctonate aldolase and UDP-glucose (heptosyl) LPS alpha-1, 3-glucosyltransferase determined by differential genome analyses are responsible for lipid A biosynthesis and hence have been assessed as essential genes responsible for the growth and survival of the pathogen *Pseudomonas aeruginosa.* The essentiality of *kdsA* and *waaG* genes was verified by targeted gene disruption experiments. Insertion duplication approach was used to demonstrate that both *kdsA* and *waaG* genes required for peptidoglycan synthesis are essential for growth and survival in *Pseudomonas aeruginosa.* To construct the mutant, the essential genes were cloned into a suicide vector and the recombinant plasmid was integrated to recombine with the wild type *Pseudomonas aeruginosa* by site directed mutagenesis. Selectable antibiotic resistance markers introduced through suicide vector confirmed whether the genes have survived the disruption suggesting non-essentiality with colony formation in the selective media or disruption resulting in lack of recovery of antibiotic resistant colonies, thereby suggesting the essentiality of these genes for survival of *Pseudomonas aeruginosa.* Through the experimental validation it was observed that both the *kdsA* and *waaG* genes form no recombinants thus confirming their essentiality for growth and survival of *Pseudomonas aeruginosa* (50).

2.3 Conclusions

Beyond applications for general target discovery, validation and assay development, genomics will also improve the clinical use of narrow spectrum antibiotics. Genomic-based analytical tools will be required for rapid early diagnosis of the pathological agent. While the task ahead in developing novel antibacterial drugs is daunting, the situation with the continued growth of antibiotic resistant pathogens makes it imperative that the search for novel antimicrobials continues. It is clear, that there is large and growing role for many new technologies in this endeavor, most notably microbial genomics. Only with the passage of time, we will we be able to see the way genomics revolutionized drug discovery.

References

[1] Fournier PE, Drancourt M, Colson P, Rolain JM, La Scola B, Raoult D. Modern clinical_microbiology: new challenges and solutions. Nat Rev Microbiol. 2013 Aug;11(8): 574–85.

[2] Lewis K. Platforms for antibiotic discovery. Nat Rev Drug Discov. 2013 May;12(5): 371–87

[3] Fraser-Liggett CM. Insights on biology and evolution from microbial genome sequencing. Genome Res. 2005 Dec;15(12): 1603–10.

[4] J. Augen. The evolving role of information technology in the drug discovery process. Drug Discovery Today, 7: 315–323, 2002.

[5] Roemer T, Boone C. Systems-level antimicrobial drug and drug synergy discovery. Nat Chem Biol. 2013 Apr;9(4): 222–31

[6] Gupta R, Michalski MH, Rijsberman FR. Can an infectious disease genomics project predict and prevent the next pandemic? PLoS Biol. 2009 Oct;7(10): e1000219.

[7] Sakharkar KR, Sakharkar MK, Chow VT. A novel genomics approach for the identification of drug targets in pathogens, with special reference to Pseudomonas aeruginosa. In Silico Biol. 2004;4(3): 355–60.

[8] Sakharkar KR, Sakharkar MK, Chow VT. Biocomputational strategies for microbial drug target identification. Methods Mol Med. 2008;142: 1–9.

[9] Dammel CS, Noller HF. 1995. Suppression of a cold-sensitive mutation in 16S rRNA by overexpression of a novel ribosome-binding factor, RbfA. Genes Dev. 9: 626–637.

[10] Joyce AR, et al. 2006. Experimental and computational assessment of conditionally essential genes in Escherichia coli. J. Bacteriol. 188: 8259–8271.

[11] Zalacain M, et al. 2003. A global approach to identify novel broadspectrum antibacterial targets among proteins of unknown function. J. Mol. Microbiol. Biotechnol. 6: 109–126.

[12] Rosamond J, Allsop A. Harnessing the power of the genome in the search for new antibiotics. Science. 2000 Mar 17;287(5460): 1973–6.

[13] Jordan IK, Rogozin IB, Wolf YI, Koonin EV. Essential genes are more evolutionarily conserved than are nonessential genes in bacteria. Genome Res. 2002 Jun;12(6): 962–8.

[14] Uversky VN. Intrinsically disordered proteins and novel strategies for drug discovery. Expert Opin Drug Discov. 2012 Jun;7(6): 475–88.

[15] Mészáros B, Tóth J, Vértessy BG, Dosztányi Z, Simon I. Proteins with complex architecture as potential targets for drug design: a case study of Mycobacterium tuberculosis. PLoS Comput Biol. 2011 Jul;7(7): e1002118.

[16] Rahman SA, Schomburg D. Observing local and global properties of metabolic pathways: 'Load points' and 'choke points' in the metabolic networks. Bioinformatics. 2006;22: 1767–1774.

[17] Jeong H, Tombor B, Albert R, Oltval ZN, Barabasi AL. The large-scale organization of metabolic networks. Nature. 2000;407: 651–654.

[18] Yeh I, Hanekamp T, Tsoka S, Karp PD, Altman RB. Computational analysis of Plasmodium falciparum metabolism: Organizing genomic information to facilitate drug discovery. Genome Research. 2004;14: 917–924.

[19] Oberhardt MA, Puchalka J, Fryer KE, Martins dos Santos VA, Papin JA (2008) Genome-scale metabolic network analysis of the opportunistic pathogen Pseudomonas aeruginosa PAO1. J Bacteriol 190: 2790–2803. doi: 10.1128/JB.01583-07.

[20] Jain R, Rivera MC, Lake JA (1999) Horizontal transfer among genomes: the complexity hypothesis. Proc Natl Acad Sci USA. 181, 3801–3806.

[21] Zheng LL, Li YX, Ding J, Guo XK, Feng KY, Wang YJ, Hu LL, Cai YD, Hao P, Chou KC. A comparison of computational methods for identifying virulence factors. PLoS One. 2012;7(8): e42517.

[22] Zhang R, Lin Y. DEG 5.0, a database of essential genes in both prokaryotes and eukaryotes. Nucleic Acids Res. 2009 Jan;37(Database issue): D455–8.

[23] Normark, S., Bergstrom, S., Edlund, T., Grundstrom, T., Jaurin, B., Lindberg, F. P. & Olsson, O. (1983). Overlapping genes. Annu Rev Genet17, 499–525.

[24] Rogozin, I. B., Spiridonov, A. N., Sorokin, A. V., Wolf, Y. I., Jordan, I. K., Tatusov, R. L. & Koonin, E. V. (2002). Purifying and directional selection in overlapping prokaryotic genes. Trends Genet 18, 228–232.

[25] Yelin, R., Dahary, D., Sorek, R. & 13 other authors (2003).Widespread occurrence of antisense transcription in the human genome. Nat Biotechnol 21, 379–386.

[26] Miyata, T. & Yasunaga, T. (1978). Evolution of overlapping genes. Nature 272, 532–535.

[27] Fukuda, Y., Washio, T. & Tomita, M. (1999). Comparative study of overlapping genes in the genomes of Mycoplasma genitalium and Mycoplasma pneumoniae. Nucleic Acids Res 27, 1847–1853.

[28] Burge, C. B. & Karlin, S. (1998). Finding the genes in genomic DNA. Curr Opin Struct Biol 8, 346–354.

[29] Yourno J.D. 1972. Gene fusion. Brookhaven Symp. Biol. 23: 95–120.

[30] Isono K. and Yourno J. 1973. Mutation leading to gene fusion in the histidine operon of Salmonella typhimurium. J. Mol. Biol. 76: 455–461.

[31] Sali A. 1999. Functional links between proteins. Nature 402: 23–26

[32] Galperin M.Y. and Koonin E.V. 2000. Who's your neighbor? New computational approaches for functional genomics. Nat. Biotechnol. 18: 609–613.

[33] Tsoka S. and Ouzounis C.A. 2001. Functional versatility and molecular diversity of the metabolic map of Escherichia coli. Genome Res. 11: 1503–1510.

[34] Yanai I., Derti A. and DeLisi C. 2001. Genes linked by fusion events are generally of the same functional category: A systematic analysis of 30 microbial genomes. Proc. Natl. Acad. Sci. USA 98: 7940–7945.

[35] Long M. 2000. A new function evolved from gene fusion. Genome Res. 10: 1655–1657.

[36] Katzen F.M., Deshmukh F.D., Daldal F. and Beckwith J. 2002. Evolutionary domain fusion expanded the substrate specificity of the transmembrane electron transporter DsbD. EMBO J. 21: 3960–3969.

[37] Berthonneau E. and Mirande M. 2000. A gene fusion event in the evolution of aminoacyl-tRNA synthetases. FEBS Lett. 470: 300–304.

[38] Ma, H. & Zeng, A. P. 2003. Reconstruction of metabolic networks from genome data and analysis of their global structure for various organisms. Bioinformatics, 19, 270–277.

[39] Oltvai, Z. N. & Barabasi, A. L. 2002. Systems biology: Life's complexity pyramid. Science, 298, 763–764.

[40] Schuster, S., Fell, D. A. & Dandekar, T. 2000. A general definition of metabolic pathways useful for systematic organization and analysis of complex metabolic networks. Nature Biotechnology, 18, 326–332.

[41] Palumbo, M. C., Colosimo, A., Giuliani, A. & Farina, L. 2005. Functional essentiality from topology features in metabolic networks: A case study in yeast. FEBS Letters, 579, 4642–4646.

[42] Galperin M.Y. and Koonin E.V. 1999. Searching for drug targets in microbial genomes. Curr. Opin. Biotechnol. 10: 571–578.

[43] Zarembinski TI, Hung LIW, Mueller-Dieckmann HJ, Kim KK, Yokota M, Kim R and Kim SH. Structure-based assignment of the biochemical function of a hypothetical protein: A test case of structural genomics. Proc. Natl. Acad. Sci. USA 95: 15189–15193, 1998.

[44] Grigoriev IV and Kim SH. Detection of protein fold similarity based on correlation of amino acid properties. Proc. Natl. Acad. Sci. USA 96: 14318–14323, 1999.

[45] Schneider G and Fechner U. Computer-based de novo design of drug like molecules. Nature Rev. Drug Discov. 4: 649–663, 2005.

[46] Schmid MB. Crystallizing new approaches for antimicrobial drug discovery. Biochem. Pharmacol. 71: 1048–1056, 2006.

[47] Stover, C. K., Pham, X. Q., Erwin, A. L., Mizoguchi, S. D., Warrener, P., Hickey, M. J., Brinkman, F. S. L., Hufnagle, W. O., Kowallk, D. J., Lagrou, M., Garber, R. L., Goltry, L., Tolentino, E., Westbrock-Wadman, S., Yuan, Y., Brody, L. L., Coulter, S. N., Folger, K. R., Kas, A., Larbig, K., Lim, R., Smith, K., Spencer, D., Wong, G. K. S., Wu, Z., Paulsen, I. T., Relzer, J., Saler, M. H., Hancock, R. E. W., Lory, S. & Olson, M. V. 2000. Complete genome sequence of Pseudomonas aeruginosa PAO1, an opportunistic pathogen. Nature, 406, 959–964.

[48] Hutchison, M. L. & Govan, J. R. W. 1999. Pathogenicity of microbes associated with cystic fibrosis. Microbes and Infection, 1, 1005–1014.

[49] Perumal D, Samal A, Sakharkar KR, Sakharkar MK. Targeting multiple targets in Pseudomonas aeruginosa PAO1 using flux balance analysis of a reconstructed genome-scale metabolic network. J Drug Target. 2011 Jan; 19(1): 1–13.

[50] Perumal D, Sakharkar KR, Tang TH, Chow VT, Lim CS, Samal A, Sugiura N, Sakharkar MK. Cloning and targeted disruption of two lipopolysaccharide biosynthesis genes, kdsA and waaG, of Pseudomonas aeruginosa PAO1 by site-directed mutagenesis. J Mol Microbiol Biotechnol. 2010;19(4): 169–79.

3

A Prioritization Analysis of Disease Association by Data-mining of Functional Annotation of Human Genes

Nicki Tiffin and Junaid Gamieldien

South African National Bioinformatics Institute/South African Medical Research Council Bioinformatics Unit, University of the Western Cape

3.1 Introduction

3.1.1 Genetics underlying disease

Identifying disease-associated genes is at the hub of understanding genetic contributors to disease phenotypes; and identifying disease-causing genes can improve early disease detection and diagnosis as well as offering indicators of prognosis. Importantly, identifying disease-causing genes can also offer new therapeutic avenues to explore, by exposing fundamental biological mechanisms that are dysregulated in the disease state. Genetic factors that underlie a disease can range from single gene (Mendelian) disorders, whereby a genetic mutation in a single gene can result in a gross phenotypic effect – for example, mutations of the cystic fibrosis conductance regulator gene (CFTR) disrupt protein structure and cause cystic fibrosis; to complex, multigenic diseases in which many different genes have variations that each contribute a small part to the disease phenotype [1–3].

Historically, research into the genetic causes of disease used a top-down, *hypothesis-testing* approach, where clinicians and researchers began with a detailed characterization of the disease phenotype and then generated and experimentally tested a hypothesis about possible mechanisms that could be causing it. The genomics revolution, however, has resulted in many more *hypothesis-generating* studies that do not rely on any prior knowledge about disease mechanisms, in which large sample sets of DNA from patients are

Post-genomic Approaches in Drug and Vaccine Development, 53–70.

compared to DNA from controls without the disease, and the difference in genetics between the two groups is determined on a global scale. By characterizing genetic differences between patient and controls in this way, new hypotheses for disease mechanisms can be generated from these studies based on the genome-wide data, and validation studies can then be designed to test the empirically-generated hypotheses [4].

3.1.2 The era of genomics

Genomic technologies in common use for disease gene identification include: genome-wide association studies (GWAS) and copy number variation analysis using microarray genotyping; expression microarrays; and whole genome and exome sequencing (WGES) by next generation sequencing (NGS) approaches to identify DNA variants that may be contributing to disease. It has been estimated that on average an individual carries approximately 3.4 million single nucleotide polymorphisms (SNPs), 344 000 indels and 717 large deletions in their genomes [5], as well as ~74 *de novo* single nucleotide variants per generation [6]. A large number of these mutations are novel; and understanding the functionality of the implicated genes is crucial in effective filtering of such data.

3.1.3 Genes and their function

Investigations into the effects of DNA variation to date have focused on the protein-coding regions of the genome – the exome – with major efforts to define and annotate all human genes within the exome portion of the genome, and to characterize their protein products. With increasingly refined analyses and the use of transcriptome data as a reference, the human exome has been estimated at approximately 21 000 protein-coding genes covering some 30 Mb. This constitutes 1.22% of the complete human genome [7]. The ensuing challenge is to understand the roles of all these genes, in order to understand the biological functioning of the cellular machinery; and furthermore, to collate all this functional information into data repositories in a way that can be effectively data-mined in an automated way. The information that is assigned specifically to each gene is referred to as the gene *annotation*; and the *functional annotation* is the information for each gene pertaining to the biological functions of the gene product. These annotations are selected from a standardized set of pre-defined function descriptions – called an *ontology*, or *controlled vocabulary* [8], [9], so that genes within the database that have the same biological function can be recognized as such through automated searches. An example is the Gene Ontology (GO), a well-developed ontology for the functional annotation of genes that is in common use in the biomedical sciences [10].

Gene functional annotations may be determined in multiple ways: Experimental data about gene function are generated in laboratories, and then the gene record is manually annotated with the new functional data, by an expert 'bio-curator' – the data is said to be 'manually curated'. Given the rapid advances in genomics technology, however, manual curation can no longer keep up with the enormous amount of data being generated by genomic experiments, and automated functional annotation methods have been developed. These rely on sequence, structure, phylogenetic or co-expression relationships between known and novel sequences [11], [12]. Data-mining of functional annotation of genes facilitates the grouping of genes according to their actions in the cell, and the recognition of genes that may have similar or redundant actions. Functional annotation can also identify the regulatory networks and pathways in which the gene participates, allowing for a functional context for genes within physiological processes of the cell. Within the arena of disease gene discovery, understanding the functions of a known disease gene can be harnessed to identify further candidates that might be participating in the same cell processes – and may therefore contribute to the dysregulation seen in the phenotype of that disease. This "guilt by association" approach is harnessed in many computational data-mining methods used to identify strong genetic candidates for disease [13]

3.1.4 Disease gene prioritization

Disease gene identification can contribute significantly to improved disease outcomes, through providing biological insights into the disease mechanism, new biomarkers for diagnosis, and new targets for therapeutic interventions. Understanding a disease genotype can provide the connections between disease phenotype and the underlying physiological mechanisms that are causing the symptoms – providing avenues to explore ways to prevent or ameliorate these symptoms. In general, genomic experiments are more likely to generate groups of genes that are potentially implicated in disease, rather than single genes. From GWAS experiments, for example, any gene in an entire chromosomal haploblock containing the disease-associated marker has the equal potential to be implicated in the disease. This could be a list of several hundred genes [14]; and the challenge lies in ranking these genes from most to least likely to be the true disease-associated gene, so that empirical analysis can start with the best candidate genes with the highest likelihood of underlying the disease phenotype [14], [15]. As information about human genes continues to rapidly expand, it has become increasingly unfeasible to investigate all the information available about a set of candidate genes in order to predict which might be more suitable disease-causing candidates. For this reason, many bioinformatics approaches have been developed to automate data-mining of all the data available for a list of genes (reviewed in [14–17]). These approaches use an array of gene properties, including: intrinsic properties such as gene length and sequence composition; tissue- and cell-type specific expression profiles of genes; membership in gene pathways and protein families containing known disease genes; sequence conservation across species, and gene functional annotation. In Section 1.6.1, we present the BioOntological Relational Graph semantic database as a prototype system for effective data-mining of gene annotation data.

3.1.5 Beyond genes – functional annotation of the entire genome

Understanding the genetic variants causing human disease requires a complete understanding of the normal functioning of human DNA. Only a tiny proportion of human DNA function had been elucidated, and was restricted to protein-coding regions of the genome until September 2012, when the Encyclopedia of DNA Elements (ENCODE) Project released a large novel annotation dataset, assigning function to 99.61% of human DNA and showing that 80.4% of the genome is involved in a biochemical and/or chromatin-associated process [7]. These analyses give insight into the regions of the genome with regulatory roles in controlling the spatiotemporal synthesis of proteins. These DNA regions can regulate transcribed non-protein coding RNA molecules (including microRNA), transcription factor-binding sites, chromatin structure and methylation sites. Understanding the function of DNA variants from the entire genome will have a significant impact on understanding the global regulatory framework within human cells, leading to better understanding of gene regulation, human disease genetics and pharmacogenetics.

3.2 Definition of a Gene

Since the identification of the first disease-causing genes, our understanding and conceptualization of what constitutes a gene has changed considerably; and this has broadened the range of annotations and functions that can underlie disease phenotype. The complete sequencing of the human genome, completed in 2001 [18], was a crucial first step towards unraveling how the genetic code underlies human biology. This opened many possibilities for identifying genetic variants that cause disease or underlie disease susceptibility, although the genes defined at this stage of the human genome projected were almost exclusively protein-coding genes that were transcribed into messenger RNA (mRNA) and then translated into an expressed protein [19]. The structure of these genes generally included exons – the protein-coding portions of the gene; introns lying between the exons – including non-coding regulatory regions, enhancers and splice sites; untranslated mRNA regions at both the 3' and 5' ends of the gene, containing regulatory motifs; and a proximal promoter region lying upstream of the coding sequence and containing binding sites for protein complexing by the transcriptional machinery of the cell [20]. With further study of the human genome, however, several other types of untranslated genes have been identified, and these have significant roles in the regulation of cell physiology. These include: (i) Pseudogenes, which are paralogues of protein-coding genes but have no protein products. Pseudogenes may have unrecognized function, and some have been implicated in disease (e.g. [21]); and the definition of what constitutes a pseudogene is inconsistent [22], [23]; (ii) Regulatory RNA molecules, including microRNA molecules that are short RNA molecules that complement mRNA sequences and regulate their degradation [24], [25], and long non-coding RNA molecules that appear to be cis-regulatory elements regulating DNA transcription [26], [27]. Regulatory RNA molecules are increasingly implicated in gene dysregulation underlying disease. Finally, the ENCODE data provides us with new concepts of what constitutes a gene – and this will inform future approaches to annotating functional elements of the human genome [28].

3.3 Data-Mining Functional Annotation

Functional annotation of human genes can include several types of annotation. The Gene Ontology (GO) is a well-established ontology that defines the function of individual gene products (proteins) using consistent, standardized descriptions that fall into the three categories of: *cellular component*, or the location within the cell; *molecular function*, the intrinsic molecular activities of the gene product; and *biological process* which defines the sets of molecular reactions in which the protein participates [10], www.geneontology.org. GO annotations can be retrieved for individual genes from many of the public genomic databases including Ensembl (www.ensembl.org, [29]); the UCSC

Genome Web browser (http://genome.ucsc.edu/, [30]) and the NIH Gene Database (http://www.ncbi.nlm.nih.gov/gene). Each gene may have multiple GO annotations associated with it in the different categories, reflecting the multiple functions that a single gene product may fulfill. Many tools have been developed for the automated mining of GO annotations, for example AmiGO is a web-based browser developed by the Gene Ontology Consortium, and can identify either the GO terms associated with protein products, or conversely the protein products associated with a particular GO term [31]. Another online tool is DAVID, developed by the NIH, USA (http://david.abcc.ncifcrf.gov/, [32]), which measures enrichment for particular GO annotation in a submitted list of genes. This enrichment function can determine which functions are over-represented in a set of genes, for example all differentially regulated genes found by transcriptional microarray analysis, and can give insights into which overall physiological cell functions are perturbed by the experimental conditions under investigation. Other similar tools exist to measure GO term enrichment (e.g. GOStat – http://gostat.wehi.edu.au/, [33]).

Further types of functional annotations of genes that can be used to gain understanding of gene function include information about gene regulatory networks in which a protein might be involved; and several pathway databases contain this type of information, including the KEGG database [34] and the REACTOME database [35]. This kind of regulatory pathway and network information can inform candidate disease gene selection in several ways [36]. One of these is the selection of candidates from physiological pathways that are likely to underlie the disease phenotype or the identification of candidates that participate in the same pathways as known disease-causing genes in a "guilt by association" model of disease gene identification (for example, in [37]). When genomic experiments generate a large list of genes, enrichment of pathway membership by these genes can inform understanding of dysregulated physiological processes that may underlie disease [38]. Finally, information on protein–protein interactions, which form the basis of many gene regulatory networks, can be directly mined to identify likely disease candidates based on their involvement in cellular processes involving already-identified disease genes [37].

3.4 Experimental Sources of Functional Annotation

The Gene Ontology project is the most comprehensive source of high quality functional annotations. A team of expert curators use experimental results published in the biomedical literature to associate a specific GO term(s) to a gene product. Each annotation includes an evidence code to indicate what type of experiment provided the evidence by which the association was made, as well as a reference identifier for the scientific publication from which the evidence was obtained [10], (http://www.geneontology.org/GO.evidence .shtml). Currently, GO annotation codes for experimental evidence reflect associations that were inferred via direct assays, physical interactions, mutant phenotypes, genetic interactions or expression patterns. Each of the aforementioned categories encompasses evidence from multiple experimental technologies, which enable making associations to different parts of the Gene Ontology. For example, enzyme assays and immunofluorescence are both direct assays, but the former provides evidence on molecular function and the latter on cellular component. While manually curated annotations should be considered to be of highest quality, associations mined from the literature using automated methods are still a valuable source of experiment–based functional annotations [39]. We recommend that candidates generated using the latter be assessed on a case-by-case basis rather than discarded.

Model organism knockout experiments are another valuable source of functional annotations for implicating a gene in the development of specific phenotypes. The Mouse Genome Informatics Database [40] (MGI, http://www.informatics.jax.org/) and Rat Genome Database [41]. (RGD, http://rgd.mcw.edu) are exemplars in this regard, providing gene annotations that reflect the observable morphological, physiological and behavioral characteristics that arise in gene knockout models over the animal's lifespan. In both the databases, the Mammalian Phenotype Ontology [42] is employed as a rich source of community accepted annotation terms. We describe in 1.6 how such experimental annotations may assist in prioritizing candidate genetic variants that may be associated with a disease of interest.

3.5 Inferred Functional Annotation Through Sequence Similarity

It has become a common practice to use tools such as BLAST2GO [11] to assign functional annotations to a novel gene or protein based on its similarity to existing functionally annotated sequences. However, several considerations such as alignment percentage similarity, coverage and whether the matching sequence is an ortholog or paralog, etc. are all factors that should be used to weigh such functional assignments. Similarly, even when working with human genes or those from organisms on the Gene Ontology Consortium's priority list, care should be exercised when making hypotheses based on functional annotations derived from sequence similarity. In fact, the GO project's practices for annotation based on sequence (http://www.geneontology.org/GO.evidence.shtml#iss) can aid the decision making process. For example, when a gene can be shown to be a functional ortholog of an existing characterized gene using phylogenetic methods [43], the derived functional annotations can be more readily trusted than when using simple pairwise or multiple sequence alignment. It is also very important to bear in mind that even when two proteins are highly similar, subtle signatures or domains that determine specific functions may have been lost or acquired in the sequence of interest. This illustrates the possibility of producing an incorrect hypothesis when only alignment quality is considered and the need for manual inspection of the evidence when inferring functional annotations based on sequence.

3.6 High Throughput Candidate Disease Gene Prioritization Through Semantic Discovery

Tools for predicting functional variants assist in the preliminary filtering of the large number of genetic variants that are invariably identified in disease research utilizing high throughput technologies such as next generation sequencing. However, many more candidates than can be verified experimentally still remain and it is rarely easy to predict a variant's potential impact at a cellular or organismal level and still more difficult to implicate them in a disease of interest. This is true even in Mendelian disorders, but multi-genic disorders add an additional layer of complexity. In the latter, multiple functional variants are proposed to each contribute only a small percentage to the development of a disease phenotype [44]. While a fraction of those can be identified through statistical association, a large number of variants contributing to disease risk or progression cannot be detected even in large cohort studies and may represent the 'missing heredity' associated with many complex diseases [45]. A likely explanation is that rare and even private functional variants occur in various combinations in a variety of genes in different individuals [46]. For that reason, assessing candidates bearing functional variants in the context of existing biomedical knowledge and their biomolecular functions is an important step in producing a manageable set of variants for further exploration. The same is true for Mendelian disorders and even rare spontaneously-occurring monogenic diseases where a single causative variant would have to be identified from amongst thousands of candidates [47].

In all genomic studies, extant knowledge about a disease of interest and its general phenotypic presentation should be holistically considered alongside gene functional annotations as a way to identify candidates that fulfill multiple criteria and are therefore most likely to contribute to the disease of interest. Such a knowledge driven approach maximizes the chances of identifying a causative variant(s) while potentially simultaneously identifying previously undetermined biological mechanisms of the disease of interest. While this approach is sound in principle, the sheer volume of leads generated in modern–omics experiments, and the variety of independent knowledge sources to interrogate in an integrated manner makes it a very cumbersome task. We present here an example of a database of gene annotation relationships that integrates multiple sources of annotation data for the synthesis of novel disease gene predictions, through semantic discovery.

3.6.1 The BioOntological Relationship Graph (BORG) Semantic Database

BORG database seamlessly integrates hundreds of thousands of curated facts about genes and their known functions, disease and phenotype associations, and pathway membership into a large knowledge structure known as a semantic network [48]. In essence, information in the system is structured in the way a biologist would think and reason about them (Figure 3.1). This simplifies both the translation of biological questions into relevant database queries and the interpretation of the results obtained from semantic (meaning-driven) queries.

At the heart of the database are human genes, which are mapped to their known orthologs in the mouse and rat genomes via an annotated link that accurately describes the semantic relationship. Several bio-ontologies modeled in the database act as 'anchors' for integration and provide domain-specific terms for composing queries. Where appropriate, we also map between terms in different ontologies, e.g. phenotypes are linked to diseases, which enables the discovery of transitive associations between genes and disease based on a gene's known involvement in phenotypes relevant to that disease. This is primarily done on a case-by-case basis and involves medical geneticists and clinicians specializing in a specific disease of interest. Genes are similarly mapped to:

1. Gene Ontology (GO) terms based on annotations published by the GO consortium [31].
2. Disease Ontology terms based on curated associations mined from the NCBI's Gene Reference Into Function database [49].

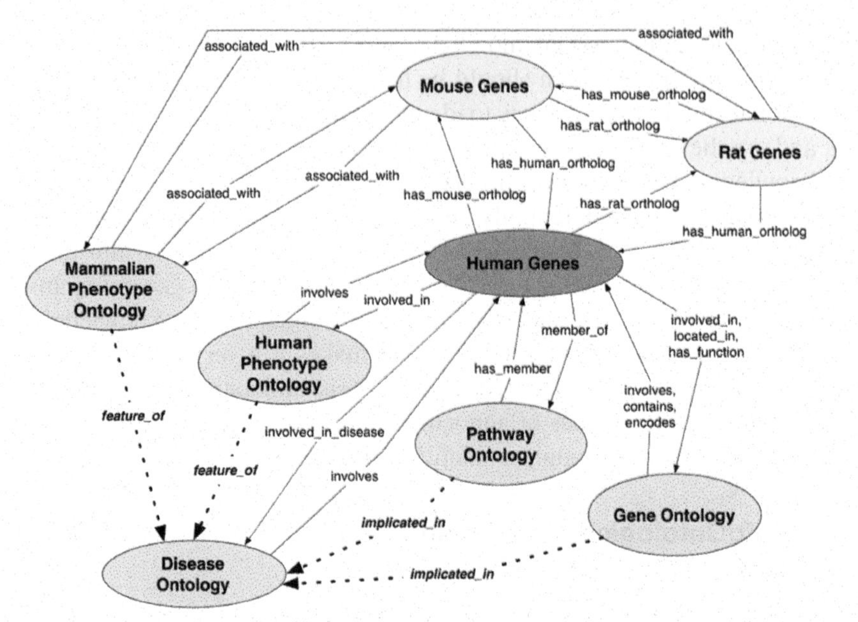

Figure 3.1 The BORG semantic database schema. Black arrows indicate semantic relations that are either derived from existing biomedical databases or are curated relationships from published text-mining projects. Red arrows indicate mapping of phenotypes, pathways and functions known or be associated with the disease of interest, resulting in a meta-ontology in which genes are transitively associated with a disease.

3. Human Phenotype Ontology [50] terms based on the phenotypes that are documented to be associated with human genes in the OMIM database.
4. Pathway Ontology [51] terms, which models gene product involvement in pathways at a conceptual rather than structural level.
5. Mammalian Phenotype Ontology [42] terms, which describe the phenotypes that arise when the gene is knocked out in mouse [40] or rat [41]. It is important to note that due to the ontological structure of the entire database model, orthologous human genes inherit these knockout phenotypes through transitive association.

3.6.2 Mining the BORG Database Through Semantic Querying

Individual facts in the BORG database are structured in the way humans naturally think about them and researchers are therefore able to ask very complex questions of the system, based on what they mean. It enables *in-silico* experimentation through complex querying, annotation retrieval and the semantic discovery of genotype-to-phenotype associations. The hierarchical structure of biological ontologies also assists in identifying transitive associations that may not always be obvious, yet may be biologically correct or plausible.

The database can be queried in three ways:

3.6.2.1 Ontology seeded queries

The BORG system has a custom natural language-like query language that enables a user to combine multiple ontology terms into a single, arbitrarily complex query spanning multiple knowledge domains if necessary. Figure 3.2 illustrates how a hypothetical user may search for human genes that are (1) not known to be involved in a disease of interest (multiple sclerosis), but (2) have been demonstrated to be involved in other diseases or phenotypes causing abnormal myelination, demyelination or nervous system degeneration, and (3) are known to play a role in functions related to normal myelination, while (4) allowing the use of biological evidence from model organisms. More information about each *in silico* generated candidate gene can easily be obtained from the BORG database by performing the query described

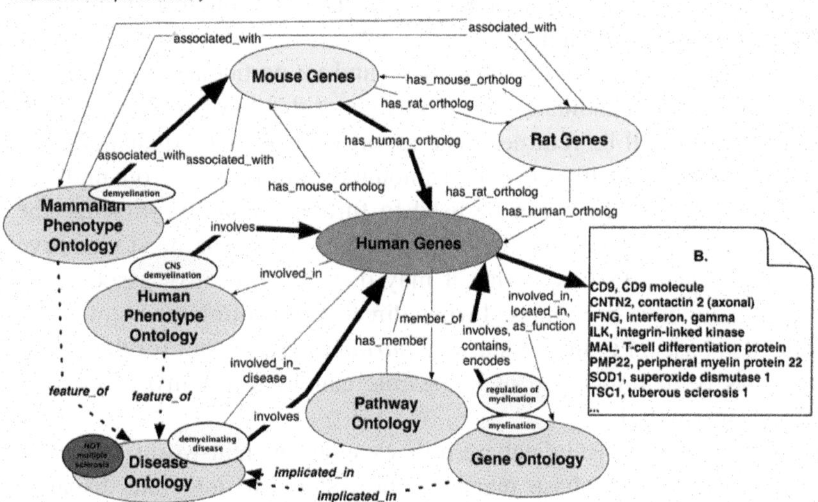

Figure 3.2 Identifying candidate genes. Genes linked to ontology seed terms or their child terms are grouped into sets and then processed according to the set theoretic operators (A), mirroring the researcher's intent and producing a final list of candidates (B).

in *1.6.2.3* below, which is particularly useful should the researcher want to manually filter the initial list.

3.6.2.2 Annotation retrieval

With this type of query, a user provides a list of genes and the BORG returns all ontology terms directly linked to it, as well as the semantic relationship between the gene and each term. When requested, the system also returns ontology terms linked to gene counterparts in model organisms, which is particularly useful when knockout phenotypes are able to shed light on a gene's potential involvement in a disease of interest.

3.6.2.3 Path-based transitive association queries

Currently, the most powerful BORG query enables researchers to discover transitive links between gene and disease. It also returns the semantic relationships between all the concepts (genes or terms) in the discovered path, thereby explaining the biological relevance of the link in an easy to understand human readable report. The main advantage of this facility is that it is able to uncover potential associations that are non-obvious, yet biologically plausible. When used for finding links between genes and disease, for example, the BORG does a directed walk on the graph to find all paths between a gene of interest and a user specified disease term. The user is able to request either the shortest or all paths between a gene of interest and the disease of interest, or all paths less than a pre-specified length.

Reports are produced on a per-gene basis and are particularly useful when filtering a large list of candidates, since only genes that have at least one path leading to disease will be returned. The report itself is self-explanatory and provides the researcher with substantial amounts of information from multiple knowledge domains, which can be used to further manually prioritize the remaining candidates based on the evidence presented. The most attractive feature of this query facility is that it may discover transitive associations that would likely have been missed when directly consulting the literature or individual databases. Figure 3.3 shows a *partial* BORG result for the *GALC* gene that has been implicated as a possible candidate for involvement in Multiple Sclerosis [52].

In summary, the BORG database utilizes superior 'real-world' modeling capabilities of graph database management systems to integrate disparate genomic and biomedical facts into a single on-disk semantic network. This simplifies the application of methods using multiple sources of functional annotations to identify candidate genes in rare diseases, rare variants

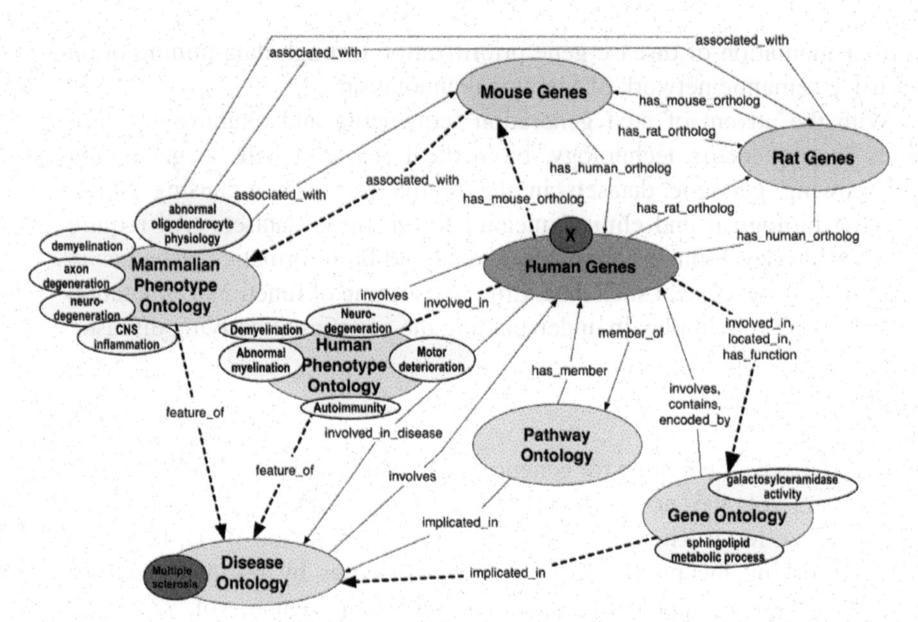

Figure 3.3 Filtering candidate genes. An example of how the BORG semantic discovery technique would use path-based searches across stored biomedical facts and its knowledge of multiple sclerosis to accumulate evidence that a gene (GALC) bearing a novel functional mutation may be involved in the disease.

of common diseases and potentially the 'missing heritability' in common diseases.

3.7 Conclusions

In this chapter, we have presented a background and set of guidelines for prioritizing disease-associated genes through data-mining of functional annotation. We have presented the concepts underlying the genetic basis of disease for Mendelian and complex genetic disorders and how the identification of these disease genes can be undertaken by harnessing the considerable existing public resources on gene function. We have introduced the concept of functional annotation, including how such annotation is derived; the annotation of genes as well as expressed RNAs; and introduced the ENCODE project in the context of annotation of the entire genome with regulatory roles that surpass the protein-coding role of the exome. We have introduced methodologies that can be used to data-mine functional annotation for disease gene prioritization, and presented the BioOntological Relationship Graph semantic database as an implementation of disease gene prioritization through data-mining of an extensive semantic network of functional annotation.

With the advent of next generation sequencing and ongoing developments in sequencing technology, biomedical research will be generating ever-growing genomic datasets in the search for disease-causing genes. Deriving biological and clinical meaning from these datasets will require considerable development of computational and bioinformatics resources in order to make best use of such data. Effective mining of functional annotation has a growing role to play in understanding the genetics underlying disease.

References

[1] N. J. Risch, "Searching for genetic determinants in the new millennium," Nature, vol. 405, no. 6788, pp. 847–856, Jun. 2000.

[2] Q. Yang, M. J. Khoury, L. Botto, J. M. Friedman, and W. D. Flanders, "Improving the prediction of complex diseases by testing for multiple disease-susceptibility genes," Am. J. Hum. Genet., vol. 72, no. 3, pp. 636–649, Mar. 2003.

[3] H. K. Tabor, N. J. Risch, and R. M. Myers, "Candidate-gene approaches for studying complex genetic traits: practical considerations," Nat. Rev. Genet., vol. 3, no. 5, pp. 391–397, May 2002.

[4] M. Dammann and F. Weber, "Personalized medicine: caught between hope, hype and the real world," Clinics (Sao Paulo), vol. 67 Suppl 1, pp. 91–97, 2012.

[5] G. R. Abecasis, A. Auton, L. D. Brooks, M. A. DePristo, R. M. Durbin, R. E. et al., "An integrated map of genetic variation from 1,092 human genomes," Nature, vol. 491, no. 7422, pp. 56–65, Nov. 2012.

[6] J. A. Veltman and H. G. Brunner, "De novo mutations in human genetic disease," Nat. Rev. Genet., vol. 13, no. 8, pp. 565–575, Aug. 2012.

[7] I. Dunham, A. Kundaje, S. F. Aldred, P. J. Collins, C. A. Davis, et al., "An integrated encyclopedia of DNA elements in the human genome," Nature, vol. 489, no. 7414, pp. 57–74, Sep. 2012.

[8] K. Liu, W. R. Hogan, and R. S. Crowley, "Natural Language Processing Methods and Systems for Biomedical Ontology Learning," J Biomed Inform, vol. 44, no. 1, pp. 163–179, Feb. 2011.

[9] C. Friedman, T. Borlawsky, L. Shagina, H. R. Xing, and Y. A. Lussier, "Bio-Ontologies and Text: Bridging the Modeling Gap Between," Bioinformatics, vol. 22, no. 19, pp. 2421–2429, Oct. 2006.

[10] M. Ashburner, C. A. Ball, J. A. Blake, D. Botstein, H. Butler, et al., "Gene ontology: tool for the unification of biology. The Gene Ontology Consortium," Nat. Genet., vol. 25, no. 1, pp. 25–29, May 2000.

[11] S. Götz, J. M. García-Gómez, J. Terol, T. D. Williams, S. H. Nagaraj, M. J. Nueda, et al. "High-throughput functional annotation and data mining with the Blast2GO suite," Nucleic Acids Res, vol. 36, no. 10, pp. 3420–3435, Jun. 2008.

[12] D. Frishman, "Protein annotation at genomic scale: the current status," Chem. Rev., vol. 107, no. 8, pp. 3448–3466, Aug. 2007.

[13] P. I. Wang and E. M. Marcotte, "It's the machine that matters: Predicting gene function and phenotype from protein networks," J Proteomics, vol. 73, no. 11, pp. 2277–2289, Oct. 2010.

[14] N. Tiffin, M. A. Andrade-Navarro, and C. Perez-Iratxeta, "Linking genes to diseases: it's all in the data," Genome Med, vol. 1, no. 8, p. 77, 2009.

[15] N. Tiffin, "Conceptual thinking for in silico prioritization of candidate disease genes," Methods Mol. Biol., vol. 760, pp. 175–187, 2011.

[16] M. Oti, S. Ballouz, and M. A. Wouters, "Web tools for the prioritization of candidate disease genes," Methods Mol. Biol., vol. 760, pp. 189–206, 2011.

[17] Y. Moreau and L.-C. Tranchevent, "Computational tools for prioritizing candidate genes: boosting disease gene discovery," Nat. Rev. Genet., vol. 13, no. 8, pp. 523–536, Aug. 2012.

[18] International Human Genome Sequencing Consortium. "Finishing the euchromatic sequence of the human genome," Nature, vol. 431, no. 7011, pp. 931–945, Oct. 2004.

[19] T. R. Gingeras, "Origin of phenotypes: genes and transcripts," Genome Res., vol. 17, no. 6, pp. 682–690, Jun. 2007.

[20] T. T. Strachan and A. P. A. P. Read, Human Molecular Genetics, 2nd ed. New York: Wiley-Liss, 1999.

[21] G. McEntee, S. Minguzzi, K. O'Brien, N. Ben Larbi, C. Loscher, et al., "The former annotated human pseudogene dihydrofolate reductase-like 1 (DHFRL1) is expressed and functional," Proc. Natl. Acad. Sci. U.S.A., vol. 108, no. 37, pp. 15157–15162, Sep. 2011.

[22] O. Svensson, L. Arvestad, and J. Lagergren, "Genome-wide survey for biologically functional pseudogenes," PLoS Comput. Biol., vol. 2, no. 5, p. e46, May 2006.

[23] K. Ohshima, M. Hattori, T. Yada, T. Gojobori, Y. Sakaki, and N. Okada, "Whole-genome screening indicates a possible burst of formation of

processed pseudogenes and Alu repeats by particular L1 subfamilies in ancestral primates," Genome Biol., vol. 4, no. 11, p. R74, 2003.

[24] J. Vera, X. Lai, U. Schmitz, and O. Wolkenhauer, "MicroRNA-Regulated Networks: The Perfect Storm for Classical Molecular Biology, the Ideal Scenario for Systems Biology," Adv. Exp. Med. Biol., vol. 774, pp. 55–76, 2013.

[25] E. S. Martens-Uzunova, M. Olvedy, and G. Jenster, "Beyond microRNA-novel RNAs derived from small non-coding RNA and their implication in cancer," Cancer Lett., Jan. 2013.

[26] J. T. Lee, "Epigenetic regulation by long noncoding RNAs," Science, vol. 338, no. 6113, pp. 1435–1439, Dec. 2012.

[27] S. Guil and M. Esteller, "Cis-acting noncoding RNAs: friends and foes," Nat. Struct. Mol. Biol., vol. 19, no. 11, pp. 1068–1075, Nov. 2012.

[28] M. B. Gerstein, C. Bruce, J. S. Rozowsky, D. Zheng, J. Du, J. O. Korbel, O. et al., "What is a gene, post-ENCODE? History and updated definition," Genome Res., vol. 17, no. 6, pp. 669–681, Jun. 2007.

[29] P. Flicek, I. Ahmed, M. R. Amode, D. Barrell, K. Beal, et al., "Ensembl 2013," Nucleic Acids Res., vol. 41, no. Database issue, pp. D48–55, Jan. 2013.

[30] D. Karolchik, A. S. Hinrichs, and W. J. Kent, "The UCSC Genome Browser," Curr Protoc Bioinformatics, vol. Chapter 1, p. Unit1.4, Dec. 2012.

[31] Gene Ontology Consortium. "The Gene Ontology in 2010: extensions and refinements," Nucleic Acids Res., vol. 38, no. Database issue, pp. D331–335, Jan. 2010.

[32] D. W. Huang, B. T. Sherman, and R. A. Lempicki, "Systematic and integrative analysis of large gene lists using DAVID bioinformatics resources," Nat Protoc, vol. 4, no. 1, pp. 44-57, 2009.

[33] T. Beißbarth and T. P. Speed, "GOstat: find statistically overrepresented Gene Ontologies within a group of genes," Bioinformatics, vol. 20, no. 9, pp. 1464–1465, Jun. 2004.

[34] M. Kanehisa, "The KEGG database," Novartis Found. Symp., vol. 247, pp. 91–101; discussion 101–103, 119–128, 244–252, 2002.

[35] R. Haw and L. Stein, "Using the reactome database," Curr Protoc Bioinformatics, vol. Chapter 8, p. Unit8.7, Jun. 2012.

[36] N. T. Doncheva, T. Kacprowski, and M. Albrecht, "Recent approaches to the prioritization of candidate disease genes," Wiley Interdiscip Rev Syst Biol Med, vol. 4, no. 5, pp. 429–442, Oct. 2012.

[37] S. Ballouz, J. Y. Liu, M. Oti, B. Gaeta, D. Fatkin, M. Bahlo, and M. A. Wouters, "Analysis of genome-wide association study data using the protein knowledge base," BMC Genet., vol. 12, p. 98, 2011.

[38] F. Zhang and R. Drabier, "IPAD: the Integrated Pathway Analysis Database for Systematic Enrichment Analysis," BMC Bioinformatics, vol. 13 Suppl 15, p. S7, 2012.

[39] N. Skunca, A. Altenhoff, and C. Dessimoz, "Quality of computationally inferred gene ontology annotations," PLoS Comput. Biol., vol. 8, no. 5, p. e1002533, May 2012.

[40] C. J. Bult, J. T. Eppig, J. A. Blake, J. A. Kadin, and J. E. Richardson, "The mouse genome database: genotypes, phenotypes, and models of human disease," Nucleic Acids Res., vol. 41, no. Database issue, pp. D885–891, Jan. 2013.

[41] S. J. F. Laulederkind, M. Tutaj, M. Shimoyama, G. T. Hayman, T. F. Lowry, et al., "Ontology searching and browsing at the Rat Genome Database," Database (Oxford), vol. 2012, p. bas016, 2012.

[42] C. L. Smith and J. T. Eppig, "The Mammalian Phenotype Ontology as a unifying standard for experimental and high-throughput phenotyping data," Mamm. Genome, vol. 23, no. 9–10, pp. 653–668, Oct. 2012.

[43] P. Gaudet, M. S. Livstone, S. E. Lewis, and P. D. Thomas, "Phylogenetic-based propagation of functional annotations within the Gene Ontology consortium," Brief. Bioinformatics, vol. 12, no. 5, pp. 449–462, Sep. 2011.

[44] X. Ke, "Presence of multiple independent effects in risk loci of common complex human diseases," Am. J. Hum. Genet., vol. 91, no. 1, pp. 185–192, Jul. 2012.

[45] T. A. Manolio, F. S. Collins, N. J. Cox, D. B. Goldstein, L. A. Hindorff, et al., "Finding the missing heritability of complex diseases," Nature, vol. 461, no. 7265, pp. 747–753, Oct. 2009.

[46] E. T. Cirulli and D. B. Goldstein, "Uncovering the roles of rare variants in common disease through whole-genome sequencing," Nat. Rev. Genet., vol. 11, no. 6, pp. 415–425, Jun. 2010.

[47] K. P. Kenna, R. L. McLaughlin, O. Hardiman, and D. G. Bradley, "Using Reference Databases of Genetic Variation to Evaluate the Potential Pathogenicity of Candidate Disease Variants," Hum. Mutat., Feb. 2013.

[48] R. Quillian, "A notation for representing conceptual information. An application to seamantics and mechanical English paraphrasing," Oct. 1963.

[49] J. D. Osborne, J. Flatow, M. Holko, S. M. Lin, W. A. Kibbe, L. J. Zhu, M. I. Danila, G. Feng, and R. L. Chisholm, "Annotating the human genome with Disease Ontology," BMC Genomics, vol. 10 Suppl 1, p. S6, 2009.
[50] P. N. Robinson, S. Köhler, S. Bauer, D. Seelow, D. Horn, and S. Mundlos, "The Human Phenotype Ontology: a tool for annotating and analyzing human hereditary disease," Am. J. Hum. Genet., vol. 83, no. 5, pp. 610–615, Nov. 2008.
[51] V. Petri, M. Shimoyama, G. T. Hayman, J. R. Smith, M. Tutaj, et al., "The Rat Genome Database pathway portal," Database (Oxford), vol. 2011, p. bar010, 2011.
[52] T. Menge, P. H. Lalive, H.-C. von Büdingen, B. Cree, S. L. Hauser, and C. P. Genain, "Antibody responses against galactocerebroside are potential stage-specific biomarkers in multiple sclerosis," J. Allergy Clin. Immunol., vol. 116, no. 2, pp. 453–459, Aug. 2005.

4

Genomics-Guided Discovery of Novel Therapeutics of Actinobacterial Origin

Dr. Janmejay Pandey and Dr. Surendra Nimesh

Department of Biotechnology, School of Life Sciences,
Central University of Rajasthan, Bandarsindri,
Kishangarh, Ajmer-305801, Rajasthan, India

4.1 Introduction

Natural products of microbial origin in general and of *Actinobacterial* origin in particular play an important role in drug discovery [1–4]. Although the precise physiological role of these molecules in microbial domain is yet to be determined, it is proposed that microorganisms produce and accumulate them for diverse physiological application e.g. interaction with important targets of cell metabolism and for using them in signaling functions that are indirectly correlated with cell growth or death [5]. From a technological point of view, these molecules constitute the single most important source of novel scaffold used for development of new drugs and drug candidates to treat life threatening infections and other human disorders [6, 7]. Of all the natural products of microbial origin, the best known examples are antibiotics. The "Golden Age of Antibiotics," (1940s to 1970s) was marked by unanticipated discovery of Penicillin by Alexander Fleming in 1929 [8] and its subsequent development by Chain and Florey in the 1940s [9]. The other noteworthy milestone in the field of microbial natural products and their derivatives was discovery of the immunosuppressants viz., Cyclosporin A, and Rapamycin [10, 11]. To date, nearly 20,000 metabolites of microbial origin have been identified, isolated and characterized to be members of different classes of natural products. Noticeably, $\sim$ 80% of these natural products have been isolated from members of phylum *Actinobacteria*

Post-genomic Approaches in Drug and Vaccine Development, 71–92.

Table 4.1 Therapeutically useful natural products of microbial origin

Acidulants	Anticancer agent	Polysaccharide
Alkaloids	Coccidiostats	Steroids
Animal growth promoter	Enzyme Inhibitor	Polysaccharides
Antibiotics	Immune suppressor	Protein
Antifungal	Immune modulator	Vitamins
Antihelminthic agents	Ionophores	
Antimetabolites	Iron transport factor	
Antioxidants	Nucleosides	
Antitumor agents	Nucleotides	
Acidulants	Anticancer agent	
Alkaloids	Coccidiostats	
Animal growth promoter	Enzyme Inhibitor	
Antibiotics	Immune suppressor	
Antifungal	Immune modulator	
Antihelminthic agents	Ionophores	

*Modified from Kurtboke (2010) [14]

[12–14]. A brief list of major classes of natural products of microbial origin is presented in Table 4.1.

Many of these natural products have been characterized for their therapeutic and biotechnologically useful properties. These properties have enabled the microbial natural products to assume a pivotal role in drug discovery and pharmaceutical industry during the later stages of the 20th century [15]. A large number of these products and their chemical derivatives have found place in government approved clinic practices [15, 16]. Despite above developments in the field of natural products-based drug industry; the beginning of 21st century experienced an unexpected withdrawal of large number of pharmaceutical companies from research programs oriented towards discovery of new microbial natural products [2]. A number of factors are regarded to be responsible for inducing this response. These factors include: (i) technical challenges associated with purification of natural products from microbial fermentations, (ii) difficulty associated with elucidation of structure and pharmacological functions of microbial natural products (iii) advent of combinatorial synthetic organic chemistry for producing an array of new compounds without the need of natural products as reaction substrate. Above all, the single most important factor for this response was discovery of redundant and previously known natural products at high frequency.

The major limitation resulting in high frequency of redundant discovery of previously known natural products has been identified as the inherent limitation of microbial culturing methodologies [17]. It has been realized for last 2–3 decades that although environmental samples contain thousands of unique bacterial species; the vast majority of these unique bacterial diversity remain recalcitrant to isolation and culturing under laboratory conditions [17, 18]. The other limitation with regard to discovery of novel microbial natural products has been the lack of scientific understanding about the culturing/incubation parameters suitable for expression and accumulation of desired natural products by the isolated *Actinobacterial* strains. To overcome these limitations, several efforts have been made in the direction of developing culture-independent genomics based approaches and progresses made with these efforts during last two decades have refueled the quest for discovery of novel microbial natural products [19]. The primary technological intervention responsible for this development has been the pioneering work of whole-genome shotgun sequencing of *Haemophilus influenzae*, published in 1995 [20], which demonstrated that microbial genome sequences could be obtained with application of large-scale microbial genome sequencing at practically applicable rapidity. At present, there are ~ 5,000 complete microbial genomes sequenced; with ~ 725 of them being those of strains belonging to phylum *Actinobacteria* (http://www.ncbi.nlm.nih.gov/genomes/MICROBES/microbial_taxtree.html).

Annotation of microbial genomes coupled with homology-guided screening have indicated vast potential of natural-product diversity which has remained underestimated; therefore, it is proposed that direct access of the genomic content of microorganisms could offer promising alternatives to existing methodologies for discovery of novel natural products. Most noticeably, genome sequencing of *Actinobacterial* strains and a few fungal strains have showed that although these strains contain an enormous variety of genes which could encode for enzymes to synthesize a range of secondary metabolites [21, 22], yet only a small fraction of these genes are expressed during standard growth conditions and fermentations [23]. In other words, annotation of genome sequences have revealed that there are many more biosynthetic gene clusters than there are currently known metabolites for culturable microorganism, suggesting that the biosynthetic potential for natural products in microorganisms has remained largely under-explored by traditional methods of natural-product discovery. Genes that are not known to be expressed are classified as 'cryptic' or 'orphan' biosynthetic gene clusters [24]. Characteristically, the physiological and ecological significance of these cryptic gene clusters remain only poorly understood. However, it is well established that they remain quiescent until suitable physicochemical signals induce their expression. At this point, it is important to comprehend that there is only very limited information available pertaining to physicochemical signals and their regulatory functions with regards to induction of the expression of cryptic gene clusters. Therefore, efforts are being directed towards: (i) identification of physicochemical inducers of cryptic gene clusters for manipulating microbial physiology and biosynthetic machinery leading to synthesis of natural-product and (ii) genetic manipulation of cryptic gene clusters for generation of genetically modified strains capable of producing previously unknown microbial natural products. These efforts have resulted in modest success in diversification of natural products of microbial origin. A summary of these efforts along with others directed towards exploitation of culture microbial isolates for diversification and identification of novel microbial natural products is presented in Figure 4.1.

Apart from efforts aligned with isolated bacterial cultures summarized above, the technological developments in the field of cloning, sequencing and annotation of DNA directly isolated from soil bypassing the need of culturing of microorganism have provided further impetus towards capturing the biosynthetic potential encoded within the genomes of enumerable microbial species in the form of metagenomic DNA libraries [25–27]. Importantly, this approach has enabled researchers to access the entire genomic diversity of the microorganism (including both culturable as well as un-culturable microbes).

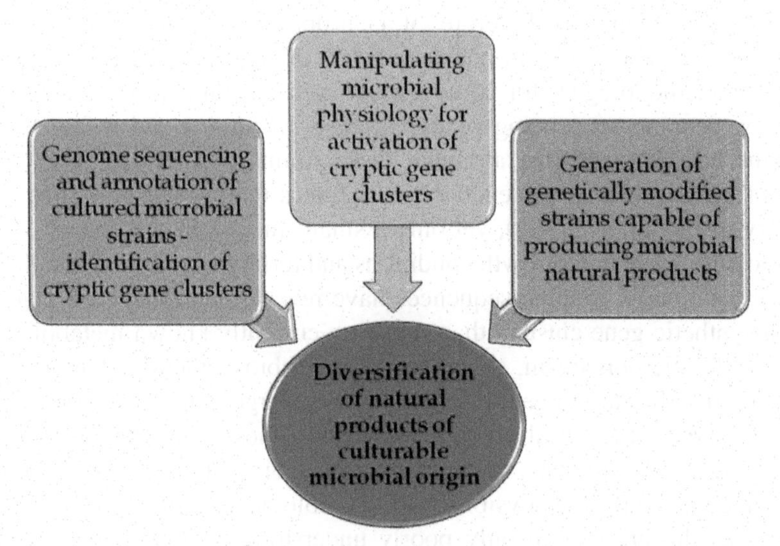

Figure 4.1 A graphical representation of genomics-guided efforts carried out for diversification of natural products obtained from culturable microorganisms.

Just like the approach is dependent upon elucidation of genome sequences of isolated microbial strains, the metagenomics-based approach for identification of novel natural products has also benefitted with homology-guided screening methodologies and resulted in identification of natural products that were not previously unforeseen [28, 29]. These efforts have been further augmented by completion of Human Genome Project, elucidation of genomes of a number of microbial human pathogens, advancements made in the directions of automation of instrument systems, robotics and development of high-throughput screening (HTS) platforms [19]. Together, these developments have provided powerful platform for screening microbial genomes for identification of natural products in a cost-effective manner. Figure 4.2 presents a graphical summary of metagenomics-based culture-independent approaches used for improved identification and diversification of novel microbial natural products.

It is pertinent to mention that the future of microbial natural products-based drug discovery largely depends upon thoughtful applications of genomics approaches combined with the complementary technologies such as HTS, proteomics, combinatorial biosynthesis, and combinatorial chemistry. In order to facilitate the expansion of microbial natural product-based therapeutic and preventive medicines, it is equally important to develop thorough understanding about biological activity and mode of action of natural products, their disease treatment efficiency, potential side effects and microbial drug resistance etc. Genomics-based approaches have already rendered significant

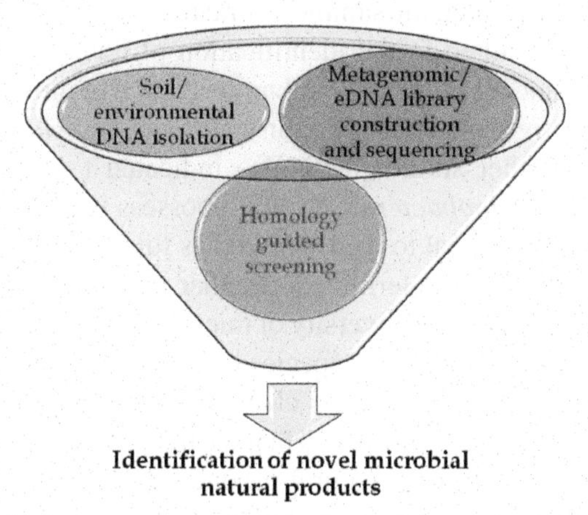

Figure 4.2 Graphical representation of the process followed for identification of microbial natural products with culture independent approach.

information relevant to these important domains related to microbial natural products-dependent drug discovery.

In light of the above prelude, the subsequent section presents a narrative with examples to reflect upon how recent developments including genomics-based approaches have guided discovery of novel microbial natural products and therapeutics.

4.2 Alternative Culturing Methodologies for Diversification of Microbial Natural Products

4.2.1 Exploring naive ecosystems for sample collections

It is well established that the majority of the microbial natural products-based drugs have been identified and extracted from microbial strains that have been cultured from mesophilic terrestrial samples [30]. However, with increasing understanding of the microbial physiology, it has now been realized that microbes are well adapted at responding to various environments [31]. Therefore, it is proposed that diversification of sample collection and thereby increasing genetic diversity of strains may have the greatest influence on the diversification of microbial natural-product extract libraries. In other words, it is critical to increase the number and diversity of sampling sites and explore underrepresented sites (e.g. deep subsurface, deep sea trenches, and sites with extreme temperature, salinity, or pH etc.). A large number of recent studies have focused on the isolation of *Actinobacterial* strains from naïve ecologically niches e.g. marine and aquatic samples, coral reef samples [32–35]. Many of these studies have resulted in the identification of novel *Actinobacterial* strains capable of producing previously unknown bioactive molecules including novel antibiotic active against methicillin-resistant *Staphylococcus aureus* (MRSA) [36]. Many other studies have also indicated towards secondary metabolites of marine *Actinobacterial* origin to possess important biological activities and have the potential to be developed as future therapeutic agents.

Although it is yet to be determined whether or not the geographical locations have direct effect on the diversity of microbial communities [31], yet reports showing presence of endemic species have prompted a hypothesis of direct relationship between geographical characteristics and microbial diversities [37]. Therefore, it could be proposed that the first step towards increasing the chance for discovery of novel natural products of microbial origin is to consider different geographical areas, including hot-spots of biodiversity for sampling. It is also pertinent to suggest that microbiologists and ecologists should work together to obtain as diverse samples and microorganisms as possible from one ecosystem to optimize the likelihood of finding novel strains and in turn novel microbial natural products.

4.2.2 Acknowledging significance of microbial symbiosis

With increasing understanding of different symbiotic associations that microorganisms are involved in under natural micro-environments [38], it is critical to carefully define the sampling strategies to ensure that many different types of matrix (e.g. soils, sediments, organic material, dead animals, dead plants, and lichens) get sampled from the selected ecosystems. Plants and lichens provide scaffold for symbiotic interaction amongst microorganisms as well as between microorganisms and eukaryotic cells [39]. These interactions are proposed to provide the physicochemical signals necessary for inducing the biosynthetic machinery involved in the production of microbial natural products. Noticeably, a number of natural products initially extracted from plant and animal materials have now been characterized to be actually produced by symbiotic microbes found as endophytes within the tissue of the corresponding plant and/or animal [40, 41]. Such microorganisms are now regarded to be difficult to culture alone in the absence of their symbiotic partner. Furthermore, the microbial genes responsible for production of active natural products may not be activated in isolated cultures. Plants samples must therefore be used judiciously as important source for isolation of microorganisms that could potentially produce novel natural products such as alkaloids. Lichens are another valuable source of microorganisms living in a unique symbiotic environment. Microorganisms present within the lichen have been reported to produce several unique secondary metabolites which make important substrates for production of valuable therapeutics [42, 43]. In light of the above description and examples, the sampling and *in vitro* strategies must be devised in such a manner that they maintain necessary structural and functional symbiotic associations and enable the symbiotic microorganisms to synthesize the natural products.

4.2.3 Culture independent sampling approaches

As it has been indicated earlier, the vast majority (~99%) of microbial diversity of environmental samples remain non-accessible to standard culturing methodologies [18]. Therefore, the major diversity of microorganisms within many samples remains unexplored. During the late stages of 20th century, the bacterial kingdom was divided into 12 groups based on the evaluation of bacterial strains available as isolated pure culture; however, at the turn of the century the number of groups expanded to 36, when the classification was carried out on the basis of molecular phylogeny [44]. Noticeably, 13 groups do not have a cultured representative yet. Similarly, till date, ~ 6,000 bacterial species have been described according to the characterization of isolated bacterial cultures; however, the number of different bacterial species present in the nature is speculated to be 100–1000 fold higher. These numbers suggest that there are diverse novel microorganisms in the natural environment that could be used as unexplored sources for drug discovery [45]. With regard to accessing the unexplored microbial diversity, there are two distinct schools of thoughts that propose (i) efforts should be directed towards exploring culturing methodologies to culture less-culturable organisms and (ii) exploring culturing methodologies for less-culturable organisms is time consuming and the methodology developed for one organism may not be applicable for the other one. Also, it is reasoned that inability to culture majority of microbial species may be as follows (i) inability to mimic cell-to-cell communication required for microbial growth within culturing setup, (ii) inhibition of microbial growth by high substrate/metabolic product concentrations, (iii) formation of viable but non-culturable (VBNC) cells, (iv) induction of lysogenic bacteriophages upon starvation of microbial cell during culturing and (v) cell damage by oxidative stress [18]. Therefore, alternate approaches for accessing the unexplored domain of microbial diversity have to be developed and implemented. Some of these approaches have already been developed and they are based upon direct isolation, cloning and sequencing of microbial DNA directly from environmental samples [46, 47]. These approaches have allowed identification of a vast number of new organisms that are significantly different from any previously cultured microorganism and also discovery of novel bioactive natural products [48, 49].

4.2.4 Manipulating culture conditions for maximizing natural product biosynthesis by cultured microorganisms

Microbial secondary metabolites, which constitute the richest source of microbial natural products, are known to be synthesized under diverse incubation conditions. However, the regulatory biochemical and molecular mechanisms of secondary metabolite biosynthesis is only poorly characterized. As a result, it is difficult to optimize incubation parameters for exploiting the entire metabolic potential of each cultured microorganism [50]. Also, the optimal incubation conditions for biosynthesis of secondary metabolites are different for different organisms and corresponding metabolite. One of the most important parameters is the 'Carbon and Nitrogen sources' [51, 52]. It has been shown that there are usually significant differences between the optimal carbon source for microbial growth and secondary metabolite synthesis [53]. It is noticeable that glucose serves as the optimal carbon source for growth of diverse microbial strains; however, it renders negative influence on production of many secondary metabolites including actinomycin, cephalosporin, alkaloids, and tylosin etc. [51, 54, 55] On the other hand, it does not interfere with the production of aflatoxin, aminoglycosides, chloramphenicol and anticapsin etc. [56]. Similarly, nitrogen sources such as ammonium salts, which help rapid microbial growth, have negative influence on secondary metabolism and inhibit the production of cephamycin, fusidin, and rifamycin [57, 58]. Sub-optimal phosphate concentration in culture medium renders positive influence on the secondary metabolite biosynthesis. Concentrations of trace elements e.g. Fe, Zn and Mn also impart significant effect on the secondary metabolite biosynthesis [59, 60]. Apart from the concentration of macro and micro-nutrients in the culture medium, the incubation parameters e.g. temperature, time and aeration etc. also have significant effect on the outcome of secondary metabolite synthesis [61]. In general, lower incubation temperatures, longer incubation time and lower-standard aeration are required for optimal synthesis of secondary metabolite based natural products. It must be comprehended that the optimal incubation parameters for secondary metabolite production vary dramatically from microbe to microbe and also from metabolite to metabolite. Hence, optimization of incubation parameters must be done very carefully in accordance with the chemical profiling and scoring of the optimization process in order to improve the incubation system.

4.2.5 Microbial co-culturing for improved synthesis of natural products

It is now rationalized that microbial communities hold greater potential for production of microbial natural products as compared to the isolated microbial strains [62]. The scientific basis of this rationale is the ever-increasing understanding about how microorganisms exist in nature. They are now considered to exist as part of micro-ecosystems, where in dynamic interactions take place by means of biochemical signaling between organisms and elicit production of microbial natural products [63]. Therefore, it is now being proposed that microbial co-culturing should be used for improving the biosynthesis of desired microbial natural products. Although some of the researchers are of the opinion that getting a stable co-culture is almost impossible, yet it may provide a means for exploiting the potential of the consortia as a whole.

4.3 Genetics/Genomics-Guided-Approaches for Diversification of Microbial Natural Products

4.3.1 Hetrologous expression of environmental DNA

As indicated in the earlier section, it has been realized that the culturable microbial diversity represents only a minor fraction ($\sim$ 1%) of the total available diversity. In order to exploit the genetic potential of the remaining majority, a number of different genetic approaches have been developed during the last 2–3 decades [64]. One of the basic approaches has been based on the isolation of DNA directly from uncultured microorganisms, followed by its digestion into large fragments with restriction enzymes, cloning into artificial vector(s) e.g. plasmid, cosmid, phasmid or artificial chromosomes. The recombinant vectors are subsequently transformed into a model expression host to induce hetrologous expression of environmental DNA [25, 64, 65]. This approach eliminates the need for culturing microorganisms in laboratory for accessing the untapped microbial genetic diversity present in various micro-environments. However, it is possible that just as in the case of their native host, the biosynthetic genes involve in production of natural products and the corresponding regulatory genes may remain dormant in the hetrologous host. Therefore screening of optimal induction conditions may be required for the production of novel natural products. On the other hand, this approach brings along an important possibility of combining multiple biosynthetic pathways leading to synthesis of novel natural products that are otherwise not possible with the conventional culture dependent approaches. With recent advancements in the high-throughput DNA sequencing and annotation, it is now also possible to identify with great precision, within the environmental DNA library (constituting of 10,000-1,00,000 clones), a recombinant clone that carries DNA fragment harbouring gene(s) or gene cluster(s) that encode for desired bioactive compounds [66]. Furthermore, the advent of chemoinformatics has now made it possible to make computer predictions of chemical structure based on gene sequence information [67].

4.3.2 Whole genome sequencing and data mining

Apart from cloning, sequencing and hetrologous expression of environmental DNA, the capability for high-throughput genome sequencing has also enabled to estimate the biosynthetic potential of a cultured microbial isolates [68]. This approach importantly complements the approach of screening of optimal incubation condition for inducing the biosynthesis of microbial natural products by cultured microbial isolates. The best and most successfully characterized examples of this approach are identification of (i) polyketide synthase (PKS) and (ii) non-ribosomal peptide synthetase (NRPS) [69]. These gene clusters are known to be involved in biosynthesis of two of the largest classes of natural products of microbial origin [70]. Noticeably, these gene clusters contain conserved features which have been the starting point of the genomics-guided discovery of several novel natural products. An interesting example of PKS and NRPS diversity has been noted with the genome sequencing of a cyanobacterium viz., *Nostoc punctiforme*. Noticeably, its genome contains more than 20 genes identified as putative PKS or NRPS [71]; however, only one of these has been related to a characterized secondary metabolite. Genome sequence of a number of *Actinobacterial* strains have also been reported to follow the similar trend. To present an example, the annotation of genome sequence of *Streptomyces avermitilis*has indicated for presence of 118 PKS and NRPS-like gene clusters [22], however, only 14 of these have been assigned to known secondary metabolites (representing ~ only 12 % of total genetic potential for biosynthesis of natural products or secondary metabolites).

4.3.3 Genome scanning with genome sequence tags

As an important alternative to the whole genome sequencing, another genetic tool viz., genome scanning has been developed and implemented as an efficient way to discover natural-product biosynthetic gene clusters [68]. This approach takes advantage of the nature of genes involved in microbial natural-product biosynthesis; as these genes exist in form of large size gene clusters (ranging from 10–200 kbps) within the microbial genomes [72]. This approach makes use of Genome Sequence Tags (GSTs), i.e. genes likely to be involved in biosynthetic pathways, as DNA or RNA probes and carries out high-throughput genome-scanning for metabolic loci involved in secondary metabolite biosynthesis, independent of their hetrologous expression [73]. This approach has been successfully implemented in isolating biosynthetic gene clusters for dynemicin and macromomycin from strains known to produce these enediyne antitumor agents [73]. Subsequently, cryptic gene clusters corresponding to enediyne biosynthesis have been identified from ∼ 70 *Actinobacterial* strains [74]. It is pertinent to mention that the success of such GST-based genome scanning largely depends upon sequence conservation among the different PKS and NRPS gene clusters. As an offshoot of the GST-based genome scanning, a PCR approach based on use of 'degenerate primers' has been developed and successfully applied for localization and cloning of biosynthetic gene clusters encoding for esperamicin and maduropeptin [75].

4.3.4 Homology-guided metagenomic screening

A recent advancement that has emerged as a complementary approach for whole genome sequencing and GST-based genome scanning is homology-guided metagenomic screening [76]. This approach relies on the PCR amplification of conserved natural product biosynthetic gene sequence motifs to identify and recover gene clusters from environmental DNA/ metagenomic DNA libraries and offers solution for targeted recovery of specific biosynthetic pathways from genomes of all bacterial species present within the environmental DNA sample. Prior bioinformatics analyses of sequences recovered through homology screening allow selective exclusion of sequences likely to be associated with gene clusters encoding with previously known natural products and thereby significantly reducing the probability for redundant isolation and increasing the odds of finding novel rare metabolites. One of the recent studies showed homology-guided metagenomic library screening, cloning, and hetrologous expression of the indolotryptoline-based *bor* biosynthetic gene cluster, which encodes for borregomycins along with several dihydroxyindolocarbazole anticancer/antibiotics [77].

4.3.5 Genomics-guided induction of cryptic gene clusters

As mentioned earlier, genome sequencing and annotation of several microbial strains in general and *Actinobacterial* strains in particular have shown presence of several gene clusters capable of synthesizing a plethora of potential secondary metabolites [73]. However, only a small fraction is expressed during the standard fermentation conditions. The remaining gene clusters are classified as 'cryptic gene clusters' which remain quiescent in the absence of adequate physico-chemical signals. Recently, a lot of interest has grown with regard to exploiting the biosynthetic potential of these cryptic gene clusters [78]. A number of different approaches (e.g. utilization of natural mutants of

RNA polymerases and ribosome engineering etc.) have been implemented for activating these silent gene clusters [78–80].

4.3.5.1 Ribosome engineering

The fundamental concept of 'ribosome engineering' has been based on the hypothesis that bacterial gene expression could be dramatically enhanced by altering transcription and translation pathways [81]. It was experimentally elucidated with a *Streptomyces lividans* strain that had a natural mutation in ribosomal protein S12 [82]. Noticeably, this mutation conferred resistance to streptomycin and also allowed production of actinorhodin which is otherwise not produced due to dormancy of corresponding gene cluster [82]. Subsequently, researchers have developed methods for selectively mutating ribosomal protein S12, as well as other ribosomal proteins and translation factors, or RNAP for activating and enhancing the production of secondary metabolites originating from cryptic gene clusters [82, 83]. The experimental outline is based on isolating such natural mutants by screening for resistantce to antibiotics, such as streptomycin, which target the ribosome. Similarly, RNA polymerase mutants may be obtained by growing bacterial strains on plates containing rifampicin that targets RNA polymerases. The genomics-guided alternatives are to generate mutants with K88E and K88R mutations in *rpsL*, which encodes the ribosomal protein S12, and the H437Y and H437R in *rpoB*, which encodes the RNAP · subunit have been effective in enhancing the yield of wide range of secondary metabolites including polyketides, macrolids, aminoglycosides and nucleosides and actinorhodin [84–86]. Importantly, combination of these mutations has further enhanced bacterial productivity of natural products.

4.3.5.2 Modulating dasR activity

N-acetylglucosamine (GlcNAc), is a major component of the cell walls of fungi. It has been shown that a high concentration and/or accumulation of GlcNAc in the culture medium may act as the major checkpoint for commencement of secondary metabolism [87]. Similar response is transmitted through *DasR*, which is identified as a global regulator of antibiotic synthesis amongst *Actinobacterial* strains [87]. *DasR* acts as pleiotropic transcriptional repressor and regulates many important cellular functions including that of GlcNAc regulon [88]. The DNA-binding activity *DasR* is inhibited in the presence of glucosamine-6-phosphate and leads to activation of antibiotic biosynthetic gene clusters. Based on this understanding, it is hypothesized that introduction of mutation(s) into *dasR* may also effectively enhance antibiotic productivity.

4.3.6 Genomics-guided combinatorial biosynthesis of novel microbial natural products

The conventional method for diversification of microbial natural products has been random mutagenesis. However, with increasing understanding of genetic basis of microbial natural product biosynthesis e.g. (i) microbial genes involved in biosynthesis of natural products are usually present in form of clusters, (ii) many of these genes are modular in nature and (iii) they produce multifunctional enzymes with relaxed substrate specificity; it is now possible to clone biosynthetic pathways gene clusters into a vector and interchange/move around genes within these clusters to generate hybrid biosynthetic machinery capable of synthesizing a wide array of novel natural products [89, 90]. An example of application of such an approach has been the pioneering work of gene transfer between strains leading to production of antibiotics actinorhodin, granaticin, and medermycin [90, 91]. In addition to the conventional PKS and NRPS gene cluster, studies with recombinant enzymes have recently resulted in the discovery of a number of novel natural products (e.g. hybrid polyketide-peptides Coronatine and Pyoluteorin) that were otherwise unexpected from predictions drawn from annotation of genome sequences [92, 93].

4.4 Conclusions

Discovery of novel bioactive compounds and chemically useful microbial natural products has been the major platform for discovery and diversification of therapeutic agents during the last century. However, the last 2–3 decades have experienced modest decline with regard to identification of natural products with novel chemical core and scaffolds. On the other hand, emergence of microbial antibiotic resistance has highlighted the need for discovery of novel therapeutics. To meet the requirement, a number of different approaches including genomics-based methodologies have been developed in recent past. Many of these approaches have resulted in the identification of novel natural products that are being used for the development of next generation therapeutics. Figure 4.3 represents list of novel microbial natural products discovered with application of different genomics-based approaches.

To summarize, it is certainly noteworthy that genomics-guided approaches had provided significant momentum to various programs focusing on discovery and development of microbial natural products-based therapeutics. Still, further efforts are required to further develop of different methodologies

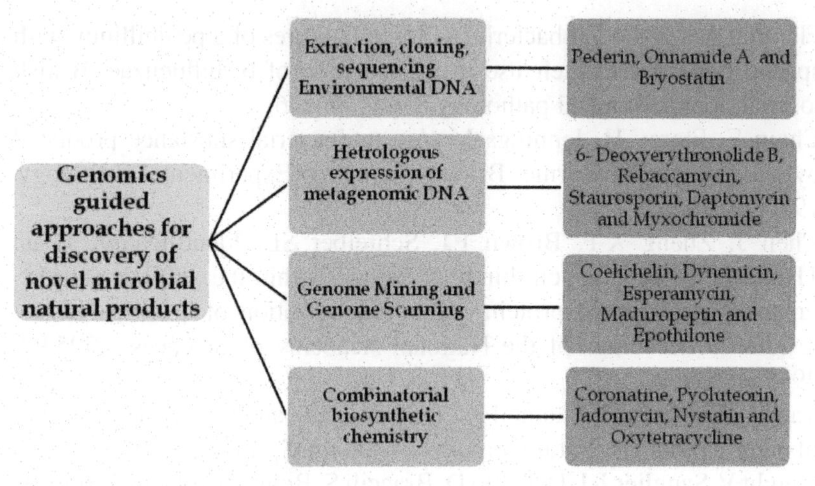

Figure 4.3 Graphical representation showing identification of novel microbial natural products with use of different genomics based approaches.

and implement them as complementary platforms to take full advantage of microbial genetic diversity for discovery of novel therapeutics.

References

[1] Grabley S, Thiericke R. The impact of natural products on drug discovery. Drug discovery from nature. Springer, New York Berlin Heidelberg. 1999: 3–37.

[2] Koehn FE, Carter GT. The evolving role of natural products in drug discovery. Nature Reviews Drug Discovery. 2005;4: 206–220.

[3] Newman DJ, Cragg GM, Snader KM. The influence of natural products upon drug discovery. Natural product reports. 2000;17: 215–234.

[4] Choudhary MI. Bioactive natural products as a potential source of new pharmacophores. A theory of memory. Pure and Applied Chemistry. 2001;73: 555–560.

[5] Williams DH, Stone MJ, Hauck PR, Rahman SK. Why are secondary metabolites (natural products) biosynthesized? Journal of Natural Products. 1989;52: 1189–1208.

[6] Harvey AL. Natural products in drug discovery. Drug discovery today. 2008;13: 894–901.

[7] Newman DJ, Cragg GM. Natural products as sources of new drugs over the last 25 years?. Journal of natural products. 2007;70: 461–477.

[8] Fleming A. On the antibacterial action of cultures of a penicillium, with special reference to their use in the isolation of b. Influenzae. British journal of experimental pathology. 1929;10: 226.
[9] Chain E, Florey H, Jennings M. An antibacterial substance produced by penicillium claviforme. British Journal of Experimental Pathology. 1942;23: 202.
[10] Chen J, Zheng X-F, Brown EJ, Schreiber SL. Identification of an 11-kda fkbp12-rapamycin-binding domain within the 289-kda fkbp12-rapamycin-associated protein and characterization of a critical serine residue. Proceedings of the National Academy of Sciences. 1995;92: 4947–4951.
[11] Van Middlesworth F, Cannell RJ. Dereplication and partial identification of natural products. Natural products isolation. Springer; 1998: 279–327.
[12] Knight V, Sanglier J-J, DiTullio D, Braccili S, Bonner P, Waters J, Hughes D, Zhang L. Diversifying microbial natural products for drug discovery. Applied microbiology and biotechnology. 2003;62: 446–458.
[13] Kurtböke I. From actinomycin onwards: Actinomycete success stories. MICROBIOLOGY. 2012;109.
[14] Kurtböke I. Biodiscovery from microbial resources: Actinomycetes leading the way. J. Austr. socie. microbial. Inc. 2010;31.
[15] Newman DJ, Cragg GM, Snader KM. Natural products as sources of new drugs over the period 1981–2002. Journal of natural products. 2003;66: 1022–1037.
[16] Butler MS. Natural products to drugs: Natural product-derived compounds in clinical trials. Natural product reports. 2008;25: 475–516.
[17] Colwell RR, Grimes DJ. Nonculturable microorganisms in the environment. ASM press; 2000.
[18] Barer MR, Harwood CR. Bacterial viability and culturability. Advances in microbial physiology. 1999;41: 93–137.
[19] Zhang L. Integrated approaches for discovering novel drugs from microbial natural products. Natural Products: Drug Discovery and Therapeutic Medicine. 2005: 33–55.
[20] Fleischmann RD, Adams MD, White O, Clayton RA, Kirkness EF, Kerlavage AR, Bult CJ, Tomb J-F, Dougherty BA, Merrick JM. Whole-genome random sequencing and assembly of haemophilus influenzae rd. Science. 1995;269: 496–512.
[21] Bentley S, Chater K, Cerdeno-Tarraga A-M, Challis G, Thomson N, James K, Harris D, Quail M, Kieser H, Harper D. Complete genome

sequence of the model actinomycete streptomyces coelicolor a3 (2). Nature. 2002;417: 141–147.

[22] Ikeda H, Ishikawa J, Hanamoto A, Shinose M, Kikuchi H, Shiba T, Sakaki Y, Hattori M. Complete genome sequence and comparative analysis of the industrial microorganism streptomyces avermitilis. Nature biotechnology. 2003;21: 526–531.

[23] Pawlik K, Kotowska M, Chater KF, Kuczek K, Takano E. A cryptic type i polyketide synthase (cpk) gene cluster in streptomyces coelicolor a3 (2). Archives of microbiology. 2007;187: 87–99.

[24] Holden MT, McGowan SJ, Bycroft BW, Stewart GS, Williams P, Salmond GP. Cryptic carbapenem antibiotic production genes are widespread in erwinia carotovora: Facile trans activation by the carr transcriptional regulator. Microbiology. 1998;144: 1495–1508.

[25] Rondon MR, August PR, Bettermann AD, Brady SF, Grossman TH, Liles MR, Loiacono KA, Lynch BA, MacNeil IA, Minor C. Cloning the soil metagenome: A strategy for accessing the genetic and functional diversity of uncultured microorganisms. Applied and environmental microbiology. 2000;66: 2541–2547.

[26] Handelsman J. Metagenomics: Application of genomics to uncultured microorganisms. Microbiology and Molecular Biology Reviews. 2004;68: 669–685.

[27] Schloss PD, Handelsman J. Biotechnological prospects from metagenomics. Current opinion in biotechnology. 2003;14: 303–310.

[28] Gillespie DE, Brady SF, Bettermann AD, Cianciotto NP, Liles MR, Rondon MR, Clardy J, Goodman RM, Handelsman J. Isolation of antibiotics turbomycin a and b from a metagenomic library of soil microbial DNA. Applied and environmental microbiology. 2002;68: 4301–4306.

[29] Elend C, Schmeisser C, Leggewie C, Babiak P, Carballeira J, Steele H, Reymond J-L, Jaeger K-E, Streit W. Isolation and biochemical characterization of two novel metagenome-derived esterases. Applied and environmental microbiology. 2006;72: 3637–3645.

[30] Lam KS. New aspects of natural products in drug discovery. Trends in microbiology. 2007;15: 279–289.

[31] Lozupone CA, Knight R. Global patterns in bacterial diversity. Proceedings of the National Academy of Sciences. 2007;104: 11436–11440.

[32] Bull AT, Stach JE. Marine actinobacteria: New opportunities for natural product search and discovery. Trends in microbiology. 2007;15: 491–499.

[33] Fiedler H-P, Bruntner C, Bull AT, Ward AC, Goodfellow M, Potterat O, Puder C, Mihm G. Marine actinomycetes as a source of novel secondary metabolites. Antonie van Leeuwenhoek. 2005;87: 37–42.
[34] Fenical W, Jensen PR. Developing a new resource for drug discovery: Marine actinomycete bacteria. Nature chemical biology. 2006;2: 666–673.
[35] Burkholder PR. The ecology of marine antibiotics and coral reefs. Biology and geology of coral reefs. 1973;2: 117–182.
[36] Asolkar RN, Kirkland TN, Jensen PR, Fenical W. Arenimycin, an antibiotic effective against rifampin-and methicillin-resistant staphylococcus aureus from the marine actinomycete salinispora arenicola. The Journal of antibiotics. 2009;63: 37–39.
[37] Konstantinidis KT, Ramette A, Tiedje JM. The bacterial species definition in the genomic era. Philosophical Transactions of the Royal Society B: Biological Sciences. 2006;361: 1929–1940.
[38] Haygood MG, Schmidt EW, Davidson SK, Faulkner DJ. Microbial symbionts of marine invertebrates: Opportunities for microbial biotechnology. Journal of molecular microbiology and biotechnology. 1999;1: 33–43.
[39] Bonfante P, Anca I-A. Plants, mycorrhizal fungi, and bacteria: A network of interactions. Annual review of microbiology. 2009;63: 363–383.
[40] Strobel G, Daisy B. Bioprospecting for microbial endophytes and their natural products. Microbiology and Molecular Biology Reviews. 2003;67: 491–502.
[41] Strobel GA. Endophytes as sources of bioactive products. Microbes and infection. 2003;5: 535–544.
[42] Boustie J, Grube M. Lichens-a promising source of bioactive secondary metabolites. Plant Genetic Resources. 2005;3: 273–287.
[43] Shukla V, Joshi GP, Rawat M. Lichens as a potential natural source of bioactive compounds: A review. Phytochemistry Reviews. 2010;9: 303–314.
[44] Pace NR. A molecular view of microbial diversity and the biosphere. Science. 1997;276: 734–740.
[45] Handelsman J, Rondon MR, Brady SF, Clardy J, Goodman RM. Molecular biological access to the chemistry of unknown soil microbes: A new frontier for natural products. Chemistry & biology. 1998;5:R245-R249.
[46] Tringe SG, Rubin EM. Metagenomics: DNA sequencing of environmental samples. Nature reviews genetics. 2005;6: 805–814.

[47] Daniel R. The metagenomics of soil. Nature Reviews Microbiology. 2005;3: 470–478.
[48] Schirmer A, Gadkari R, Reeves CD, Ibrahim F, DeLong EF, Hutchinson CR. Metagenomic analysis reveals diverse polyketide synthase gene clusters in microorganisms associated with the marine sponge discodermia dissoluta. Applied and environmental microbiology. 2005; 71: 4840–4849.
[49] Banik JJ, Brady SF. Recent application of metagenomic approaches toward the discovery of antimicrobials and other bioactive small molecules. Current opinion in microbiology. 2010;13: 603–609.
[50] Singh SB, Barrett JF. Empirical antibacterial drug discovery—foundation in natural products. Biochemical pharmacology. 2006;71: 1006–1015.
[51] Gallo M, Katz E. Regulation of secondary metabolite biosynthesis: Catabolite repression of phenoxazinone synthase and actinomycin formation by glucose. Journal of bacteriology. 1972;109: 659–667.
[52] Aharonowitz Y. Nitrogen metabolite regulation of antibiotic biosynthesis. Annual Reviews in Microbiology. 1980;34: 209–233.
[53] Drew SW, Demain AL. Effect of primary metabolites on secondary metabolism. Annual Reviews in Microbiology. 1977;31: 343–356.
[54] James D, Gutterson NI. Multiple antibiotics produced by pseudomonas fluorescens hv37a and their differential regulation by glucose. Applied and environmental microbiology. 1986;52: 1183–1189.
[55] Küenzi MT. Regulation of cephalosporin synthesis in cephalosporium acremonium by phosphate and glucose. Archives of microbiology. 1980;128: 78–83.
[56] Boeck L, Christy K, Shah R. Production of anticapsin by streptomyces griseoplanus. Applied microbiology. 1971;21: 1075–1079.
[57] El-Tayeb O, Salama A, Hussein M, El-Sedawy H. Optimization of industrial production of rifamycin b by amycolatopsis mediterranei. I. The role of colony morphology and nitrogen sources in productivity. African Journal of Biotechnology. 2004;3: 266–272.
[58] Aharonowitz Y, Demain AL. Nitrogen nutrition and regulation of cephalosporin production in streptomyces clavuligerus. Canadian journal of microbiology. 1979;25: 61–67.
[59] Weinberg D. Secondary metabolism: Regulation by phosphate and trace elements. Folia microbiologica. 1978;23: 496–504.
[60] Parra R, Aldred D, Magan N. Medium optimization for the production of the secondary metabolite squalestatin s1 by a < i > phoma < /i > sp.

Combining orthogonal design and response surface methodology. Enzyme and microbial technology. 2005;37: 704–711.
[61] Williams RP. Biosynthesis of prodigiosin, a secondary metabolite of serratia marcescens. Applied microbiology. 1973;25: 396–402.
[62] Pettit RK. Mixed fermentation for natural product drug discovery. Applied microbiology and biotechnology. 2009;83: 19–25.
[63] Shank EA, Kolter R. New developments in microbial interspecies signaling. Current opinion in microbiology. 2009;12: 205–214.
[64] Brady SF, Clardy J. Long-chain n-acyl amino acid antibiotics isolated from heterologously expressed environmental DNA. Journal of the American Chemical Society. 2000;122: 12903–12904.
[65] Brady SF, Clardy J. Cloning and heterologous expression of isocyanide biosynthetic genes from environmental DNA. Angewandte Chemie. 2005;117: 7225–7227.
[66] Brady SF. Construction of soil environmental DNA cosmid libraries and screening for clones that produce biologically active small molecules. Nature protocols. 2007;2: 1297–1305.
[67] Brandt W, Haupt VJ, Wessjohann LA. Chemoinformatic analysis of biologically active macrocycles. Current topics in medicinal chemistry. 2010;10: 1361–1379.
[68] Winter JM, Behnken S, Hertweck C. Genomics-inspired discovery of natural products. Current opinion in chemical biology. 2011;15: 22–31.
[69] Bergmann S, Schümann J, Scherlach K, Lange C, Brakhage AA, Hertweck C. Genomics-driven discovery of pks-nrps hybrid metabolites from aspergillus nidulans. Nature chemical biology. 2007;3: 213–217.
[70] Du L, Lou L. Pks and nrps release mechanisms. Natural product reports. 2010;27: 255–278.
[71] Meeks JC, Elhai J, Thiel T, Potts M, Larimer F, Lamerdin J, Predki P, Atlas R. An overview of the genome of nostoc punctiforme, a multicellular, symbiotic cyanobacterium. Photosynthesis research. 2001;70: 85–106.
[72] Schwecke T, Aparicio JF, Molnar I, König A, Khaw LE, Haydock S, Oliynyk M, Caffrey P, Cortes J, Lester JB. The biosynthetic gene cluster for the polyketide immunosuppressant rapamycin. Proceedings of the National Academy of Sciences. 1995;92: 7839–7843.
[73] Zazopoulos E, Huang K, Staffa A, Liu W, Bachmann BO, Nonaka K, Ahlert J, Thorson JS, Shen B, Farnet CM. A genomics-guided approach for discovering and expressing cryptic metabolic pathways. Nature biotechnology. 2003;21: 187–190.

[74] Van Lanen SG, Oh T-j, Liu W, Wendt-Pienkowski E, Shen B. Characterization of the maduropeptin biosynthetic gene cluster from actinomadura madurae atcc 39144 supporting a unifying paradigm for enediyne biosynthesis. Journal of the American Chemical Society. 2007;129: 13082–13094.

[75] Gao Q, Zhang C, Blanchard S, Thorson JS. Deciphering indolocarbazole and enediyne aminodideoxypentose biosynthesis through comparative genomics: Insights from the at2433 biosynthetic locus. Chemistry & biology. 2006;13: 733–743.

[76] Kalyanaraman C, Imker HJ, Fedorov AA, Fedorov EV, Glasner ME, Babbitt PC, Almo SC, Gerlt JA, Jacobson MP. Discovery of a dipeptide epimerase enzymatic function guided by homology modeling and virtual screening. Structure. 2008;16: 1668–1677.

[77] Chang F-Y, Brady SF. Discovery of indolotryptoline antiproliferative agents by homology-guided metagenomic screening. Proceedings of the National Academy of Sciences. 2013;110: 2478–2483.

[78] Chiang Y-M, Chang S-L, Oakley BR, Wang CC. Recent advances in awakening silent biosynthetic gene clusters and linking orphan clusters to natural products in microorganisms. Current opinion in chemical biology. 2011;15: 137–143.

[79] Bergmann S, Funk AN, Scherlach K, Schroeckh V, Shelest E, Horn U, Hertweck C, Brakhage AA. Activation of a silent fungal polyketide biosynthesis pathway through regulatory cross talk with a cryptic non-ribosomal peptide synthetase gene cluster. Applied and environmental microbiology. 2010;76: 8143–8149.

[80] Barker MM, Gaal T, Gourse RL. Mechanism of regulation of transcription initiation by ppgpp. Ii. Models for positive control based on properties of rnap mutants and competition for rnap. Journal of molecular biology. 2001;305: 689–702.

[81] Ochi K, Okamoto S, Tozawa Y, Inaoka T, Hosaka T, Xu J, Kurosawa K. Ribosome engineering and secondary metabolite production. Advances in applied microbiology. 2004;56: 155–184.

[82] Shima J, Hesketh A, Okamoto S, Kawamoto S, Ochi K. Induction of actinorhodin production by rpsl (encoding ribosomal protein s12) mutations that confer streptomycin resistance in streptomyces lividans and streptomyces coelicolor a3 (2). Journal of bacteriology. 1996;178: 7276–7284.

[83] Powers T, Noller HF. Selective perturbation of g530 of 16 s rrna by translational miscoding agents and a streptomycin-dependence

mutation in protein s12. Journal of molecular biology. 1994;235: 156–172.

[84] Hosaka T, Xu J, Ochi K. Increased expression of ribosome recycling factor is responsible for the enhanced protein synthesis during the late growth phase in an antibiotic-overproducing streptomyces coelicolor ribosomal rpsl mutant. Molecular microbiology. 2006;61: 883–897.

[85] Tanaka Y, Kasahara K, Hirose Y, Murakami K, Kugimiya R, Ochi K. Activation and products of the cryptic secondary metabolite biosynthetic gene clusters by rifampin resistance (rpob) mutations in actinomycetes. Journal of bacteriology. 2013;195: 2959–2970.

[86] Ochi K, Tanaka Y, Tojo S. Activating the expression of bacterial cryptic genes by rpob mutations in rna polymerase or by rare earth elements. Journal of industrial microbiology & biotechnology. 2013: 1–12.

[87] Rigali S, Titgemeyer F, Barends S, Mulder S, Thomae AW, Hopwood DA, van Wezel GP. Feast or famine: The global regulator dasr links nutrient stress to antibiotic production by streptomyces. EMBO reports. 2008;9: 670–675.

[88] Rigali S, Nothaft H, Noens EE, Schlicht M, Colson S, Müller M, Joris B, Koerten HK, Hopwood DA, Titgemeyer F. The sugar phosphotransferase system of streptomyces coelicolor is regulated by the gntr-family regulator dasr and links n-acetylglucosamine metabolism to the control of development. Molecular microbiology. 2006;61: 1237–1251.

[89] Silakowski B, Kunze B, Müller R. Multiple hybrid polyketide synthase/non-ribosomal peptide synthetase gene clusters in the myxobacterium < i > stigmatella aurantiaca < /i > . Gene. 2001;275: 233–240.

[90] Oliynyk M, Brown MJ, Cortés J, Staunton J, Leadlay PF. A hybrid modular polyketide synthase obtained by domain swapping. Chemistry & biology. 1996;3: 833–839.

[91] Omura S, Ikeda H, Malpartida F, Kieser H, Hopwood D. Production of new hybrid antibiotics, mederrhodins a and b, by a genetically engineered strain. Antimicrobial agents and chemotherapy. 1986;29: 13–19.

[92] Du L, Sánchez C, Shen B. Hybrid peptide–polyketide natural products: Biosynthesis and prospects toward engineering novel molecules. Metabolic engineering. 2001;3: 78–95.

[93] Molnar I, Schupp T, Ono M, Zirkle R, Milnamow M, Nowak-Thompson B, Engel N, Toupet C, Stratmann A, Cyr D. The biosynthetic gene cluster for the microtubule-stabilizing agents epothilones a and b from < i > sorangium cellulosum < /i > so ce90. Chemistry & biology. 2000;7: 97–109.

5

Chemogenomics Approach to Computer Aided Drug Discovery

Varun Khanna[1] and Shoba Ranganathan[2]

[1]Institute of Life Sciences, Ahmedabad University, Ahmedabad, Gujarat, India
[2]Department of Chemistry and Biomolecular Sciences & ARC Centre of Excellence in Bioinformatics, Macquarie University, Sydney, Australia

5.1 Introduction

The availability of a large number of potential new drug targets as a result of human genome sequencing and the high-throughput chemical screening of multiple compounds on gene/protein expression arrays has led to the emergence of new research disciplines like 'chemogenomics' [1], chemical biology and chemical genetics [2]. The definitions in the literature are somewhat diffuse as the three fields are inter-related because all the approaches investigate the perturbation of biological systems by small molecules. Chemical biology is defined as the study of biological systems, e.g. whole cells under the influence of chemical compounds while chemical genetics is dedicated to the study of protein function. Chemogenomics, on the other hand, is an inter-disciplinary, target-family-based approach to, drug discovery that can best be described as the study of the biological effects of screening congeneric chemical libraries against protein target families, such as, nuclear receptors, ion channels, G protein coupled receptors (GPCRs), proteases or kinases [1]. The term Chemogenomics was coined by Caron *et al.* [3] in 2001 and is rapidly replacing traditional screening methods that are limited to assessing the response of a small set of chemical compounds on a single target. In chemogenomics small molecule leads identified by the virtue of their interaction with a single member of a protein family are used to study the

Post-genomic Approaches in Drug and Vaccine Development, 93–114.

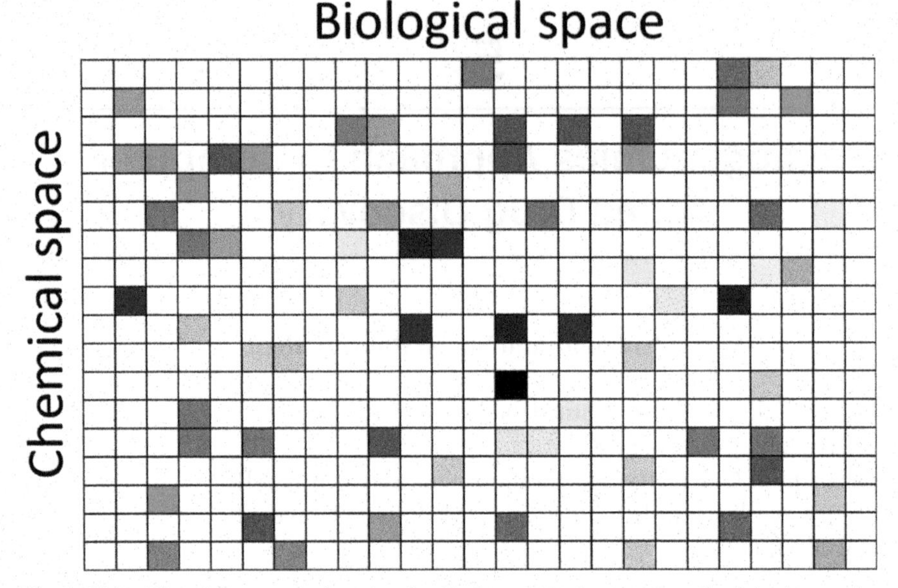

Figure 5.1 The chemogenomics matrix. Each matrix element describes the interaction between a particular compound in chemical space (rows) and a specific target in biological space (columns).

biological role of other member of the same family, usually clustered together using sequence homology. The advantages of the chemogenomic analysis include easy discovery of specific analogs within a target family, cross-target results exploration in related target families and a more through coverage of the patent space. The ultimate goal of chemogenomics is to study the biological effects of the entire chemical space, which corresponds to the superset of all small molecules on the entire protein space (all possible drug targets). Therefore, chemogenomics analysis is a never-ending learning experience aimed at developing a two-dimensional (2D) matrix (Figure 5.1) where gene or targets are reported as columns, compounds as rows and the data usually corresponds to the binding constants [K_i(binding affinity), IC_{50}(half maximal inhibitory concentration)] or functional effects [EC_{50}(half maximal effective concentration)].

Currently, this matrix is sparse as only a small number of compounds out of all the possible compounds have been experimentally tested on known gene or protein targets. Given the efforts in terms of cost and time to run biological assays, this matrix will remain sparse with respect to experimental data in the foreseeable future.

Predictive chemogenomics thus attempts to fill the gaps as the enormous size of the protein-ligand space renders large-scale systematic experimental analysis prohibitively expensive. Even the current size of existing datasets (compounds, targets and assays) and the information generated on gene or protein expression levels and binding constants is too large for manual analysis. Therefore, informatics techniques will play a crucial role in any chemogenomics analyses. Large, well annotated ligand-target-assay datasets are at the very heart of chemogenomics studies. In recent years, information about bioactive small molecules and protein targets has become available in the form of large public databases. Table 5.1 tabulates common databases that are relevant to chemogenomics drug discovery. Each of these databases has been designed for a different purpose. Some are mainly focused on describing chemical entities, while others are focused on specific aspects of proteins that make them possible drugs targets, such as position in a metabolic pathway or some unique structural features. Examples of databases that are primarily focused on describing chemical entities include Chem-Bank [4], which describes biological activities of small molecules derived from various screening projects and Chemical Entities of Biological Interest (ChEBI) [5] which contain chemical information and properties of small molecules. Databases such as UniProtKb [6], Protein Data Bank (PDB) [7]

Table 5.1 Database of relevance in chemogenomics drug discovery

Database	Number of small molecules	Number of proteins	Availability
PubChem Bioassay [12]	~71,700,000	~5,000	Public
ChEMBL [13]	1,295,510	9,844	Public
ChemBank [4]	~1,200,000	~1,000	Public
ChemPort [15]	1,150,000	15,290	Public
TDR Targets [16]	826,487	–	Public
PDB [7]	~300,000	~97,000	Public
SuperTarget [17]	195,770	6,219	Public
ChEBI [5]	~34,720	–	Public
TTD [11]	17,816	2,025	Public
BindingDB [10]	12,394	110	Public
DrugBank [18]	6,675	4141	Public
PharmGKB [19]	3,090	–	Public
iPHACE [20]	739	181	Public
GoSTAR (https://gostardb.com /gostar/index.jsp)	~6,000,000	4,900	Commercial
WOMBAT [21]	163,102	677	Commercial

and Entrez Proteins [8], Transporter Classification Database (TCDB) [9] exclusively contain protein information while others like BindingDB [10], Therapeutic Target Database (TTD) [11], PubChem Bioassay [12], ChEMBL [13] catalogue protein-ligand interactions. Apart from these databases which link chemical structure to protein targets, pharmacogenomics databases such as those providing important information of occurrence and consequences of variant genes encoding drug metabolizing enzymes, drug targets and other proteins important for drug activity are now becoming publicly available [14]. In this chapter, we outline the application of chemogenomics approach to drug discovery, discuss future prospects and unique challenges it faces. Such challenges include a better integration of bio- and cheminformatics data, development of computational infrastructure and tools to discover biomedically relevant information from existing biological screening data in public domain and the ability to build more focused screening libraries.

At the outset, the outcomes of recent studies involving chemogenomic datasets are presented to provide a starting point for further discussion on chemogenomics. Yongye and Franco [22] analyzed a dataset of 15,000 compounds containing natural products (NP) and synthetic molecules (from academic and commercial sources) screened across 100 diverse proteins. The authors analyzed the structural changes which lead to promiscuous behavior in compound binding patterns and concluded that NPs showed the least change in promiscuity due to minor structural modifications. On the other hand, synthetic compounds derived from academic groups showed greater promiscuity than the compounds obtained from commercial sources. Milletti and Hermann [23] developed a method for identifying chemical transformations that achieve selectivity against a specific "unwanted" kinase while maintaining activity for a target kinase. Blower *et al.* [24] mapped gene expression patterns to substructures and other chemical features of drugs. The authors were able to identify two subclasses of quinones whose activity pattern in the National Cancer Institute's 60-cell lines (NCI-60) correlated strongly with the expression pattern of particular genes. Strombergsson and Kleywegt [1] compared protein-ligand interactions in the PDB [7] with approved drugs and drug targets stored in the DrugBank database [18]. Principal component analysis (PCA) of the multidimensional data and nearest neighbors provided the overlap between two datasets. Receveur *et al.* [25] identified surprising binding site similarity of melanin-concentrating hormone receptor 1 (MHC1) to an unrelated target called D2/D3 receptor based on physicochemical features (hydrophobicity, aromaticity, positive or negative charge and polarity) of the amino acid residues in the putative ligand binding sites. Based on the insights of the binding site similarity, the authors were able to find a new class of benzamide ligands. Li *et al.* [26] analyzed a set of 189,907 bioactive compounds from PubChem and found that 62% of the compounds bind to more than one target while half of the remaining 38% were highly selective.

5.2 Description of Protein Ligand Spaces

The basic assumptions of any chemogenomic study are twofold: (i) similar compounds would share the protein targets and (ii) target-sharing similar ligands would share similar binding sites.

5.2.1 Protein space

It is essential to identify protein targets as knowledge of the mode of action of a drug or ligand can help enhance desirable features thus allowing researchers to design more potent drug variants. Further, a greater understanding of the biological pathways involving the target can be obtained if protein targets are known. Currently, there are over 41 million sequence entries in the UniProt [6] database, and over 97,000 3D protein structures deposited in the PDB [7]. However, how many of these known proteins are drug target is debatable. A number of reviews have been published to address this question [27–29]. Overington *et al.* [28] listed 324 drug targets for all classes of approved therapeutic drugs.

Table 5.2 Structural classification of proteins

Dimension	Classification scheme	Databases
1 Dimensional	Sequences, Patterns	UniProt [6], NCBI Protein database [8], Pfam [30], PRINTS [31], PROSITE [32]
2 Dimensional	Secondary structures	CATH [33], SCOP [34], TOPS [35]
3 Dimensional	Tertiary structure, atom coordinates	PDB [7], MODBASE [36], MMDB [37], PMDB [38]

Due to the rapid increase in protein information, many approaches for the classification of proteins have been developed. Using functional criteria a target protein might be classified into an enzyme or receptor, transporter, ion-channel or other protein. Further, an enzyme falls into one of the six major classes: oxidoreductases, hydrolases, lyases, transferases, isomerases, whereas receptors could be GPCRs, nuclear hormone receptors etc. In addition, proteins can be classified according to their sequence and structure as described in Table 5.2. The simplest representation of a protein which consists of full amino acid sequence is the first interesting information which can be used for reliable clustering of protein targets. However, the length of proteins can vary considerably (e.g. GPCRs) making it hard to analyze similarities and differences within a protein family, without using 2D or 3D representations.

5.2.2 Ligand space

The chemical space even for reasonably sized molecules i.e., those up to about molecular weight 600 Da is very large with the conservative estimate to be the order of 10^{40}[39]. Therefore, only a small fraction of compounds describing the current chemical space have been tested on known target space. To efficiently navigate the ligand space, it is necessary to represent molecules by descriptors that capture their structural characteristics and properties. The growing interest within the scientific community in selecting molecular characteristics has resulted in over 5,000 descriptors being proposed [40]. These descriptors can be computed using dedicated software tools such as Dragon (www.talete.mi.it). For a more comprehensive review on widely applicable descriptors, users are referred to [41–43]. There are numerous ways to classify molecular descriptors that vary in complexity of the encoded information and compute time. Two major types of descriptors that are relevant are structure-based (topological and geometrical) and property-based (experimentally derived or predicted). Descriptors are also classified according to their dimensionality ranging from one-dimensional (1D) to three-dimensional (3D) [44]. Table 5.3 lists examples of the major ligand descriptors reported in the literature.

1D descriptors are also known as whole molecule descriptors and are the easiest to compute. They describe global properties such as the count of bonds, molecular weight, and the number of atoms. Despite their low dimensionality, these descriptors in the past have been linked to drug-like characteristics [39], and were used to predict properties like aqueous solubility [45], partition-coefficient [46], bioavailability [47], distinguish drugs from non-drugs [48] and drugs from toxic compounds [49]. To hasten up the comparison and

Table 5.3 Ligand descriptors reported in literature

Dimension	Nature	Examples
1 Dimensional	Whole molecule	Molecular weight, Atom and bond counts, Log P, Molecular refractivity
2 Dimensional	Topological	Topological and connectivity indices (Wiener, Randic, Chi, Balaban, Kappa Shape indices, Hosoya Z index, Kier shape descriptors, BCUT, Burden eigenvalues, 2D autocorrelation descriptors), fingerprints (dictionary-dependent and dictionary-independent)
3 Dimensional	Conformational	Pharmacophores, shapes, fields, spectra, quantum-chemical (charges, electronegativity, superdelocalizibility), GETAWAY, WHIM, 3D-Morse, EVA, EEVA descriptors, gravitational indices, CoMFA

for easy interchange of chemical data, various linear 1D representations [Simplified Molecular Line Entry System (SMILES), Smiles Arbitrary Target Specification (SMARTS), SYBYL Line Notation (SLN), Wiswesser Line Notation (WLN) and InChI keys] have also been proposed in the literature. The most common of these is SMILES and its variants [50].

Most small molecule descriptors, however, belong to the family of 2D descriptors which are computed from topological representation of molecules such as the molecular graph [51]. Graph-based methods which transform the molecule into a set of nodes (atoms) connected by edges (chemical bonds) despite being slow are popular in substructure searching and clustering of molecules [52]. A number of 2D descriptors have been proposed (the Wiener Index [53], the Zagreb index [54], connectivity indices [40] and BCUT descriptors [55]). These descriptors are divided into two main categories *viz* topostructural (Wiener, Zagreb and Randic connectivity index) and topochemical (BCUT, Burden eigenvalues). Other popular 2D descriptors are fingerprint-based descriptors where a presence or absence of a particular 2D substructure or motif can be encoded in a fingerprint pattern [56].

3D descriptors, on the other hand, require the generation of 3D conformations thus can be computationally expensive with large datasets. 3D descriptors are mainly divided into quantum-chemical based (electronegativities, charges, LUMO, HOMO, superdelocalizibility), volume based (van der Waals, geometric volume) and interaction energy based (CoMFA, GRID, G-WHIM descriptors). Like 2D fingerprints 3D pharmacophoric fingerprints are based on atoms, substructures, ring centroids and planes that are thought to be relevant for receptor binding. The pharmacophoric features that are typically included for calculating 3D descriptors are hydrogen bond donors and acceptors, charged centers, hydrophobic centers and aromatic ring centers. Other 3D descriptors are 3D topographical indices, geometric atom pairs, GETAWAY, EVA, 3D-Morse descriptors. For a general review on all the above descriptors, readers are referred to comprehensive handbook of molecular descriptors [57].

5.3 Applications

Chemogenomic strategies are increasingly being used by various fields of medical research – for example cancer, inflammatory and hormone diseases. In general, chemogenomic approaches can be used for three different purposes in disease research. Chemogenomics can be used primarily to identify potential drug targets (target fishing). It is also used for predicting bioprofiles of drugs and drug repurposing. These three applications are briefly described below.

5.3.1 Target fishing

In silico target fishing is an emerging technique where the biological target is predicted on the basis of the information available in biologically annotated chemical databases. Given that a prohibitively huge effort is required in terms of cost, time and resources to experimentally determine all possible chemical-target associations, computational approaches are actively being pursued. Various approaches for computational target prediction have been developed with the help of advanced data-mining algorithms. Detailed description of the technologies involved has been reviewed by number of researchers [58–60]. Bender *et al.* [61], have grouped the current computational approaches for predicting targets or the mode of action into four classes: pharmacophore searching or similarity searching, data-mining, analysis of bioactivity spectra and molecular docking. Traditionally, these techniques have been used for searching novel compounds of known targets; however, it was recently realized that it is also possible to predict novel targets for known compounds, using the same approaches.

5.3.1.1 Similarity searching in chemical databases

Chemical similarity searching is a well-established technique in ligand-based virtual screening and has been used traditionally to screen a database of compounds for structural similarity to the query chemical structure. Interested readers are referred to a general review on chemical similarity [62]. Chemical similarity searching in the context of target prediction compares a compound whose target is not known to the database of compounds with known targets. Thus, conventional similarity searching is inverted when a new target for a group of known ligands is desired. The small molecules are usually represented as chemical fingerprints (2D descriptors) and the similarity between two compounds is measured by the Tanimoto similarity metric. However, any chemical descriptor or chemical similarity metric can be used to measure similarity between compounds. Nettles *et al.* [63] suggested that while 2D descriptors are powerful for similarity searching in annotated datasets, 3D descriptors are often more appropriate for searching orphan compounds having low 2D similarity to the database molecules. An obvious disadvantage of similarity-based searching is that prior knowledge of target is not incorporated during the search. However, there have been reports where search performance was improved by weighting molecular fingerprints with the target class knowledge [64, 65]. In addition to chemical similarity searching, target-based ontologies are also being exploited to predict novel targets for old compounds. The goal in this case is to assign compounds to a new target, based on the sequence identity of known targets [66].

5.3.1.2 Data-mining in annotated databases

Data-mining, in general, refers to finding correlations or patterns automatically from large relational databases [67]. It is a somewhat more sophisticated method than similarity searching for predicting biological targets in small molecules. This method tries to build statistical models by analyzing the properties of the known actives for a target. The generated model is then used to predict the probability of a query compound being active against the target. In a pioneering effort Poroikov, Filimonov and others predicted *in silico* activity spectra of substances (PASS) by training models of chemical features of activity classes [68]. An example of this method was reported by Nidhi *et al.* [69] where the authors used WOMBAT datasets to build multiple-category Laplacian-modified naïve Bayesian models using extended connectivity fingerprints from 964 target classes. The model was used to predict the top three most likely protein targets for all MDL Drug Report database compounds. The algorithm was subsequently used to predict adverse drug reactions and off-target effects [70]. However, one caveat is that, a target may have an array of structurally diverse active compounds; therefore, one model may not able to capture all features resulting in lower performance.

5.3.1.3 Molecular docking

The approaches described above are based on small molecule information alone. An alternative method is to include 3D information of the biological target and perform a ligand-target docking analysis. Traditionally, molecular docking has been the most frequently used method for structure-based drug design. However, in the chemogenomics context, a small molecule can be docked to an array of related proteins in order to determine the most likely interaction partner. Compared to the other approaches, molecular docking puts tremendous strain on computational resources since a given small molecule needs to be docked to a large number of proteins. The scoring functions, which quantitatively measure the interaction between the protein and the ligand, lie at the heart of docking-based target identification approaches. Despite over a decade of development, current scoring functions might not be fully accurate. Nonetheless, there are few reports on successful applications of docking in target identification. Chen *et al.* [71] used an inverse docking procedure termed as INVDOCK to identify targets for 4H-tamoxifen and vitamin E. For 4H-tamoxifen, many known receptors like protein kinase C, alcohol dehydrogenase, prostaglandin synthase and alcohol dehydrogenase were predicted. Overall, it was shown that 50% of the computer-identified potential protein targets were implicated or confirmed by experiments. TarFisDock [72] has been launched to automate the searching for small molecules-proteins interactions. It also contains a potential drug target database (PDTD), containing 841 known and potential drug targets categorized into 15 therapeutic areas. Zeng *et al.* [73] employed a reverse docking strategy (one ligand, multiple targets, instead of one target, multiple ligands) to predict potential antineoplastic targets of main functional components of green tea such as epigallocatechingallate (EGCG), epigallocatechin (EGC), epicatechingallate (ECG) and epicatechin (EC). For EGCG, several known targets such as HIV protease, glutathione reductase, catalase, eEF1-α and cholesterol oxidase with experimental evidence were predicted. Experiments also confirm the predicted involvement of DNA methyltransferase, ZAP-70 and 67kD laminin receptors. Novel targets identified include dihydrofolatereducatse, farnesyl protein transferase and histone deacetylase. However, more experiments are needed to verify EGCG binding targets found in their study. Lee *et al.* [74] generated a 2D matrix of docking scores among all X-ray protein structures in humans and yeast with 35 famous drugs. They were able to improve the sensitivity of reverse docking by including a large number of protein structures and concluded that as many protein structures as possible were important in finding real binding targets. In recent years, web-based search engines such as idTarget [75] have become available. These servers can predict possible binding targets of small molecules *via* a divide and conquer approach, in combination with scoring functions based on robust regression analysis and quantum chemical charge models.

5.3.2 Prediction of The Bioprofiles of Drugs

The prediction of the activity profile of compounds for a set of targets is a fundamental role of chemogenomics. The activity profiling of bioactive compounds can be carried out using chemical structures (ligand-based approaches) or using structure of the target (structures-based approaches). Specific methods and application of both approaches have been reviewed extensively [60, 76, 77]. In addition, protein-ligand-based approaches employ interaction information of proteins and ligands to build machine learning models (proteochemometrics) [78]. However, due to the limited availability of 3D information on protein targets the use of explicit target information in computational chemogenomics is restricted to the exactly known X-ray structures or structures that can be estimated in good approximation e.g. homology models. Therefore, most of the computational chemogenomics studies are ligand-based and are centered on the *similarity principle* [79, 80] which states that "structurally similar compounds have similar biological activities". However, this is not always true and several structure-activity relationships studies have demonstrated that chemically similar compounds may have significantly different biological activities while molecules with different structures can be very similar in their biological actions. These limitations are well documented in the literature [81–83]. Nevertheless, the similarities in ligand sets indicate to some extent the complementarity in binding pocket of the target. The pair-wise comparison of all the ligands generates a similarity matrix where the (i, j)th element of the matrix is the similarity of a ligand i to a ligand j. Chemical similarity between a pair of ligand is calculated using structural descriptors, weighting schemes and one of the similarity coefficients [43]. Likewise, wherever possible a target similarity matrix can be separately constructed using the same mathematical framework and protein descriptors. Machine learning models can be trained if a large number of ligands are known for a target [84–87]. The prediction quality of the model will depend on the number and diversity of ligands and on the selection of false positives used for the training purpose.

5.3.3 Drug Repurposing

Drug repurposing or repositioning is a primary example of beneficial application of chemogenomics to speed up drug discovery by identifying new clinical targets for existing approved drugs [88, 89]. This approach has the clear advantage of rapid clinical deployment, without going through the lengthy and costly procedures involved in drug discovery. The concept behind drug repositioning is based on promiscuous nature of small molecules [90] and interconnectedness of cellular pathways [91]. In this regard, compounds that have passed through phase II or III clinical trials but were unable to make it to the market due to other issues are suitable candidates for drug repositioning approaches. In the past, serendipity has played a major in drug repositioning [92]. For example, sildenafil was originally developed to treat heart disease however, it was later repositioned for male erectile dysfunction. Similarly through repositioning, thalidomide was approved to treat multiple myeloma and erythema nodosum in leprosy, despite being withdrawn from the market earlier due to carcinogenic effects [93]. The Rare Disease Repurposing Database described by Xu and Cote [94] is a promising new resource for drug repurposing efforts. The database contains 236 products that have the potential to be repurposed and offer new tools to develop niche therapies for rare disease patients. Haupt and Schroeder [95] review recent developments, tools and success stories in computational drug repurposing.

5.4 Conclusions

Chemogenomics approaches are becoming very popular with the rise of high-throughput data for both ligands and protein targets of pharmaceutical importance. The fact that several drugs act by interacting with multiple targets is shifting the drug discovery paradigm from one-target one-drug model to multiple-target model. In this context, instead of pursuing a highly selective compound for a unique target, i.e., “single key for a specific lock” the goal is to select an ideal “master key” which acts on a battery of targets. The “master key” however, should not operate on any possible lock, in order to avoid any adverse effects (e.g. “promiscuous binders”). The challenge is to identify the set of targets that are associated with a desired clinical effect. Further, for chemogenomics to create an appropriate increase in the efficiency of drug discovery process it will require a better integration of chemical and biological data by new tools and databases. In addition, the success of chemogenomics will not only depend upon the progress made to create desired links between relevant datasets and novel methodologies developed to overcome current limitations but also on what can realistically be done within budgeting constraints.

References

[1] H. Strombergsson and G. J. Kleywegt: "A chemogenomics view on protein-ligand spaces". BMC bioinformatics,Vol. 10 Suppl 6, No., 2009, S13.

[2] D. R. Spring: "Chemical genetics to chemical genomics: small molecules offer big insights". Chemical Society reviews, Vol. 34, No. 6, 2005, 472–482.

[3] P. R. Caron, M. D. Mullican, R. D. Mashal, K. P. Wilson, M. S. Su and M. A. Murcko: "Chemogenomic approaches to drug discovery". Current opinion in chemical biology, Vol. 5, No. 4, 2001, 464–470.

[4] K. P. Seiler, G. A. George, M. P. Happ, N. E. Bodycombe, H. A. Carrinski, S. Norton, S. Brudz, J. P. Sullivan, J. Muhlich, M. Serrano, P. Ferraiolo, N. J. Tolliday, S. L. Schreiber and P. A. Clemons: "ChemBank: a small-molecule screening and cheminformatics resource database". Nucleic acids research, Vol. 36, No. Database issue, 2008, D351–359.

[5] P. de Matos, N. Adams, J. Hastings, P. Moreno and C. Steinbeck: "A database for chemical proteomics: ChEBI". Methods in molecular biology, Vol. 803, No., 2012, 273–296.

[6] T. U. Consortium: "Update on activities at the Universal Protein Resource (UniProt) in 2013". Nucleic acids research, Vol. 41, No. D1, 2013, D43–D47.

[7] P. W. Rose, C. Bi, W. F. Bluhm, C. H. Christie, D. Dimitropoulos, S. Dutta, R. K. Green, D. S. Goodsell, A. Prliæ, M. Quesada, G. B. Quinn, A. G. Ramos, J. D. Westbrook, J. Young, C. Zardecki, H. M. Berman and P. E. Bourne: "The RCSB Protein Data Bank: new resources for research and education". Nucleic acids research, Vol. 41, No. D1, 2013, D475–D482.

[8] D. L. Wheeler, T. Barrett, D. A. Benson, S. H. Bryant, K. Canese, V. Chetvernin, D. M. Church, M. DiCuccio, R. Edgar, S. Federhen, L. Y. Geer, Y. Kapustin, O. Khovayko, D. Landsman, D. J. Lipman, T. L. Madden, D. R. Maglott, J. Ostell, V. Miller, K. D. Pruitt, G. D. Schuler, E. Sequeira, S. T. Sherry, K. Sirotkin, A. Souvorov, G. Starchenko, R. L. Tatusov, T. A. Tatusova, L. Wagner and E. Yaschenko: "Database

resources of the National Center for Biotechnology Information". Nucleic acids research, Vol. 35, No. suppl 1, 2007, D5–D12.

[9] M. H. Saier, Jr., M. R. Yen, K. Noto, D. G. Tamang and C. Elkan: "The Transporter Classification Database: recent advances". Nucleic acids research, Vol. 37, No. Database issue, 2009, D274–278.

[10] T. Liu, Y. Lin, X. Wen, R. N. Jorissen and M. K. Gilson: "BindingDB: a web-accessible database of experimentally determined protein-ligand binding affinities". Nucleic acids research, Vol. 35, No. Database issue, 2007, D198–201.

[11] F. Zhu, B. Han, P. Kumar, X. Liu, X. Ma, X. Wei, L. Huang, Y. Guo, L. Han, C. Zheng and Y. Chen: "Update of TTD: Therapeutic Target Database". Nucleic acids research, Vol. 38, No. suppl 1, 2010, D787–D791.

[12] Y. Wang, E. Bolton, S. Dracheva, K. Karapetyan, B. A. Shoemaker, T. O. Suzek, J. Wang, J. Xiao, J. Zhang and S. H. Bryant: "An overview of the PubChem BioAssay resource". Nucleic acids research, Vol. 38, No. Database issue, 2010, D255–266.

[13] A. Gaulton, L. J. Bellis, A. P. Bento, J. Chambers, M. Davies, A. Hersey, Y. Light, S. McGlinchey, D. Michalovich, B. Al-Lazikani and J. P. Overington: "ChEMBL: a large-scale bioactivity database for drug discovery". Nucleic acids research, Vol. 40, No. Database issue, 2012, D1100–1107.

[14] S. C. Sim, R. B. Altman and M. Ingelman-Sundberg: "Databases in the area of pharmacogenetics". Human mutation, Vol. 32, No. 5, 2011, 526–531.

[15] S. Kim Kjærulff, L. Wich, J. Kringelum, U. P. Jacobsen, I. Kouskoumvekaki, K. Audouze, O. Lund, S. Brunak, T. I. Oprea and O. Taboureau: "ChemProt-2.0: visual navigation in a disease chemical biology database". Nucleic acids research, Vol. 41, No. D1, 2013, D464–D469.

[16] M. P. Magarinos, S. J. Carmona, G. J. Crowther, S. A. Ralph, D. S. Roos, D. Shanmugam, W. C. Van Voorhis and F. Aguero: "TDR Targets: a chemogenomics resource for neglected diseases". Nucleic acids research, Vol. 40, No. Database issue, 2012, D1118–1127.

[17] N. Hecker, J. Ahmed, J. von Eichborn, M. Dunkel, K. Macha, A. Eckert, M. K. Gilson, P. E. Bourne and R. Preissner: "SuperTarget goes quantitative: update on drug–target interactions". Nucleic acids research, Vol. 40, No. D1, 2012, D1113–D1117.

[18] C. Knox, V. Law, T. Jewison, P. Liu, S. Ly, A. Frolkis, A. Pon, K. Banco, C. Mak, V. Neveu, Y. Djoumbou, R. Eisner, A. C. Guo and D. S. Wishart: "DrugBank 3.0: a comprehensive resource for 'omics' research on drugs". Nucleic acids research, Vol. 39, No. Database issue, 2011, D1035–1041.

[19] C. F. Thorn, T. E. Klein and R. B. Altman: "PharmGKB: the Pharmacogenomics Knowledge Base". Methods in molecular biology, Vol. 1015, No., 2013, 311–320.

[20] R. Garcia-Serna, O. Ursu, T. I. Oprea and J. Mestres: "iPHACE: integrative navigation in pharmacological space". Bioinformatics, Vol. 26, No. 7, 2010, 985–986.

[21] M. Olah, R. Rad, L. Ostopovici, A. Bora, N. Hadaruga, D. Hadaruga, R. Moldovan, A. Fulias, M. Mractc and T. I. Oprea: WOMBAT and WOMBAT-PK: Bioactivity Databases for Lead and Drug Discovery. In: Chemical Biology. Wiley-VCH Verlag GmbH; 2008: 760–786

[22] A. B. Yongye and J. L. Medina-Franco: "Toward an Efficient Approach to Identify Molecular Scaffolds Possessing Selective or Promiscuous Compounds". Chemical Biology & Drug Design, Vol., No., 2013, n/a-n/a.

[23] F. Milletti and J. C. Hermann: "Targeted Kinase Selectivity from Kinase Profiling Data". ACS Medicinal Chemistry Letters, Vol. 3, No. 5, 2012, 383–386.

[24] P. E. Blower, C. Yang, M. A. Fligner, J. S. Verducci, L. Yu, S. Richman and J. N. Weinstein: "Pharmacogenomic analysis: correlating molecular substructure classes with microarray gene expression data". The pharmacogenomics journal, Vol. 2, No. 4, 2002, 259–271.

[25] J.-M. Receveur, E. Bjurling, T. Ulven, P. B. Little, P. K. Nørregaard and T. Högberg: "4-Acylamino-and 4-ureidobenzamides as melanin-concentrating hormone (MCH) receptor 1 antagonists". Bioorganic & medicinal chemistry letters, Vol. 14, No. 20, 2004, 5075–5080.

[26] Q. Li, T. Cheng, Y. Wang and S. H. Bryant: "PubChem as a public resource for drug discovery". Drug Discovery Today, Vol. 15, No. 23–24, 2010, 1052–1057.

[27] P. Imming, C. Sinning and A. Meyer: "Drugs, their targets and the nature and number of drug targets". Nature Reviews Drug Discovery, Vol. 5, No. 10, 2006, 821–834.

[28] J. P. Overington, B. Al-Lazikani and A. L. Hopkins: "How many drug targets are there?". Nature Reviews Drug Discovery, Vol. 5, No. 12, 2006, 993–996.

[29] A. L. Hopkins and C. R. Groom: "The druggable genome". Nature Reviews Drug Discovery, Vol. 1, No. 9, 2002, 727–730.
[30] R. D. Finn, J. Tate, J. Mistry, P. C. Coggill, S. J. Sammut, H.-R. Hotz, G. Ceric, K. Forslund, S. R. Eddy, E. L. L. Sonnhammer and A. Bateman: "The Pfam protein families database". Nucleic acids research, Vol. 36, No. suppl 1, 2008, D281–D288.
[31] T. K. Attwood, P. Bradley, D. R. Flower, A. Gaulton, N. Maudling, A. L. Mitchell, G. Moulton, A. Nordle, K. Paine, P. Taylor, A. Uddin and C. Zygouri: "PRINTS and its automatic supplement, prePRINTS". Nucleic acids research, Vol. 31, No. 1, 2003, 400–402.
[32] C. J. A. Sigrist, E. de Castro, L. Cerutti, B. A. Cuche, N. Hulo, A. Bridge, L. Bougueleret and I. Xenarios: "New and continuing developments at PROSITE". Nucleic acids research, Vol. 41, No. D1, 2013, D344–D347.
[33] A. L. Cuff, I. Sillitoe, T. Lewis, A. B. Clegg, R. Rentzsch, N. Furnham, M. Pellegrini-Calace, D. Jones, J. Thornton and C. A. Orengo: "Extending CATH: increasing coverage of the protein structure universe and linking structure with function". Nucleic acids research, Vol. 39, No. suppl 1, 2011, D420–D426.
[34] A. Andreeva, D. Howorth, J.-M. Chandonia, S. E. Brenner, T. J. P. Hubbard, C. Chothia and A. G. Murzin: "Data growth and its impact on the SCOP database: new developments". Nucleic acids research, Vol. 36, No. suppl 1, 2008, D419–D425.
[35] I. Michalopoulos, G. M. Torrance, D. R. Gilbert and D. R. Westhead: "TOPS: an enhanced database of protein structural topology". Nucleic acids research, Vol. 32, No. suppl 1, 2004, D251–D254.
[36] U. Pieper, N. Eswar, B. M. Webb, D. Eramian, L. Kelly, D. T. Barkan, H. Carter, P. Mankoo, R. Karchin, M. A. Marti-Renom, F. P. Davis and A. Sali: "modbase, a database of annotated comparative protein structure models and associated resources". Nucleic acids research, Vol. 37, No. suppl 1, 2009, D347–D354.
[37] T. Madej, K. J. Addess, J. H. Fong, L. Y. Geer, R. C. Geer, C. J. Lanczycki, C. Liu, S. Lu, A. Marchler-Bauer, A. R. Panchenko, J. Chen, P. A. Thiessen, Y. Wang, D. Zhang and S. H. Bryant: "MMDB: 3D structures and macromolecular interactions". Nucleic acids research, Vol. 40, No. D1, 2012, D461–D464.
[38] T. Castrignanò, P. D. O. De Meo, D. Cozzetto, I. G. Talamo and A. Tramontano: "The PMDB Protein Model Database". Nucleic acids research, Vol. 34, No. suppl 1, 2006, D306–D309.

[39] C. A. Lipinski: "Drug-like properties and the causes of poor solubility and poor permeability". Journal of pharmacological and toxicological methods, Vol. 44, No. 1, 2000, 235–249.

[40] R. Todeschini and V. Consonni: Molecular Descriptors for Chemoinformatics. In.: Wiley-VCH; 2009

[41] P. Labute: "A widely applicable set of descriptors". Journal of Molecular Graphics and Modelling, Vol. 18, No. 4–5, 2000, 464–477.

[42] V. Khanna and S. Ranganathan: "In Silico Methods for the Analysis of Metabolites and Drug Molecules". Algorithms in Computational Molecular Biology: Techniques, Approaches and Applications, Vol., No., 361–381.

[43] V. Khanna and S. Ranganathan: "Molecular similarity and diversity approaches in chemoinformatics". Drug Development Research, Vol. 72, No. 1, 2011, 74–84.

[44] A. Bender and R. C. Glen: "Molecular similarity: a key technique in molecular informatics". Organic & biomolecular chemistry, Vol. 2, No. 22, 2004, 3204–3218.

[45] J. R. Votano, M. Parham, L. H. Hall, L. B. Kier and L. M. Hall: "Prediction of aqueous solubility based on large datasets using several QSPR models utilizing topological structure representation". Chemistry & biodiversity, Vol. 1, No. 11, 2004, 1829–1841.

[46] M. Clark: "Generalized fragment-substructure based property prediction method". Journal of chemical information and modeling, Vol. 45, No. 1, 2005, 30–38.

[47] J. Wang, G. Krudy, X. Q. Xie, C. Wu and G. Holland: "Genetic algorithm-optimized QSPR models for bioavailability, protein binding, and urinary excretion". Journal of chemical information and modeling, Vol. 46, No. 6, 2006, 2674–2683.

[48] J. Sadowski and H. Kubinyi: "A scoring scheme for discriminating between drugs and nondrugs". Journal of medicinal chemistry, Vol. 41, No. 18, 1998, 3325–3329.

[49] V. Khanna and S. Ranganathan: "Physicochemical property space distribution among human metabolites, drugs and toxins". BMC bioinformatics, Vol. 10, No. Suppl 15, 2009, S10.

[50] N. M. O'Boyle: "Towards a Universal SMILES representation - A standard method to generate canonical SMILES based on the InChI". Journal of cheminformatics, Vol. 4, No. 1, 2012, 22.

[51] R. Gozalbes, J. P. Doucet and F. Derouin: "Application of topological descriptors in QSAR and drug design: history and new trends". Current drug targets. Infectious disorders, Vol. 2, No. 1, 2002, 93–102.
[52] J. W. Raymond, C. J. Blankley and P. Willett: "Comparison of chemical clustering methods using graph- and fingerprint-based similarity measures". Journal of molecular graphics & modelling, Vol. 21, No. 5, 2003, 421–433.
[53] D. H. Rouvray: The rich legacy of half a century of the Wiener index. In: Topology in Chemistry: Discrete Mathematics of Molecules. Horwood Publishing; 2002: 16–37
[54] K. Das, I. Gutman and B. Zhou: "New upper bounds on Zagreb indices". Journal of Mathematical Chemistry, Vol. 46, No. 2, 2009, 514–521.
[55] R. Pearlman and K. M. Smith: "Novel software tools for chemical diversity". Perspectives in Drug Discovery and Design, Vol. 9–11, No. 0, 1998, 339–353.
[56] P. Willett: "Similarity-based virtual screening using 2D fingerprints". Drug Discovery Today, Vol. 11, No. 23–24, 2006, 1046–1053.
[57] R. Todeschini and V. Consonni: Handbook of molecular descriptors. In.: Wiley. com; 2008
[58] J. L. Jenkins, A. Bender and J. W. Davies: "In silico target fishing: Predicting biological targets from chemical structure". Drug Discovery Today: Technologies, Vol. 3, No. 4, 2006, 413–421.
[59] A. Koutsoukas, B. Simms, J. Kirchmair, P. J. Bond, A. V. Whitmore, S. Zimmer, M. P. Young, J. L. Jenkins, M. Glick and R. C. Glen: "Target prediction to multi-target drug design: Current databases, methods and applications". Journal of proteomics, Vol. 74, No. 12, 2011, 2554–2574.
[60] D. Rognan: "Structure-Based Approaches to Target Fishing and Ligand Profiling". Molecular Informatics, Vol. 29, No. 3, 2010, 176–187.
[61] A. Bender, D. W. Young, J. L. Jenkins, M. Serrano, D. Mikhailov, P. A. Clemons and J. W. Davies: "Chemogenomic data analysis: prediction of small-molecule targets and the advent of biological fingerprint". Combinatorial chemistry & high throughput screening, Vol. 10, No. 8, 2007, 719–731.
[62] P. Willett, J. Barnard and G. Downs: "Chemical Similarity Searching". Journal of chemical information and computer sciences, Vol. 38, No. 6, 1998, 983–996.
[63] J. H. Nettles, J. L. Jenkins, A. Bender, Z. Deng, J. W. Davies and M. Glick: "Bridging Chemical and Biological Space:? "Target Fishing" Using 2D

and 3D Molecular Descriptors". Journal of medicinal chemistry, Vol. 49, No. 23, 2006, 6802–6810.

[64] N. Stiefl and A. Zaliani: "A knowledge-based weighting approach to ligand-based virtual screening". Journal of chemical information and modeling, Vol. 46, No. 2, 2006, 587–596.

[65] K. Birchall, V. J. Gillet, G. Harper and S. D. Pickett: "Training similarity measures for specific activities: application to reduced graphs". Journal of chemical information and modeling, Vol. 46, No. 2, 2006, 577–586.

[66] U. Visser, S. Abeyruwan, U. Vempati, R. P. Smith, V. Lemmon and S. C. Schürer: "BioAssay Ontology (BAO): a semantic description of bioassays and high-throughput screening results". BMC bioinformatics, Vol. 12, No. 1, 2011, 257.

[67] M. Kantardzic: Data-mining: concepts, models, methods, and algorithms. In.: John Wiley & Sons; 2011

[68] V. V. Poroikov, D. A. Filimonov, Y. V. Borodina, A. A. Lagunin and A. Kos: "Robustness of biological activity spectra predicting by computer program PASS for noncongeneric sets of chemical compounds". Journal of chemical information and computer sciences, Vol. 40, No. 6, 2000, 1349–1355.

[69] Nidhi, M. Glick, J. W. Davies and J. L. Jenkins: "Prediction of Biological Targets for Compounds Using Multiple-Category Bayesian Models Trained on Chemogenomics Databases". Journal of chemical information and modeling, Vol. 46, No. 3, 2006, 1124–1133.

[70] A. Bender, J. Scheiber, M. Glick, J. W. Davies, K. Azzaoui, J. Hamon, L. Urban, S. Whitebread and J. L. Jenkins: "Analysis of Pharmacology Data and the Prediction of Adverse Drug Reactions and Off-Target Effects from Chemical Structure". ChemMedChem, Vol. 2, No. 6, 2007, 861–873.

[71] Y. Z. Chen and D. G. Zhi: "Ligand-protein inverse docking and its potential use in the computer search of protein targets of a small molecule". Proteins, Vol. 43, No. 2, 2001, 217–226.

[72] H. Li, Z. Gao, L. Kang, H. Zhang, K. Yang, K. Yu, X. Luo, W. Zhu, K. Chen, J. Shen, X. Wang and H. Jiang: "TarFisDock: a web server for identifying drug targets with docking approach". Nucleic acids research, Vol. 34, No. Web Server issue, 2006, W219–224.

[73] R. Zheng, T. S. Chen and T. Lu: "A comparative reverse docking strategy to identify potential antineoplastic targets of tea functional components and binding mode". International journal of molecular sciences, Vol. 12, No. 8, 2011, 5200–5212.

[74] M. Lee and D. Kim: "Large-scale reverse docking profiles and their applications". BMC bioinformatics, Vol. 13 Suppl 17, No., 2012, S6.

[75] J. C. Wang, P. Y. Chu, C. M. Chen and J. H. Lin: "idTarget: a web server for identifying protein targets of small chemical molecules with robust scoring functions and a divide-and-conquer docking approach". Nucleic acids research, Vol. 40, No. Web Server issue, 2012, W393–399.

[76] E. Jacoby: "Computational chemogenomics". Wiley Interdisciplinary Reviews: Computational Molecular Science, Vol. 1, No. 1, 2011, 57–67.

[77] J. L. Jenkins: "Large-Scale QSAR in Target Prediction and Phenotypic HTS Assessment". Molecular Informatics, Vol. 31, No. 6–7, 2012, 508–514.

[78] C. R. Andersson, M. G. Gustafsson and H. Strombergsson: "Quantitative chemogenomics: machine-learning models of protein-ligand interaction". Current topics in medicinal chemistry, Vol. 11, No. 15, 2011, 1978–1993.

[79] V. Monev: "Introduction to similarity searching in chemistry". MATCH Communication in Mathematical and in Computational Chemistry, Vol. 51, No., 2004, 7–38.

[80] F. Barbosa and D. Horvath: "Molecular similarity and property similarity". Current topics in medicinal chemistry, Vol. 4, No. 6, 2004, 589–600.

[81] P. M. Petrone, B. Simms, F. Nigsch, E. Lounkine, P. Kutchukian, A. Cornett, Z. Deng, J. W. Davies, J. L. Jenkins and M. Glick: "Rethinking Molecular Similarity: Comparing Compounds on the Basis of Biological Activity". ACS Chemical Biology, Vol. 7, No. 8, 2012, 1399–1409.

[82] H. Kubinyi: "Chemical similarity and biological activities". Journal of the Brazilian Chemical Society, Vol. 13, No. 6, 2002, 717–726.

[83] Y. C. Martin, J. L. Kofron and L. M. Traphagen: "Do Structurally Similar Molecules Have Similar Biological Activity?". Journal of medicinal chemistry, Vol. 45, No. 19, 2002, 4350–4358.

[84] Y. Yamanishi, M. Araki, A. Gutteridge, W. Honda and M. Kanehisa: "Prediction of drug–target interaction networks from the integration of chemical and genomic spaces". Bioinformatics, Vol. 24, No. 13, 2008, i232–i240.

[85] K. Bleakley and Y. Yamanishi: "Supervised prediction of drug–target interactions using bipartite local models". Bioinformatics, Vol. 25, No. 18, 2009, 2397–2403.

[86] X. Ning and G. Karypis: "In silico structure-activity-relationship (SAR) models from machine learning: a review". Drug Development Research, Vol. 72, No. 2, 2011, 138–146.
[87] M. Gönen: "Predicting drug–target interactions from chemical and genomic kernels using Bayesian matrix factorization". Bioinformatics, Vol. 28, No. 18, 2012, 2304–2310.
[88] J. Aubé: "Drug Repurposing and the Medicinal Chemist". ACS Medicinal Chemistry Letters, Vol. 3, No. 6, 2012, 442–444.
[89] D.-L. Ma, D. S.-H. Chan and C.-H. Leung: "Drug repositioning by structure-based virtual screening". Chemical Society reviews, Vol. 42, No. 5, 2013, 2130–2141.
[90] A. L. Hopkins: "Drug discovery: Predicting promiscuity". Nature, Vol. 462, No. 7270, 2009, 167–168.
[91] A. L. Hopkins: "Network pharmacology: the next paradigm in drug discovery". Nature chemical biology, Vol. 4, No. 11, 2008, 682–690.
[92] T. T. Ashburn and K. B. Thor: "Drug repositioning: identifying and developing new uses for existing drugs". Nature reviews. Drug discovery, Vol. 3, No. 8, 2004, 673–683.
[93] B. Ladizinski, E. J. Shannon, M. R. Sanchez and W. R. Levis: "Thalidomide and analogues: potential for immunomodulation of inflammatory and neoplastic dermatologic disorders". Journal of drugs in dermatology: JDD, Vol. 9, No. 7, 2010, 814–826.
[94] K. Xu and T. R. Coté: "Database identifies FDA-approved drugs with potential to be repurposed for treatment of orphan diseases". Briefings in Bioinformatics, Vol. 12, No. 4, 2011, 341–345.
[95] V. J. Haupt and M. Schroeder: "Old friends in new guise: repositioning of known drugs with structural bioinformatics". Briefings in Bioinformatics, Vol. 12, No. 4, 2011, 312–326.

6

Network Biology Methods for Drug Repositioning

Cheng Zhu[1], Chao Wu[1] and Anil G Jegga[2,3]

[1]Department of Computer Science, College of Engineering and Applied Science, University of Cincinnati, Cincinnati, OH Division of Biomedical Informatics, Cincinnati Children's Hospital Medical Center, Cincinnati, OH 45229, USA
[2]Division of Biomedical Informatics, Cincinnati Children's Hospital Medical Center, Cincinnati, OH 45229, USA
Department of Pediatrics, University of Cincinnati College of Medicine, Cincinnati, OH 45229, USA
[3]Department of Computer Science, College of Engineering and Applied Science, University of Cincinnati, Cincinnati, OH
Anil.Jegga@cchmc.org

6.1 Introduction

Rapid and cost-effective drug development is a vision that contrasts sharply with the current state of drug discovery. The current costly and time-consuming paradigm of drug discovery is ill-equipped to handle rapidly emerging diseases (e.g., swine flu, drug-resistant pathogens) and diseases that have a small financial market (e.g., rare disorders). One solution is to identify new uses for existing drugs commonly referred to as Drug Repositioning or Drug Repurposing. Drug repositioning or repurposing is nothing but discovering novel therapeutic indications for existing approved drugs, which provides an alternative and cost-efficient strategy to boost the discovery of disease therapeutics [1]. Because existing drugs have known pharmacokinetics and safety profiles, and are often approved by regulatory agencies for human use, any newly identified use can be rapidly evaluated in phase II clinical trials, which last two years and cost much less. Maximizing the indications

Post-genomic Approaches in Drug and Vaccine Development, 115–132.

potential and revenue from drugs that are already marketed thus offers a new take on the famous mantra of the Nobel Prize-winning pharmacologist, Sir James Black, "The most fruitful basis for the discovery of a new drug is to start with an old drug". Drug repositioning has many advantages, for instance, repositioned drugs can enter late-stage clinical phases and be approved with a shorter time and lower cost than novel compounds, because the starting points are usually approved compounds with known bioavailability and safety profiles, proven formulations and manufacturing routes, and well-characterized pharmacology [2]. The risks and failures that are typically accompanied with drug development can thus be minimized. It is therefore not surprising that in recent years, of the new drugs that reach their first markets, repositioned drugs account for ~30%! Drug repositioning in general is based on the hypothesis that by reshuffling disease and approved drug related knowledge in novel and interesting ways, potential new indications could be discovered [3].

Rational design of drug mixtures poses formidable challenges because of the lack of information about *in vivo* cell regulation, mechanisms of genetic pathway activation, and *in vivo* pathway interactions. Most of the repositioned drugs therefore are the result of "serendipity" – based on late phase clinical studies of unexpected findings. To convert this exercise into "systematic serendipity" requires a more rational way that involves, but not necessarily limited to, integrating and mining the enormous amounts of data generated by various techniques, in different formats and from diverse domains. In other words, the computational task is to identify potential repositioning candidates systematically. To this effect several network-based approaches have been developed and implemented to identify and rank drug repositioning candidates. Network biology, an emerging field, integrates heterogeneous -omics data into a biologically meaningful framework suitable for knowledge extraction and hypothesis generation.

6.2 Principles of Drug Repositioning Strategies

Conceptually, drug repositioning is motivated by two fundamental scientific principles. First, drugs by nature are "promiscuous", i.e., a single drug can bind multiple targets and/or impact several pathways. Second, drug targets relevant to a specific disease or pathway can also play critical roles in other related or unrelated diseases or pathways [3, 4]. Some commonly used drug repositioning strategies are listed in the following sections.

6.2.1 Screening of Related but Heterogeneous Knowledgebases

The fundamental approach in this strategy is that various accumulated heterogeneous data sets, such as pharmacological, biomedical, genomic, and chemical data, will be integrated and mined by novel computational and analytical algorithms. The "virtual screening" is performed to discover hidden or non-explicit relations between a drug, target gene, and disease. Under the case studies section we present some recent examples that are based on mining integrated heterogeneous data to identify repositioning candidates.

6.2.2 Pharmacopeia Scanning

This unbiased strategy re-screens the existing compounds against a variety of targets by a semi-blind approach to identify possible therapeutic benefits or side-effects, the recent advanced screening technologies are adopted.

6.2.3 Phenotype-centric Screening

The strategy starts from a point of an interested phenotype and the existing drugs that are screened to discover drugs that may generate an unexpected (either as on- and/or off-target effects of compounds), but yet desired phenotypic result.

6.3 Computational Network Approaches for Drug Repositioning

As part of *in silico* drug repositioning strategies, computational approaches have been employed as efficient, fast and economical alternatives. Most of these are built around integration and mining of large and heterogeneous data sets from different but related domains [5]. Some of the successful approaches are based on similarity between drugs [6], target proteins [7], or side effects and phenotypes [8, 9]. The underlying hypothesis is that "similarity" between drugs can be exploited to extrapolate or discover novel indications. Other successful approaches are based on gene expression signatures (e.g., connectivity map) [10, 11] and literature mining [12]. The constant flux of huge amounts of data because of the high throughput analytical pipelines necessitates development of novel, robust, and efficient computational approaches to integrate, store, mine, and extract knowledge from terabytes of data to and effectively formulate testable hypotheses.

6.4 Network Analysis Concepts

A network in general comprises a set of nodes and edges connecting the nodes. To characterize the properties of a network, several measurements ranging from centrality measures to their modular or community structure are used. These network measurements have been extensively employed to analyze the properties of large-scale networks in different domains including biomedical systems to extract meaningful patterns or make recommendations. In the following sections, we outline some of the common network measures used to analyze the networks. Table 6.1 summarizes several commonly used network measurements along with their equations to calculate.

6.4.1 Measures of Node Centrality

Centrality in general is used as a measure to calculate the "importance" of a node in a network. Four common centrality measures that are commonly used are: degree, betweenness centrality, closeness centrality and eigenvector centrality.

6.4.1.1 Degree

Node degree indicates the number of direct connections (edges) that a node has. When a network is directed, it has two respective measures of degree, namely in-degree and out-degree. In-degree represents the number of edges that are directed to a node while out-degree, as the name indicates, represents the number of outgoing connections or edges from a node.

6.4.1.2 Betweenness centrality

A node's betweenness centrality, also referred to as the shortest path betweenness, is the ratio of the number of the shortest paths that pass through a node to the number of the shortest path between any pair of nodes in the network [13].

6.4.1.3 Closeness centrality

Closeness centrality is a measurement on how close a node is connected to all other nodes of the network [14, 15]. A node's closeness is defined as the inverse of sum of the node's distances to all other nodes in the network. Closeness can be defined as a measurement of how fast the information originating from a node can spread to all other nodes sequentially.

6.4.1.4 Eigenvector centrality

Eigenvector centrality is a measurement of the influence of a node in a network. It assigns relative scores to all nodes in the network based on the concept that connections to high-scoring nodes contribute more to the score of the node than connections to low-scoring nodes. Google's PageRank [16] is a variant of the Eigenvector centrality measure.

6.4.2 Shortest Path Length

The shortest path length, sometimes also called geodesic distance, is the smallest number of edges or "hops" that are required to pass from one node to another.

6.4.3 Network Clustering Coefficient

Clustering coefficient is a measurement of degree to nodes that tend to cluster together in a network [17]. Clustering coefficient indicates how densely the neighbors of a node are connected. The average clustering coefficient for nodes in a network represents the tendency to form clusters. It is also an indicator of the modularity in a network.

6.4.4 Network Density

The density measures the extent of the contacts between pairs of nodes of the network [18]. The density is computed as the proportion of contacts that could possibly occur in the network compared with those that are actually observed in the network.

6.4.5 Sub-networks, Modules and Communities

Sub-network, or connected component, or loosely connected network, is a way to interconnect the components in a system or network so that those components depend on each other to the least extent. A loosely connected network can be easily decomposed into definable modules. A module, on the

Table 6.1 Summary of network measurements

Indicator	Calculation
Degree(with In-degree and Out-degree)	•Undirected networks: D = the number of edges connected to a node. •Directed networks: Din = the number of directed-in edges; $Dout$ = the number of directed-out edges of a node. •Average degree $<D> = 2E/N$, E is the total number of edges, N is the total number of nodes, and $< >$ denotes average.
Betweenness centrality	$Bci = \sum \frac{\#SP\_through\_node\_i}{\#SP}$ #SP number shortest paths
Closeness centrality	$Cc\,i = \frac{1}{\sum_{j=1}^{N} d(n_i, n_j)}$ N is the number of nodes, d is distance from node ni to nj
Eigenvector centrality	$E_i = \frac{1}{\lambda} \sum_{t \in M(i)} E_t = \frac{1}{\lambda} \sum_{t \in G} a_{i,t} E_t$ $AE=\lambda E$ $M(i)$ is a set of the neighbors of node i and λ is a constant, A = (ai,t) is the adjacent matrix of node i
Shortest path length	Lij = the smallest number of edges between nodes i and j.
Clustering coefficient	$Cc = \frac{2n}{E(E-1)}$ n is the number of a node's neighbors in the network, and E is the total number of edges.
Density for undirected network	$D = \frac{2E}{N(N-1)}$ N is the number of nodes, and E is the number of edges in the network
Density for directed network	$Bce = \sum \frac{\#SP\_through\_edge\_e}{\#SP}$ #SP number shortest paths

other hand, is a group of nodes that are more densely connected to each other than to the nodes outside the group in a network [19]. In biomedical domain, networks are composed of modules [20]. A functional module for instance represents a group of highly interconnected molecules with relatedness in one or more biological functions [21]. In a biomedical network, functional modules are believed to be comparable to topologically densely connected modular structures that are relatively isolated from other parts of the network [22]. A community, similarly, is a densely connected sub-network within a large network, such as a close-knit group of genes in a protein interaction network or a group of close friends in a social network. Often, the terms network modularity and communities are used interchangeably.

Finding modules or clusters in a network is an active field of research and is widely used in several domains, including machine learning, information retrieval, pattern recognition and bioinformatics. The extraction and analysis of clusters in large-scale biomedical networks have been extensively studied in protein interaction networks, functional enrichment networks, and regulatory networks [19, 23, 24]. Network clustering approaches can be broadly categorized as: (a) connectivity-based clustering (hierarchical clustering), (b) centroid-based clustering, (c) density-based clustering, and (d) centrality based clustering.

Connectivity-based clustering, also known as hierarchical clustering, is based on the principle that objects are more related to nearby objects than to objects that are farther away. These algorithms connect "objects" to form "clusters" based on the pair-wise distance [25–27].

A typical centroid-based clustering algorithm is K-means, where the number of clusters in the network is fixed to k. K-means clustering tries to find the k cluster centers and allocate the objects to their nearest cluster center, such that the squared distances of each objects to their cluster center are minimized [28].

A density-based clustering approach tries to find higher densely connected components than the remainder of the network. Sparsely distributed objects in the areas are usually considered to be noise or border points [29]. The most popular density–based clustering method is DBSCAN [30].

Centrality-based clustering methods start by removing an edge that has the highest centrality measure (e.g., the edge betweenness centrality), the edge removal procedure will be repeated iteratively until a network has been partitioned to several clusters after a fraction of edge have been removed [22, 31].

Besides these main categories, other clustering approaches have also been developed and applied in network analyses. For example, the Louvain method for community detection [32], employs a greedy optimization method which attempts to iteratively optimize the "modularity" of a partition from the network, the procedure stops until a maximum of modularity is attained and a hierarchy of communities is generated.

6.5 Network Biomedicine

The various types of the "omics" data, such as protein interactions, gene expressions, gene-disease, drug-target, and disease–drug relations, etc., can be viewed from a network perspective, where a node represents a biological entity and an edge its annotations and/or associations. Because networks in biomedical domain have been found to be comparable to communication and social networks [33] through commonalities such as scale-freeness and small-world properties, the algorithms used for social and Web networks should be equally applicable to biomedical networks.

Network-based approaches have been proven useful in biomedical domains to tackle the challenges for two primary reasons. First, various types of the "omics" data, such as protein interactions, gene expressions, gene to disease relations and disease to drug relations, can be viewed from a network perspective, where a node represents a biological entity and an edge its annotations and/or associations. Second, networks in biomedical domain have been found to have some in common properties with communication and social networks such as scale-freeness and small-world properties [33], the algorithms used for social and web networks are therefore regarded equally applicable to biomedical networks as they provide a systems-level view of the corresponding biological systems. Retrieving useful patterns and knowledge from biomedical systems are important because such patterns in the networks may uncover the underlying biomedical characteristics and provide possible solutions for medical treatment. In addition to representing as another paradigm shift in drug discovery science, network-based approaches provide a fresh perspective in understanding important proteins in the context of their cellular environments. This can provide a rational basis for deriving useful strategies in drug design [34]. Apart from identifying novel drug targets and infer mechanism of action, networks enable us to formulate novel, testable hypotheses facilitating novel drug discovery, drug repositioning candidates, drug combinatorials and ultimately realize the goal of personalized medicine [34, 35].

Network analysis approaches have an inherent ability to represent the diverse interactions found in higher organisms mathematically and graphically. As a result, network representations of biological systems have greatly facilitated the analysis of the complexity and the diversity of different types of data. These approaches generally focus on extracting modules/patterns comprising nodes and edges representing functional relatedness from a large, global biological network structure. The analysis can also be extended to extract network connections that appear to be critical in integrated heterogeneous networks. The underlying reasoning, as discussed earlier, is that modularization of biological networks provides deep insight into living systems and human diseases [36]. The discovery of high connectivity among different diseases, drugs or genes can be used not only to infer the common mechanism and targeted pathways, but also help to find possible drug repositioning candidates, especially when one or more diseases in the same module has an approved drug. Therefore, by integrating drug, gene, disease and other biological entity relationships to represent them as different types of networks, and applying appropriate network algorithms, interesting drug and disease combinations for drug targets may be discovered [37–40]. To this effect, there have been several studies published. In a protein–protein interaction (PPI) network, hubs and bottlenecks which are topologically important and essential may serve as potential drug targets [31, 41]. Predicting a PPI network for the pathogenic trypanosome *Leishmania major* Florez *et al.* [42] identified 142 hubs and bottleneck proteins that have no orthologs in humans and may serve as potential drug targets. In another study, combining important proteins/genes from both the interactome and the reactome of *Mycobacterium tuberculosis*, Raman *et al.* proposed a drug target identification pipeline ("targetTB"), to predict and refine drug targets for tuberculosis [43]. On the other hand, potential drug repositioning candidates can be inferred from similar known drug targets. This requires the knowledge of known drug-target relationships for drug similarity networks, as well as the measures of drug similarity and target similarity [3]. Yang *et al.* identified potential drug targets for Alzheimer's disease by applying a two-directional Z-transformation on a score matrix that characterizes the docking strength between chemicals and targets [44]. Zhao and Li on the other hand combined the measures of drug therapeutic (or phenotypic) similarity and drug chemical structural similarity to predict links between drugs and target proteins by a regression model called drug CIPHER [45, 46]. Additional approaches based on the analysis of integrated heterogeneous networks include a study by Lee *et al.* [47] wherein a tripartite network was constructed by integrating known disease, drug, and protein interactions. The authors apply a shared neighborhood scoring (SNS) algorithm on this tripartite network to identify potential novel indications for a drug used to treat hypertension. In a related study, through a co-module analysis across herb-biomolecule-disease multilayer networks, Li *et al.* [48] constructed a Chinese herb network to measure the strongly connected herbs and herb pairs.

In the following sections, we present three case studies which have successfully applied network-based approaches to identify and prioritize drug repositioning candidates.

6.5.1 Case Studies

6.5.1.1 Mantra: mode of action by network analysis

Iorio *et al.* [49] developed MANTRA (Mode of Action by Network Analysis), a computational approach to identify potential drug repositioning candidates. The analysis is based on a novel similarity measure among the cellular responses elicited by a large set of compounds in humans. For each of the compound, the cellular response is summarized by a prototype genome-wide ranked list (PRL) of genes. This PRL is built *in silico* by combining genome-wide profiles of differential expression, following treatments with the compound on a number of different human cell lines. The PRLs are then compared to each other with a novel approach based on Gene Set Enrichment Analysis (GSEA) ending up with an exhaustive set of drug pairwise distance values. Using these distance measures, a drug–drug network is built and analyzed to identify modules or highly interconnected sub-networks or communities. Communities enriched for drugs with similar mechanism of action are then identified using known knowledge and literature searches. These communities are used to make novel hypotheses on known drugs and to classify novel drugs when they are integrated in the network. Using this framework, the authors discovered that fasudil (a Rho-kinase inhibitor) could potentially be repositioned as treatment for neurodegenerative disorders.

6.5.1.2 Disease profiles and protein interactions mining

Suthram *et al.* [50] integrated molecular profiles of diseases with protein–protein interaction data, to infer protein functional modules and networks that were shared among many diseases. The molecular profiles of diseases were obtained from the NCBI Gene Expression Omnibus (GEO) [51] while the protein–protein interaction (PPI) data for humans was obtained from the Human Protein Reference Database (HPRD) [52]. Using this human PPI data that was organized into 'modules' of functionally interacting proteins, a statistical approach was used to evaluate the molecular signatures of diseases for gene functional module activity, whereby the module activity was determined as the mean normalized transcriptional activity of its component genes in the disease molecular profile. A disease–disease network was then formed on the basis of functional module activity shared between diseases and mined to identify sub-networks with multiple known drug targets.

In another interesting study, Chiang and Butte [53] used a "guilt by association" approach to discover alternative uses for drugs by systematically evaluating a drug treatment-based view of diseases. The authors used the DRUGDEX System, as a gold-standard comprehensive pharmacopeia resource. The DRUGDEX system provides literature-backed drug information on both (FDA) approved indications and off-label uses and is an U.S. government approved source for medical insurance reimbursable off-label uses. Using this system, the authors systematically captured the FDA-approved (FDA approved drug and indications) and practiced (FDA-approved drugs, but off-label drug uses) views into a Drug-Disease Knowledge Base consisting of 726 diseases and 2,022 drugs. Given that the treatment profiles between a small subset of disease pairs can be quite different between the FDA-approved and practiced views, the authors reasoned that these discrepancies can be mined to suggest novel drug uses. This was based on the hypothesis that if two diseases share similar therapies, then drugs that are currently used for only one of the two diseases may also be extrapolated to the other.

6.5.1.3 Heterogeneous network clustering

In a recent study (Wu *et al.*, 2013, Unpublished), we built a gene and feature-based (shared biological processes, pathways, phenotype) disease and drug heterogeneous network and applied network clustering to identify drug repositioning candidates. We used two state-of-art network clustering approaches [54, 55] to identify the modules of diseases–drugs. We downloaded known disease–gene and drug-target associations from KEGG Medicus (Feb, 2013), [56]. There were a total of 1301 diseases and 3613 drugs with at least one known gene association along with 1976 known indications (representing 364 diseases and 1066 drugs).

We first generated disease–disease, drug–drug, and disease–drug pairs based on shared genes or features. Thus, the nodes in our heterogeneous network were diseases and drugs while the edges represented either a shared gene or a shared feature (enriched biological process, pathway or phenotype). We first built a gene-based network where two nodes (disease or drug) are connected if they share a gene. We used *Jaccard* coefficient to measure the similarity between two nodes. Since a disease or drug can be related to other disease or drug even if they do not share a gene, we further enhanced our network by adding edges that represent shared features (biological processes, pathways, and mouse phenotype). To do this, we first performed an enrichment analyses of each of the disease and drug using ToppFun application of the ToppGene Suite [57]. For each of disease and drug, we first computed the enriched biological processes, pathways, and mouse phenotype. We then built a feature-based network where nodes represent disease or drug while the edges represent shared enriched features (biological process, mouse phenotype and pathways; p-value = 0.05 *Bonferroni* correction). We used *Jaccard* score to measure the feature similarity between each pair of the nodes. We thereby generated a list of disease–disease, drug–drug, and disease–drug pairs based on shared genes and/or enriched features. Based on whether a pair of nodes (disease–disease, disease–drug, and drug–drug) shares genes or enriched features or both, we assigned weights to all the edges in the filtered pairs. For instance, a pair of nodes with a weighted edge of 1 indicates that they share either a gene or one of the three features whereas a weight of 4 indicates that the two nodes showed significant associations (sharing not only a gene but also the three features, namely, biological process, pathway, and phenotype). The resulting weighted heterogeneous network consisted of 657 disease nodes and 3489 drug nodes. The total number of edges in this network is 116493; 680 edges were between two diseases, 1626 were between a disease–drug and 114187 between two drugs

We applied graph clustering to the weighted drug–disease heterogeneous network to extract densely connected clusters of diseases and drugs and mined them to extract potential candidates for drug repositioning. We used Cluster ONE [55] and Louvain's modularity [54] for the module detection. Using the Cluster ONE and Louvain detected communities we generated all possible disease–drug combinations on a per cluster basis. We call these the "drug repositioning candidates". To test the robustness of these novel drug repositioning candidate pairs, we removed 10% of the edges at a time and calculated the recovery rate of our predictions in a repetitive manner. Briefly, in each run, we randomly removed 10% of edges from the heterogeneous weighted disease–drug network and performed graph clustering (both ClusterONE and Louvain methods) to detect the communities and extract drug repositioning candidate pairs. We repeated this for ten times and compared the drug repositioning candidates with those from the original network (before randomly removing the 10% edges). The average recovery rate in case of drug repositioning candidates generated by ClusterONE was ~95% while in case of Louvain clustering it was ~85%. This demonstrates that the drug repositioning candidates we have discovered are robust and that additional edge removal or addition will not affect the output significantly.

Two of the drug repositioning candidates in our results that overlapped with the literature reports and clinical trials were derived from a cluster with drugs vismodegib and erismodegib and diseases basal cell carcinoma (BCC) and Gorlin syndrome. Interestingly, vismodegib, an oral inhibitor of the hedgehog pathway, is the first drug approved by the US Food and Drug Administration (FDA) for the treatment of locally advanced and metastatic BCC [58, 59]. Additionally, another study reported the efficacy of vismodegib on patients with Gorlin syndrome (basal cell nevus syndrome), a rare autosomal dominant disorder in which those with the disease are prone to developing multiple BCCs at an early age [60] (clinical trial NCT00957229). In our analyses, vismodegib and Gorlin syndrome do not share a common gene but are still clustered together because of the pathway-based connectivity (hedgehog signaling pathway). This demonstrates the utility of our approach in using feature-based heterogeneous networks to identify drug repositioning candidates.

6.6 Conclusions

In this chapter, we have outlined and shown how network-based computational approaches are used to address the drug repositioning problem. With the premise of deciphering mechanistic relationships underlying disease progress and therapeutic intervention, network-based methods are appealing because the focus is on the interactions or associations of various factors, While promising network-based approaches for drug repositioning have certain limitations. First, most of the current network-based computational approaches for drug discovery including drug repositioning tend to be gene-centric. As a result, diseases and drugs that do not have gene annotations are often ignored. Second, currently there is no systematic and efficient ranking strategy for the discovered disease–drug pairs or repositioning candidates. Most of them use either literature or clinical trials-based searches to "validate" or prioritize their predictions. However, this strategy may miss true positives simply because the records related to a novel association have not yet been added to the literature database or they are not yet in any clinical trials. Lastly, the lack of a "real", ready-to-use gold standard for drug repositioning makes it hard to validate most of these computational approaches in a more systematic way.

References

[1] Li, Y.Y. and S.J. Jones, Drug repositioning for personalized medicine. Genome Med, 2012. **4**(3): p. 27.

[2] Boguski, M.S., K.D. Mandl, and V.P. Sukhatme, Drug discovery. Repurposing with a difference. Science, 2009. **324**(5933): p. 1394–5.

[3] Sardana, D., et al., Drug repositioning for orphan diseases. Brief Bioinform, 2011. **12**(4): p. 346–56.
[4] Pujol, A., et al., Unveiling the role of network and systems biology in drug discovery. Trends Pharmacol Sci, 2010. **31**(3): p. 115–23.
[5] Wishart, D.S., Discovering drug targets through the web. Comp Biochem Physiol Part D Genomics Proteomics, 2007. **2**(1): p. 9–17.
[6] Keiser, M.J., et al., Predicting new molecular targets for known drugs. Nature, 2009. **462**(7270): p. 175–81.
[7] Kinnings, S.L., et al., Drug discovery using chemical systems biology: repositioning the safe medicine Comtan to treat multi-drug and extensively drug resistant tuberculosis. PLoS Comput Biol, 2009. **5**(7): p. e1000423.
[8] Campillos, M., et al., Drug target identification using side-effect similarity. Science, 2008. **321**(5886): p. 263–6.
[9] Yang, L. and P. Agarwal, Systematic drug repositioning based on clinical side-effects. PLoS One, 2011. **6**(12): p. e28025.
[10] Lamb, J., The Connectivity Map: a new tool for biomedical research. Nat Rev Cancer, 2007. **7**(1): p. 54–60.
[11] Lamb, J., et al., The Connectivity Map: using gene-expression signatures to connect small molecules, genes, and disease. Science, 2006. **313**(5795): p. 1929–35.
[12] Agarwal, P. and D.B. Searls, Literature mining in support of drug discovery. Brief Bioinform, 2008. **9**(6): p. 479–92.
[13] Freeman, L., A set of measures of centrality based upon betweenness. Sociometry, 1977. **40**: p. 35–41.
[14] Beauchamp, M.A., An improved index of centrality. Behav. Sci., 1965(10): p. 161–163.
[15] Sabidussi, G., The centrality index of a graph. Psychometrika, 1966. **31**: p. 581–603.
[16] Brin, S. and L. Page, The anatomy of a large-scale hypertextual Web search engine. Computer Networks and ISDN Systems, 1998(30): p. 107–117.
[17] Watts, D.J., and S. H. Strogatz, Collective dynamics of ‘small-world’ networks. Nature, 1998(393): p. 440–442.
[18] Friedkin, N.E., Structural cohesion and equivalence explanations of social homogeneity. Sociol. Method. Res., 1984. **12**: p. 235–261.
[19] Palla, G., et al., Uncovering the overlapping community structure of complex networks in nature and society. Nature, 2005. **435**(7043): p. 814–8.

[20] Hartwell, L.H., et al., From molecular to modular cell biology. Nature, 1999. **402**(6761 Suppl): p. C47–52.

[21] Eisenberg, D., et al., Protein function in the post-genomic era. Nature, 2000. **405**(6788): p. 823–6.

[22] Girvan, M. and M.E. Newman, Community structure in social and biological networks. Proc Natl Acad Sci U S A, 2002. **99**(12): p. 7821–6.

[23] Guimera, R. and L.A. Nunes Amaral, Functional cartography of complex metabolic networks. Nature, 2005. **433**(7028): p. 895–900.

[24] Rives, A.W. and T. Galitski, Modular organization of cellular networks. Proc Natl Acad Sci U S A, 2003. **100**(3): p. 1128–33.

[25] Székely, G.J.a.R., M. L., Hierarchical clustering via Joint Between-Within Distances: Extending Ward's Minimum Variance Method. Journal of Classification, 2005. **22**: p. 151–183.

[26] Samanta, M.P. and S. Liang, Predicting protein functions from redundancies in large-scale protein interaction networks. Proc Natl Acad Sci U S A, 2003. **100**(22): p. 12579–83.

[27] Xenarios, I., et al., DIP, the Database of Interacting Proteins: a research tool for studying cellular networks of protein interactions. Nucleic Acids Res, 2002. **30**(1): p. 303–5.

[28] Kanungo, T.M., D. M.; Netanyahu, N. S.; Piatko, C. D.; Silverman, R.; Wu, A. Y., An efficient k-means clustering algorithm: Analysis and implementation. EEE Trans. Pattern Analysis and Machine Intelligence, 2002. **24**: p. 881–892.

[29] Hans-Peter Kriegel, P.K., Jörg Sander, Arthur Zimek Density-based Clustering. WIREs Data Mining and Knowledge Discovery, 2011. **1**(3): p. 231–240.

[30] Martin Ester, H.-P.K., Jörg Sander, Xiaowei Xu A density-based algorithm for discovering clusters in large spatial databases with noise, in Proceedings of the Second International Conference on Knowledge Discovery and Data Mining (KDD-96). 1996. p. 226–231.

[31] Yu, H., et al., The importance of bottlenecks in protein networks: correlation with gene essentiality and expression dynamics. PLoS Comput Biol, 2007. **3**(4): p. e59.

[32] Vincent D Blondel, J.-L.G., Renaud Lambiotte, Etienne Lefebvre, Fast unfolding of communities in large networks. Journal of Statistical Mechanics: Theory and Experiment, 2008. **10**.

[33] Junker, B.H., D. Koschutzki, and F. Schreiber, Exploration of biological network centralities with CentiBiN. BMC Bioinformatics, 2006. **7**: p. 219.

[34] Chandra, N. and J. Padiadpu, Network approaches to drug discovery. Expert Opin Drug Discov, 2013. **8**(1): p. 7–20.
[35] Wu, Z., Y. Wang, and L. Chen, Network-based drug repositioning. Mol Biosyst, 2013. **9**(6): p. 1268–81.
[36] Barabasi, A.L. and Z.N. Oltvai, Network biology: understanding the cell's functional organization. Nat Rev Genet, 2004. **5**(2): p. 101–13.
[37] Li, Y. and P. Agarwal, A pathway-based view of human diseases and disease relationships. PLoS One, 2009. **4**(2): p. e4346.
[38] Mei, H., et al., Opportunities in systems biology to discover mechanisms and repurpose drugs for CNS diseases. Drug Discov Today, 2012. **17**(21–22): p. 1208–16.
[39] Hu, G. and P. Agarwal, Human disease-drug network based on genomic expression profiles. PLoS One, 2009. **4**(8): p. e6536.
[40] Wu, M., X. Yang, and C. Chan, A dynamic analysis of IRS-PKR signaling in liver cells: a discrete modeling approach. PLoS One, 2009. **4**(12): p. e8040.
[41] Jeong, H., et al., Lethality and centrality in protein networks. Nature, 2001. **411**(6833): p. 41–2.
[42] Florez, A.F., et al., Protein network prediction and topological analysis in Leishmania major as a tool for drug target selection. BMC Bioinformatics, 2010. **11**: p. 484.
[43] Raman, K., K. Yeturu, and N. Chandra, targetTB: a target identification pipeline for Mycobacterium tuberculosis through an interactome, reactome and genome-scale structural analysis. BMC Syst Biol, 2008. **2**: p. 109.
[44] Yang, L., et al., Identifying unexpected therapeutic targets via chemical-protein interactome. PLoS One, 2010. **5**(3): p. e9568.
[45] Zhao, S. and S. Li, Network-based relating pharmacological and genomic spaces for drug target identification. PLoS One, 2010. **5**(7): p. e11764.
[46] Wishart, D.S., et al., DrugBank: a knowledgebase for drugs, drug actions and drug targets. Nucleic Acids Res, 2008. **36**(Database issue): p. D901–6.
[47] Lee, H.S., et al., Rational drug repositioning guided by an integrated pharmacological network of protein, disease and drug. BMC Syst Biol, 2012. **6**: p. 80.
[48] Li, S., et al., Herb network construction and co-module analysis for uncovering the combination rule of traditional Chinese herbal formulae. BMC Bioinformatics, 2010. **11 Suppl 11**: p. S6.

[49] Iorio, F., et al., Discovery of drug mode of action and drug repositioning from transcriptional responses. Proc Natl Acad Sci U S A, 2010. **107**(33): p. 14621–6.

[50] Suthram, S., et al., Network-based elucidation of human disease similarities reveals common functional modules enriched for pluripotent drug targets. PLoS Comput Biol, 2010. **6**(2): p. e1000662.

[51] Barrett, T., et al., NCBI GEO: mining tens of millions of expression profiles–database and tools update. Nucleic Acids Res, 2007. **35**(Database issue): p. D760–5.

[52] Goel, R., et al., Human Protein Reference Database and Human Proteinpedia as resources for phosphoproteome analysis. Mol Biosyst, 2012. **8**(2): p. 453–63.

[53] Chiang, A.P. and A.J. Butte, Systematic evaluation of drug-disease relationships to identify leads for novel drug uses. Clin Pharmacol Ther, 2009. **86**(5): p. 507–10.

[54] Vincent, D.B., et al., Fast unfolding of communities in large networks. Journal of Statistical Mechanics: Theory and Experiment, 2008. **2008**(10): p. P10008.

[55] Nepusz, T., H. Yu, and A. Paccanaro, Detecting overlapping protein complexes in protein-protein interaction networks. Nat Meth, 2012. **9**(5): p. 471–472.

[56] Kanehisa, M., et al., KEGG for representation and analysis of molecular networks involving diseases and drugs. Nucleic Acids Research, 2010. **38**(suppl 1): p. D355–D360.

[57] Chen, J., et al., ToppGene Suite for gene list enrichment analysis and candidate gene prioritization. Nucleic Acids Research, 2009. **37** (suppl 2): p. W305–W311.

[58] Sekulic, A., et al., Efficacy and safety of vismodegib in advanced basal-cell carcinoma. N Engl J Med, 2012. **366**(23): p. 2171–9.

[59] Von Hoff, D.D., et al., Inhibition of the hedgehog pathway in advanced basal-cell carcinoma. N Engl J Med, 2009. **361**(12): p. 1164–72.

[60] Tang, J.Y., et al., Inhibiting the hedgehog pathway in patients with the basal-cell nevus syndrome. N Engl J Med, 2012. **366**(23): p. 2180–8.

7

Hypothesis Driven Multi-target Drug Design

Prem kumar Jayaraman[1], Dr. Mohammad Imran Siddiqi[2], Meena K Sakharkar[3], Ramesh Chandra[4] and Kishore R Sakharkar[5]

[1]Biomedical Engineering Research Center
Nanyang Technologies University Singapore
[2]CDRI, Lucknow India
[3]Department of Pharmacy and Nutrition,
University of Saskatchewan, SK, Canada
[4]Department of Chemistry, Delhi University, India,
B. R. Ambedkar Center for Biomedical Research,
University of Delhi, India
[5]OmicsVista, Singapore

7.1 Introduction

The discovery and development of antibiotics in the early 1900s is undoubtedly one of the greatest achievements of medicine and must rank as one of the most important events in the history of humankind. More than half a century of antibiotic use has led to the emergence of bacterial resistant strains able to resist many commercially available antibiotics in the market [1, 2]. There are many reasons associated with the emergence of bacterial resistance, including unnecessary, inappropriate use of antimicrobials, drug toxicity, increased morbidity and healthcare costs [3]. Certainly, the introduction of chemotherapeutic to combat the bacterial infections is one part of the success story but the management of antimicrobial agents concerning the consumption and overuse imposed by selection pressure of different antibiotics has contributed to forming a reservoir of resistance genes among bacterial pathogens [4]. The proper dosage and administration of available antibacterial compounds needs to be updated considering the patient's clinical response and it is a major

Post-genomic Approaches in Drug and Vaccine Development, 133–178.

challenge constant among the clinical settings due to the appearance and spread of multi-drug and pan-drug resistant pathogens such as imipenem-resistant *P. aeruginosa* and/or *A. baumannii*, carbapenems-resistant Gram-negative bacteria harboring carbapenemases and vancomycin-resistant *Enterococcus spp.* [5]. Overall, the total consumption/dosage regimes of antimicrobials are not alone to blame for the rapid development of resistance, in addition many different forms of antibiotic resistance mechanisms exemplified by both Gram-positive and Gram-negative pathogens also play a larger part [6].

Especially, Gram-negative pathogens such as *P. aeruginosna* have developed sophisticated resistance strategies through multiple mechanisms that has led to the selection of multi-drug resistant phenotypes, which have extremely limited the therapeutic options for *Pseudomonas* infections [7, 8]. *P. aeruginosa* is a highly drug resistant and opportunistic pathogen. Due to the permeability barrier in the outer membrane it is naturally resistant to many antibiotics. Multi-drug resistant phenotypes in *P. aeruginosa* occur through the acquisition of multiple imported resistance mechanisms coupled with chromosomally encoded resistance mechanisms, accumulation of multiple chromosomal changes over time and/or a single mutational event leading to over expression of one or more efflux pump(s) [9, 10]. This opportunistic "ESKAPE" pathogen is currently causing the majority of the nosocomial and community-acquired infections and effectively "escapes" the effects of many commercially available antibacterial drugs, thereby joining the ranks of "superbugs" [11, 8]. Alarmingly, the prospects for novel anti-pseudomonal agents in subsequent years are quite poor with most of the drugs in the pipeline being simply more or less new derivatives of existing families of antimicrobial agents [12, 13]. Hence, there is a pressing need to identify novel therapeutic strategies and more potent drugs to combat the constant evolution of resistance and circumvent the spread of multi-drug and pan-drug resistance development in *P. aeruginosa*.

Despite the growing concerns with the rise in resistance rates amongst important bacterial pathogens, for the past few decades the discovery of novel antimicrobial agents has declined steeply [14, 15]. In spite of increasing levels of investments in Pharmaceutical R & D, the question raised here is why is there a high clinical attrition rate of drugs in early/late stage of preclinical or clinical development? Many reasons have been proposed for this decline in academic/pharmaceutical research focused on the discovery and production of new effective antimicrobials.

The most fundamental reason is that most of the important targets in main metabolic pathways which allow selective toxicity have already been discovered and all the classes of antimicrobial agents for those obvious targets were also being developed and marketed [16, 15]. In addition, the common causes for shortage of the antimicrobial compounds in the late-stage pre-clinical development are due to lack of efficacy, clinical safety and toxic side effects accounting each for 30% failures [17]. However, in the recent years there is mounting evidence that the decline in productivity has been to a larger extent concurrently with the introduction of target-based drug discovery [18]. This drug discovery paradigm for decades was largely based upon discovery and development of therapeutics that exert their activities by modulating individual targets. Also, this dominant paradigm of 'one drug, one target, one disease' demonstrates the therapeutic effect based on Ehrlich's philosophy [26] of 'magic bullets' through identification of disease-associated gene with *in vivo* validation and subsequent screening of drugs that modulates the specific gene or gene products [18]. Although this approach has led to the discovery of many successful antimicrobials, recent studies revealed that the existence of inherent redundancy and robustness in many biological networks leading to the poor efficacy of monotherapeutics implies that inhibiting a single target might fail to produce the desired therapeutic effect [18, 19].

Given the crisis of a continuing trend in new resistance mechanisms along with increase in clinical attrition figures, the future of medicine without any doubt looks bleak. This means that the present drug discovery process is no longer sustainable [16] in the long run hence a definite change in direction is absolutely necessary to keep our hopes alive.

It is now widely recognized that, in contrast to 'one drug, one target, one disease', a single chemical entity able to target several nodes in a disease-causing network is considered to be highly beneficial for treating complex diseases [20, 19]. In comparison to single-strong affinity in monotherapy, multi-target inhibitors have low affinities towards their targets, since it expands the number of target molecules with smaller side effects and toxicity [21, 22]. In particular, multi-targeting offers an indirect relation of many existing targets of pharmacologically important pathways easing the constraints of druggability and the multiplied low affinities has more intense effects [21]. Thus, the underlying importance in the development of drugs with a desired polypharmacology profile are identification of targets to be perturbed in the disease-causing network and lead compounds with the desired biological activity against the identified target combinations [19]. The three complementary methods listed for the comparative analyses of disease-causing networks are systematic screening (drug–drug combination assays), knowledge-based combinations and network analyses [19]. The rational quest of cell-based *in vitro* drug combination assays between unknown compounds and drugs with known biomolecular targets can identify interactions between disease relevant pathways and multi-target mechanisms providing a new path for drug discovery in the coming decades [23].

Many specific lead generation strategies have been used earlier for the rational design of inhibitors with predefined multi-target profile, of which most frequently reported is the framework combination strategy. In this approach, two separate pharmacophores that inhibit different targets are synthetically linked or merged into a single molecule [24, 31]. The major complexity that is involved in the design of multiple ligands lies in balancing both drug-like physicochemical properties and desired multi-target profiles with unwanted off-target effects. However, if the starting compounds are small and the fragments are well integrated the framework combination strategy can be successful in generating multiple ligands with good oral bioavailability [25]. Previously, several studies have proposed an alternative standard antibacterial therapy for broader empirical coverage through the simultaneous administration of two drugs each with a separate action leading to a synergistic combination effect [27, 28]. Though the use of antibiotic combinations for serious infections is a recommended therapy in many hospitals, there is no direct evidence to support its clinical effectiveness against *P. aeruginosa* infections and in addition there are many potential shortcomings using combination therapy of which more concern are the effects of increased drug toxicity, higher treatment costs and enhanced risk of cross-infections due to the creation of more resistant organisms [29, 30]. In comparison to the drug combinations, a single multi-target inhibitor has greater predictable pharmacokinetic and pharmacodynamic relationships due to the administration of a single drug, increased penetration capacity due to additional pharmacophore, lower toxicity and delayed onset, thus overcoming thus the existing bacterial resistance mechanisms due to activity at multiple sites [31].

Recently, the hybrid antibiotic approach in targeting multiple enzymes has been quite successful, such as trimethoprim (benzyl pyrimidine) linked fluoroquinolone (BP-4Q-002) hybrid compounds [32], aminoglycoside-fluoroquinolone [33] and peptide-aminoglycosides hybrids [34] have been shown to be potent antimicrobial agents against both Gram-negative and Gram-positive bacteria. In addition, three hybrid compounds TD-1792, MCB-3837 and CBR-2092 currently in human clinical trials are reported to have significantly higher activity against multi-drug resistant strains in comparison to both the drugs alone and in combination [31]. Therefore, from the recent progress in the design and development of multi-target hybrid compounds, a promising strategy for fresh research avenues has emerged.

As a first step in the discovery of multi-target therapeutics the most crucial part is to identify or prioritize possible combination of targets that can be targeted by a single agent to achieve an optimal therapeutic benefit [19]. The network-based knowledge of scientific literature is increasingly gaining interest for the insights it offers into the identification of relevant target sets and development of multi-target therapeutics [19]. The knowledge-based methods employ text mining of biological data from the scientific literature i.e. exploit information from literature to generate novel hypotheses and connecting concepts by new and indirect associative techniques to contribute to drug discovery [35, 36]. Natural secondary metabolites from plants are considered as privileged scaffolds that display multiple biological effects by interacting with multiple proteins [37]. Also, botanical mixtures that contain various bioactive natural products shutdown the whole process either by network pharmacology or biochemical synergism by weakly binding to multiple proteins within the same signaling network [38].

In light of the meager antibiotic pipeline for Gram-negative bacteria and threat of multi-drug resistance, the motivation for rational design of multi-target inhibitors came from one of the best studied and investigated plant polyphenol, epigallocatechin gallate (EGCG, from green tea). EGCG and its derivatives have numerous biological activities including interactions with multiple enzymes, such as DNA gyrase subunit B (GyrB)/topoisomerase IV subunit B (ParE) and dihydrofolatereductase (DHFR) [39–41]. It is reported that ligands with exact and similar structures are more likely to have similar properties and therefore often bind proteins with similar active sites and based on this, proteins can be related for multi-target drug design [42, 43]. Among these well-established targets of antibacterial agents, GyrB and ParE belong to DNA replication pathway and enzyme DHFR belongs to folate biosynthesis pathway (DNA precursor's synthesis pathway). Both GyrB and ParE are highly homologous enzymes and the inhibitors (coumarins and cyclothia-lidines) bind to the ATP binding pocket in the N-terminal domain of GyrB/ParE and block the access of ATP [44]. DHFR enzyme is another successful target for the bacterial folate competitive inhibitor trimethoprim and is often given in combination with sulfonamides (competitive inhibitors of DHPS) for a broader coverage in the clinical settings [45]. The use of hybrid-linked molecules incorporating both sulfonamide groups and pyrimidine subunits (trimethoprim) have earlier been shown to have simultaneous inhibition of both DHPS and DHFR enzymes leading to autosynergism similar to the combinations of sulfonamides and trimethoprim [46]. Recently, Bennett *et al.* [47] reported an inhibitor (MANIC compound, isocytosine derivative) that binds to both the pterin site of DHPS and folate binding site of DHFR inhibiting both the enzymes and has also recommended the use of this inhibitor as a fragment to build compounds that are more potent against both the enzymes [47] (Figure 7.1A). Using these data the biomedical literature was surveyed to establish a common connecting bridge between the two most important pathways: DNA replication pathway enzymes (GyrB and TopoIV) and folate biosynthesis pathway enzymes (DHPS and DHFR) and it was found out that the derivatives of benzene sulfonamide groups are reported to bind to all the four enzymes [48, 49] (Figure 7.1A). This approach of discovering new hidden relationships is known as co-occurrence-based methods, which establishes relationships between biomedical concepts such as genes and pathways [50].

Previously, using drug-phytochemical combination assays with network analyses it is shown that phytochemicals (protocatechuic acid, gallic acid, quercetin and myricetin) which are structural analogues and fragments of epigallocatechin gallate, bind to *P. aeruginosa* folate binding site of DHFR and ATP-binding site of GyrB/ParE enzymes and are synergistic in combination with sulfonamides (sulfamethoxazole and sulfadiazine) [51–53]. However, are more importantly, the major concern in at study was higher minimum inhibitory concentration (MIC) values of the chosen phytochemicals compared to that of the antibiotics used. In support of this, several studies have reported that phytochemicals have lesser stability and selectivity, adverse phytochemical-antibiotic interactions and limited bioavailability. They are also less potent due to their MIC ranges (greater than 1, 000 μg/ml) in comparison to the compounds from microbial origin (antibiotics) [54, 55]. Interestingly, phytochemicals can be converted to pharmacologically acceptable antimicrobial agents by optimization through structural modifications [56].

Thus, using three dimensional-structure-based pharmacophore, information for all the four enzymes (GyrB, TopoIV, DHPS and DHFR) from published literature coupled with a framework combination strategy (Figure 7.1B), design and validation of a novel class of hybrid compounds (structural analogue of MANIC fragment-benzene sulfonamide groups conjugates) were investigated for simultaneous inhibition of these quadri-enzymes using various *in silico* methods. Important interactions for the inhibition of the enzymes (GyrB/TopoIV) and DHFR of *E. coli* have been identified based on their individual pharmacophore models from molecular dynamics simulations [57, 58]. Similarly, for DHPS of *B. anthracis*, pharmacophore model based on

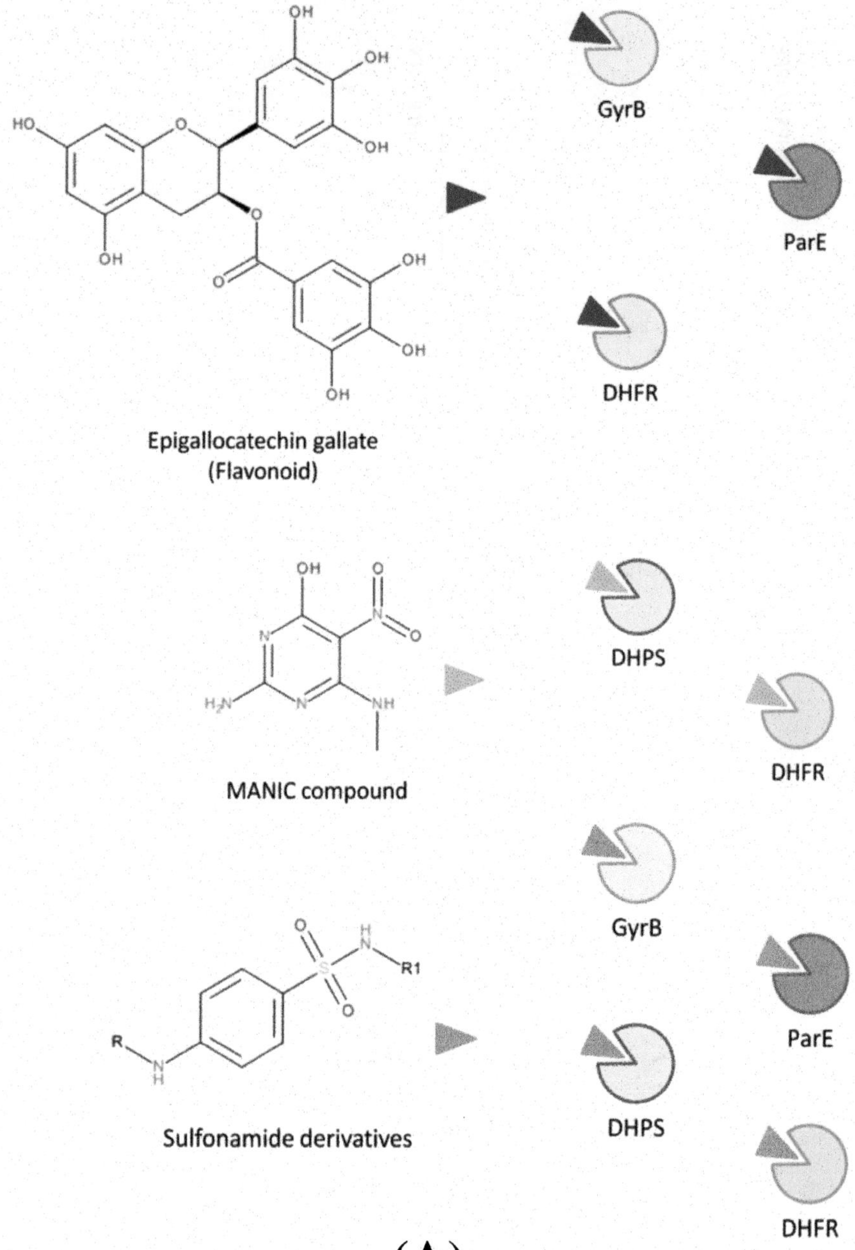

GyrB

ParE

DHFR

Epigallocatechin gallate
(Flavonoid)

DHPS

DHFR

MANIC compound

GyrB

ParE

DHPS

Sulfonamide derivatives

DHFR

(A)

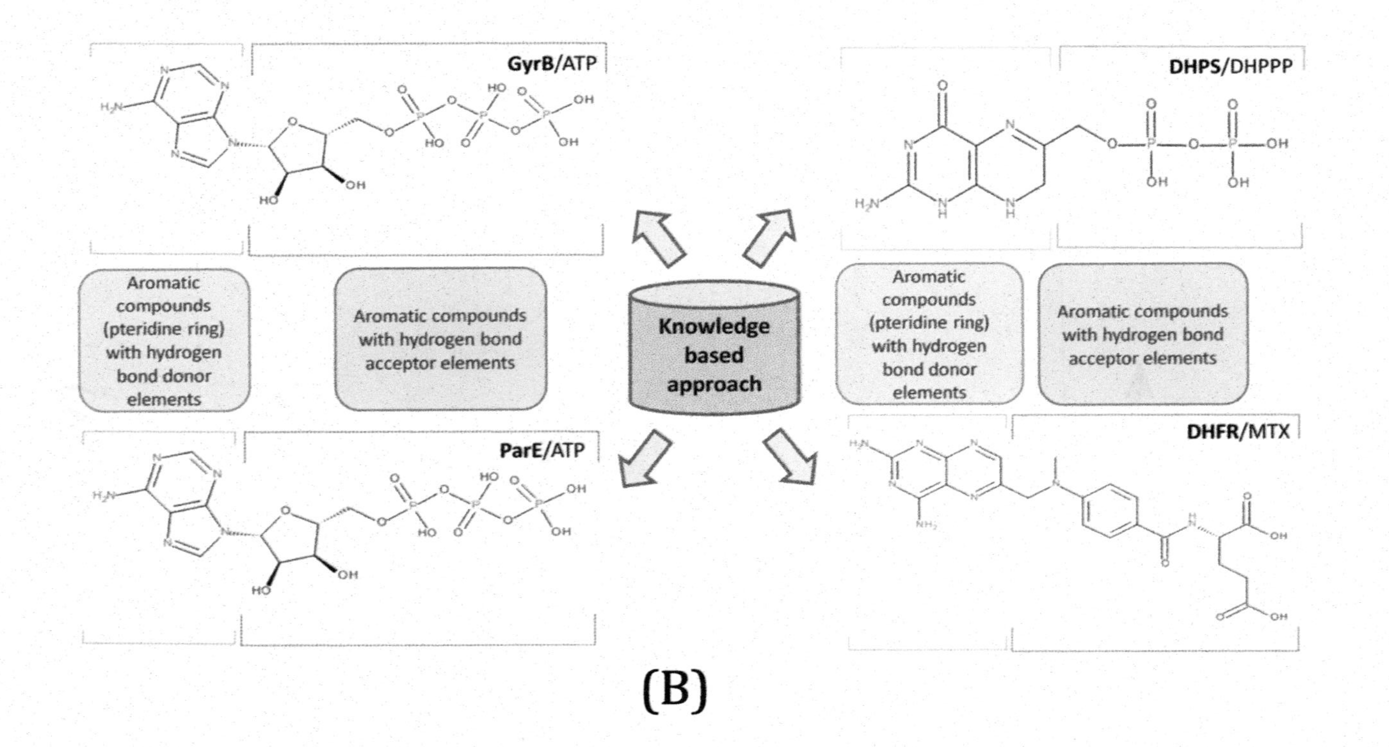
GyrB/ATP
DHPS/DHPPP
Aromatic compounds (pteridine ring) with hydrogen bond donor elements
Aromatic compounds with hydrogen bond acceptor elements
Knowledge based approach
Aromatic compounds (pteridine ring) with hydrogen bond donor elements
Aromatic compounds with hydrogen bond acceptor elements
ParE/ATP
DHFR/MTX
(B)

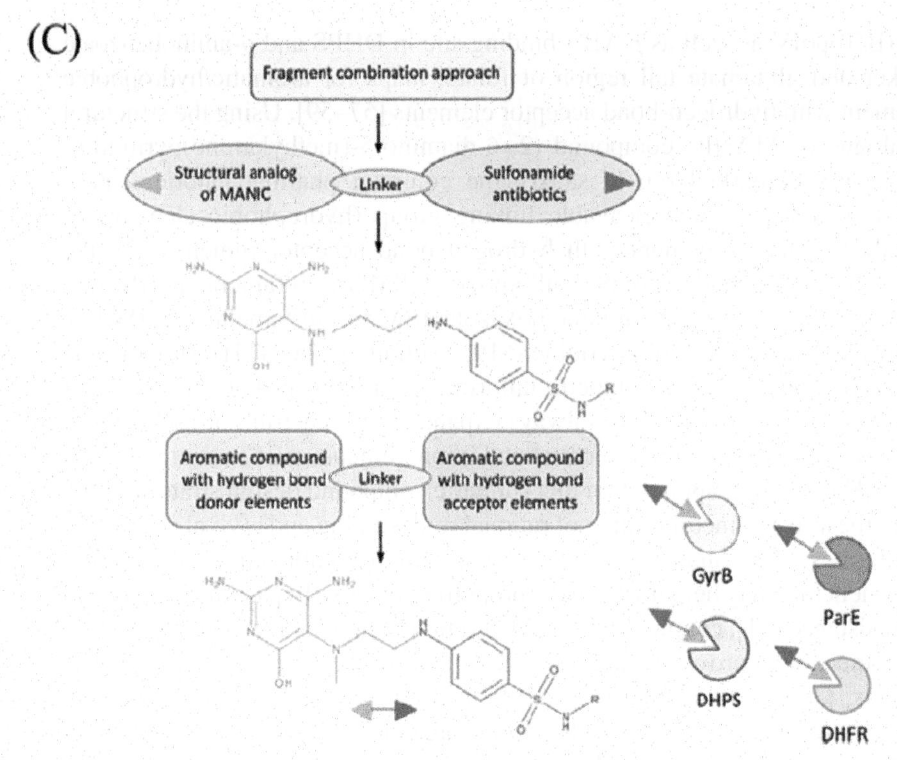

Figure 7.1 How we identified the targets? - Hypothesis (III) and design of multi-action hybrid compounds. (A) Mechanism of action of epigallocatechin gallate (plant flavonoid, red arrow head) targeting multiple enzymes GyrB, ParE & DHFR, MANIC (blue arrow head), new class of pterin-based dual inhibitors of DHPS and DHFR, sulfonamide derivatives (violet arrow head) that binds to all the four reported enzymes; (B) Knowledge-based approach was used to establish relationship between therapeutic targets (GyrB, ParE, DHPS and DHFR). Common pharmacophoric features derived for multi-inhibition of GyrB, ParE, DHPS and DHFR enzymes from previously reported individual pharmacophore models and known pterin/pteridine-based inhibitors; (C) Schematic illustration of the fragment combination method used to derive the hybrid compounds (structural analogue of MANIC compound (blue arrow head) linked with antibiotics (violet arrow head) using non-cleavable linker (orange line) that can simultaneously bind to all the four enzymes.

the pterin two-ring structure of the substrate has been derived [59]. Combining all these models a common pharmacophoric feature model (Figure 7.1B) was derived for multiple inhibition of the quadri enzymes are (1) an aromatic element with hydrogen-bond donor elements mapping the adenine moiety of ATP in GyrB/TopoIV, pterin moieties of dihydropterin pyrophosphate in DHPS and dihydrofolate in DHFR [57–59], (2) ribose and phosphate groups of ATP in

GyrB/TopoIV as well as *p* ABA binding site in DHPS and *p*-aminobenzoate linker and glutamate tail region of folate, maps for aromatic/hydrophobic element with hydrogen-bond acceptor elements [57–59]. Using the structural analogue of MANIC compound (2, 6-diamino-5-(methylamino)pyrimidin-4-ol) was embarked on, to satisfy the common pharmacophoric feature (1) linked with a non-cleavable linking group (hydrophobic element) to another aromatic element with hydrogen bond acceptor moieties (2), the benzene sulfonamide groups (sulfamoxole (SMO), sulfamerazine (SMR), sulfamethazine (SMT), sulfamethoxazole (SMX), sulfamethizole (SMZ), sulfisoxazole (SOX), sulfapyridine (SPR), sulfathiazole (STH), sulfadiazine (SUD)) (Figure 7.1C). Synthetic tailoring of previously known classes of antibiotics over the past have always expanded the spectrum of use of those agents and has also offered incremental improvements over current therapies [60]. The winning approach in this chimeric compound design strategy in the hypothesis is the incorporation of rational re-designing and revival of existing antibiotics (sulfonamides) through the structural combination with another fragment that may help improve the properties and also bypass the mechanism of resistance [61], coupled with the integration of multi-target specificity. This hypothesis also provides the most crucial link for the current accessible drug targets, GyrB/ParE and DHPS/DHFR in the biological network. It is also significant that when designed multiple ligands have activity against four different enzymes from two different pharmacologically important pathways, it possibly will serve to mitigate the development and spread of antibiotic resistance. Clearly, this hypothesis forms a base for multi-target drug discovery and reveals the enhancement of already approved drugs through a hybrid design framework for the treatment of complex diseases. The two-dimensional representations of the designed hybrid compounds are shown in Figure 7.2.

Designing new drug scaffolds, containing pyrimidine moiety conjugated with benzene-sulfonamide derivatives (sulfonamide antibiotics), is an alternative novel drug-design strategy for inhibition of quadri-enzymes DNA gyrase subunit B (GyrB) and topoisomerase IV subunit B (ParE) dihydropteroate synthase (DHPS), dihydrofolatereductase (DHFR) of two different pathways.

Further on, the proposed hybrid-drug design hypothesis was also successfully demonstrated using various integrated computer-aided molecular modeling methodologies such as docking, dynamics simulations and molecular/electronic properties analyses. The *in silico*-based methods such as molecular docking, combined molecular docking and common pharmacophore identification and fragment combination have earlier been shown to

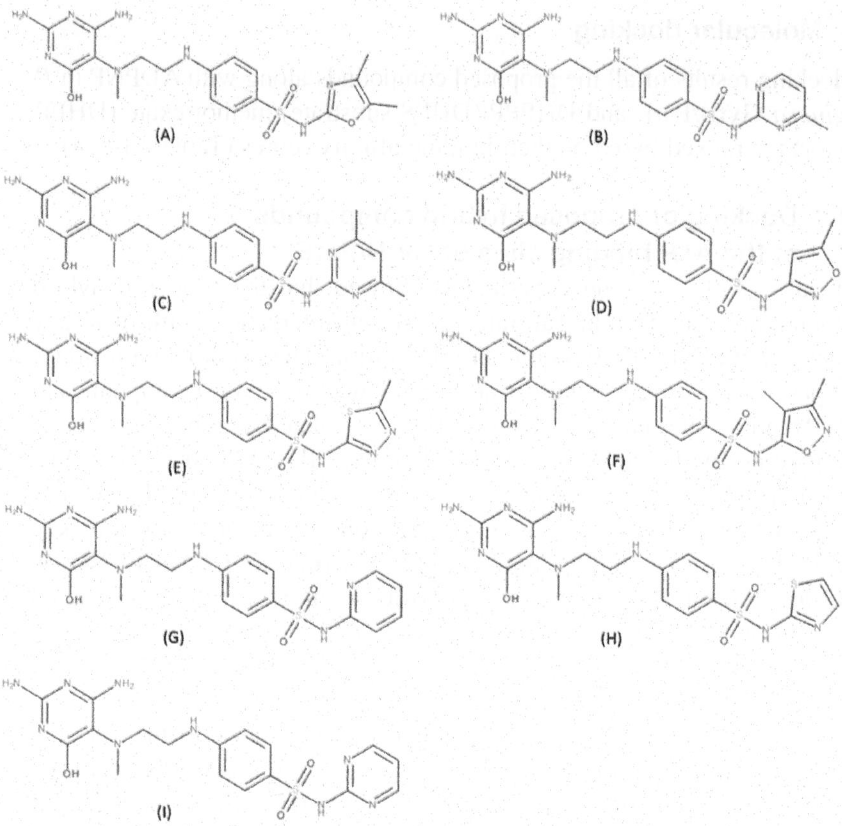

Figure 7.2 Structures of the designed hybrid compounds. (A) MD_SMO; (B) MD_SMR; (C) MD_SMT; (D) MD_SMX; (E) MD_SMZ; (F) MD_SOX; (G) MD_SPR; (H) MD_STH; (I) MD_SUD.

be successful in the design and discovery of multi-target compounds directed at selective multiple targets [62]. The proposed hybrid compound's drug-like pharmacotherapeutic profiles was also investigated using various physico-chemical properties prediction mechanism and also the derived common pharmacophoric features were verified using stereo-electronic properties calculated by quantum chemical methods. To the best of the author's knowledge, this study is the first to successfully combine knowledge from the scientific literature with previous drug combination data and fragment combination approach towards the identification of novel hybrid compounds that can strongly provide potential model systems in the discovery of antimicrobial hybrid drugs development.

7.1.1 Molecular docking

The docking results of all the proposed compounds along with ADPNP in *P. aeruginosa* GyrB/ParE and DHPPP (DHPS substrate)/methotrexate (DHFR inhibitor) are ranked based on their binding affinity scores (Table 7.1–7.4).

7.1.1.1 Docking of proposed hybrid compounds at the ATP binding sites of GyrB/ParE

The results of the docking experiments showed that the proposed hybrid compounds have good binding affinities and strong interactions with all the key residues of both GyrB/ParE required for inhibition. The list of major residues participating in hydrogen bonding interactions with the hybrid compounds in the ATP binding sites cavity of GyrB/ParE are tabulated in Table 7.5.

Regarding the top ranked hybrid compound MD_SMX, the pyrimidine ring occupies the adenine binding region of ATP in both GyrB/ParE enzymes forming hydrogen bond interactions with the conserved water molecule, key **GyrB**/ParE residues including **Asp75**/Asp68, **Thr167**/Thr162, **Ser49**, **Gly79**, **Glu45** and the nitrogen atom at the 5-position engages in hydrogen bond interactions with **Tyr111**/Tyr104 residues in GyrB/ParE respectively (Figures 7.3A & 7.4A).

The phenyl ring of MD_SMX occupies both ribose and phosphate binding regions of ATP in both **GyrB**/ParE enzymes, while the sulfonyl groups positions are similar to the α -phosphate group of ADPNP forming hydrogen bond interactions with **Asn48**/Asn41, **Val122,** Ile115 and Gly114 residues. In addition, residues **Ile80**/Ile89, **Val122**/Ile115, **Val120**/Val113 form favourable hydrophobic contacts with the pyrimidine ring and phenyl ring of sulfamethoxazole in MD_SMX. The oxazole ring of MD_SMX occupies the binding regions in between β-phosphate and γ-phosphate groups of ADPNP in both **GyrB**/ParE and engages in hydrogen bond interactions with **Gly119**/Gly112, **His118**/His111, **Leu117**/Leu110 and Lys339/Lys333 residues in domain II region of both the enzymes. The docked poses of the entire designed hybrid compounds have similar binding conformations and showed similar interactions binding at the ATP binding sites of both GyrB/ParE (Figures 7.3B & 7.4B)

7.1.1.2 Docking of proposed hybrid compounds at the pterin binding site of DHPS

The docking results show that the proposed hybrid compounds have similar binding conformations and good binding affinities at the pterin binding pocket

Table 7.1 Docking Results Of Adpnp And Designed Hybrid Compounds (Ranked On The Basis Of Binding Affinity Scores) At The Atp Binding Site In *P. aeruginosa* Gyrb

Compounds	Scoring function		E_{inter} [c]	E_{intra} [d]	Hbond[e]	LE1[f]	LE3[g]	Binding affinity (KJ/mol)
	MolDock[a]	Rerank[b]						
MD_SMX	–192.67	–158.03	–205.32	12.65	–24.07	–6.42	–5.27	–38.42
MD_SMZ	–189.26	–154.23	–204.92	15.65	–23.78	–6.31	–5.14	–37.89
MD_SUD	–185.87	–156.51	–210.99	25.13	–21.05	–6.19	–5.22	–37.79
MD_STH	–170.94	–133.65	–191.94	21	–19.48	–5.89	–4.61	–36.68
MD_SOX	–194.895	–148.92	–208.73	13.83	–23.48	–6.29	–4.8	–36.56
MD_SPR	–181.09	–154.98	–205.67	24.59	–20.45	–6.03	–5.17	–35.16
MD_SMO	–194.13	–145.27	–202.63	8.5	–20.53	–6.26	–4.69	–34.91
MD_SMT	–186.81	–150.95	–212.78	25.97	–19.43	–5.84	–4.72	–34.33
MD_SMR	–189.74	–158.19	–213.66	23.92	–18.97	–6.12	–5.1	–34.27
ADPNP	–234.13	–183.03	–219.98	–14.15	–28.94	–7.55	–5.9	–33.13

[a]MolDock: docking score function by MolDock; [b]Rerank: reranking score function; [c]E_{Inter}: ligand – protein (DHPS) interaction energy; [d]E_{Intra}: internal energy of the ligand; [e]HBond: hydrogen bonding energy; [f]LE1: Ligand efficiency 1 - MolDock score divided by Heavy atoms count; [g]LE3: Ligand efficiency 3 – Rerank score divided by Heavy atoms count; ADPNP (non-hydrolyzable analogue of ATP) – GyrB substrate

Table 7.2 Docking results of ADPNP and designed hybrid compounds (ranked on the basis of binding affinity scores) at the ATP binding site in *P. aeruginosa* ParE.

Compounds	Scoring function		E_{inter} [c]	E_{intra} [d]	Hbond[e]	LE1[f]	LE3[g]	Binding affinity (KJ/mol)
	MolDock[a]	Rerank[b]						
MD_SMX	–183.89	–143.72	–198.27	14.37	–23.39	–6.13	–4.79	–38.11
MD_SMZ	–183.99	–143.42	–199.48	15.49	–23.47	–6.13	–4.78	–36.84
MD_STH	–177.62	–146.3	–194.78	17.16	–17.89	–6.12	–5.04	–35.56
MD_SMR	–178.87	–148.36	–204.62	25.75	–18.01	–5.77	–4.79	–35.41
MD_SMO	–187.54	–155.86	–202.92	15.38	–16.58	–6.05	–5.03	–34.47
MD_SUD	–174.01	–139.79	–188.67	14.63	–18.65	–5.8	–4.66	–34.32
MD_SPR	–177.19	–150.47	–202.29	25.09	–15.65	–5.91	–5.02	–33.86
MD_SOX	–196.31	–161.22	–206.87	10.56	–17.93	–6.33	–5.2	–33.68
MD_SMT	–186.64	–153.13	–210.97	24.33	–15.98	–5.83	–4.79	–33
ADPNP	–234.21	–184.57	–224.59	–9.62	–32.15	–7.56	–5.95	–33.86

[a]MolDock: docking score function by MolDock; [b]Rerank: reranking score function; [c]E_{Inter}: ligand – protein (DHPS) interaction energy; [d]E_{Intra}: internal energy of the ligand; [e]HBond: hydrogen bonding energy; [f]LE1: Ligand efficiency 1 - MolDock score divided by Heavy atoms count; [g]LE3: Ligand efficiency 3 – Rerank score divided by Heavy atoms count; ADPNP (non-hydrolyzable analogue of ATP) – ParE substrate

Table 7.3 Docking results of DHPPP and designed hybrid compounds (ranked on the basis of binding affinity scores) at the pterin binding site in *P. aeruginosa* DHPS

Compounds	Scoring function		E_{inter}[c]	E_{intra}[d]	Hbond[e]	LE1[f]	LE3[g]	Binding affinity (KJ/mol)
	MolDock[a]	Rerank[b]						
MD_SMX	–158.97	–120.2	–172.46	13.49	–23.54	–5.29	–4.01	–36.94
MD_SMZ	–158.74	–120.53	–171.79	13.06	–24.22	–5.29	–4.02	–36.21
MD_SOX	–163.11	–123.39	–175.5	12.39	–24.36	–5.26	–3.98	–36.19
MD_STH	–144.67	–110.55	–159.62	14.84	–20.87	–4.99	–3.81	–35.6
MD_SPR	–137.55	–93.31	–161.94	24.39	–20.84	–4.58	–3.11	–35.27
MD_SUD	–141.35	–108.42	–163.25	21.9	–20.82	–4.71	–3.61	–34.17
MD_SMO	–142.44	–44.08	–155.73	13.29	–19.89	–4.59	–1.42	–33.94
MD_SMR	–148 18	–114.27	–167.91	19.73	–20.17	–4.78	–3.69	–33.36
MD_SMT	–147 27	–99.99	–169.2	21.93	–19.03	–4.6	–3.12	–32.95
DHPPP	–116.91	–96.71	–123 89	6.982	–29.33	–5.31	–4.39	–33.33

[a]MolDock: docking score function by MolDock; [b]Rerank: reranking score function; [c]E_{Inter}: ligand – protein (DHPS) interaction energy; [d]E_{Intra}: internal energy of the ligand; [e]HBond: hydrogen bonding energy; [f]LE1: Ligand efficiency 1 - MolDock score divided by Heavy atoms count; [g]LE3: Ligand efficiency 3 – Rerank score divided by Heavy atoms count; DHPPP (dihydropterin pyrophosphate) –DHPS substrate

Table 7.4 Docking results of MTX and designed hybrid compounds (ranked on the basis of binding affinity scores) at the folate binding site in *P. aeruginosa* DHFR

Compounds	Scoring function		E_{inter}[c]	E_{intra}[d]	Hbond[e]	LE1[f]	LE3[g]	Binding affinity (KJ/mol)
	MolDock[a]	Rerank[b]						
MD_SMX	–140.09	–113.97	–146.08	5.99	–20.89	–4.67	–3.79	–34.71
MD_SMZ	–134.16	–107.69	–140.7	6.6	–20.44	–4.47	–3.59	–34 61
MD_SPR	–116.61	–93.17	–133.09	16.48	–16.87	–3.89	–3.11	–32.63
MD_STH	–123.33	–92.82	–129.31	5.98	–14.96	–4.25	–3.2	–32.52
MD_SOX	–139.51	–113.27	–141.33	1.82	–15.73	–4.5	–3.65	–31.62
MD_SMO	–120.04	–31.82	–126.29	6.25	–15.53	–3.87	–1.03	–31.51
MD_SMR	–123.49	–103.36	–137.81	14.31	–13.25	–3.98	–3.33	–30.59
MD_SUD	–111.13	–92.5	–131.23	20.09	–11.07	–3.7	–3.08	–29.93
MD_SMT	–123.95	–61.98	–134.62	10.67	–12.01	–3.87	–1.94	–29.27
MTX	–151.34	–119.61	–174.19	22.85	–18.08	–4.48	–3.54	–33.05

[a]MolDock: docking score function by MolDock; [b]Rerank: reranking score function; [c]E_{Inter}: ligand – protein (DHPS) interaction energy; [d]E_{Intra}: internal energy of the ligand; [e]HBond: hydrogen bonding energy; [f]LE1: Ligand efficiency 1 - MolDock score divided by Heavy atoms count; [g]LE3: Ligand efficiency 3 – Rerank score divided by Heavy atoms count; MTX (methotrexate) – DHFR inhibitor

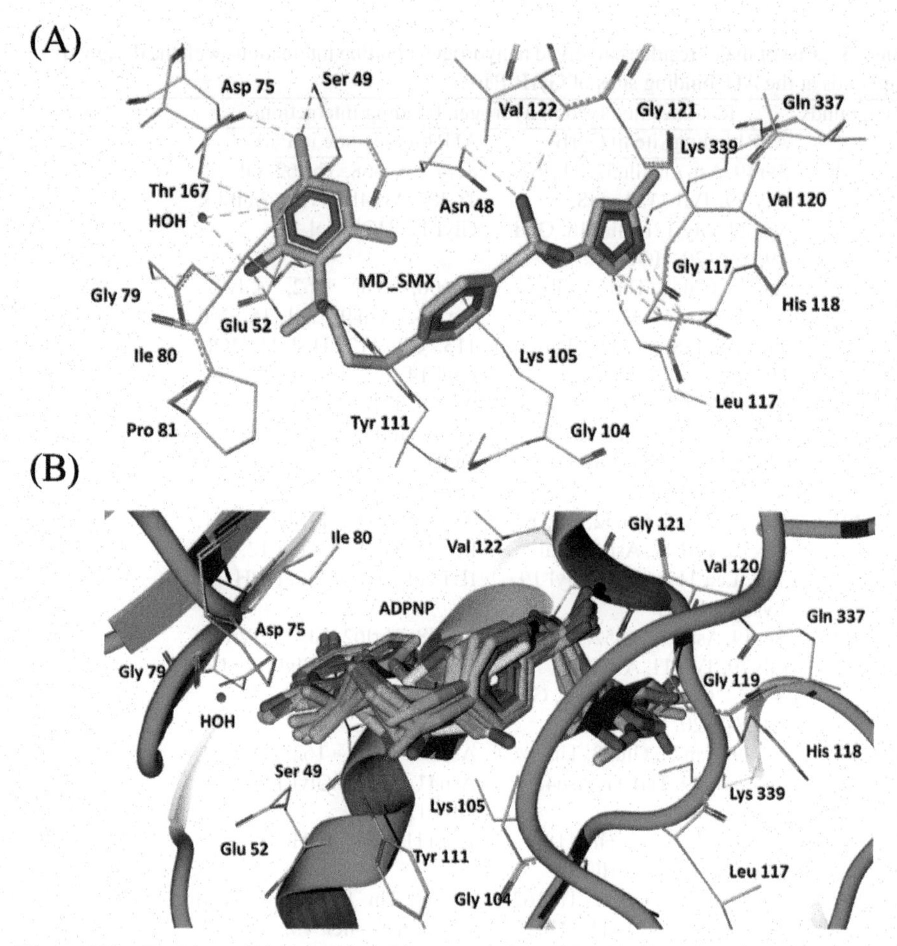

Figure 7.3 Docking of designed hybrid compounds at ATP binding site of *P. aeruginosa* GyrB. (A) Binding pose of the top-scored compound MD_SMX (grey stick model); (B) Binding poses of all the designed hybrid compounds (grey stick models) superimposed with ADPNP (yellow stick model). The residues are shown as wireframe coloured by their element and the hydrogen bonds are illustrated as dotted green lines.

of DHPS maintaining contacts with all the key conserved residues required for inhibition. The list of major residues participating in hydrogen bond interactions with the designed inhibitors in the DHPS/DHFR active sites are tabulated in Table 7.6 based on their respective binding affinity scores.Considering the top ranked compound MD_SMX, the pyrimidine ring engages the pterin binding pocket and the hydroxyl moiety form hydrogen bonds with Gly224 and Lys228, the unsubstituted amine at the 2-position forms hydrogen bond

Table 7.5 List of major residues involved in hydrogen bonding interactions with the designed compounds at the ATP binding sites of GyrB/ParE

Compounds	Residues involved in hydrogen bonding interactions	
	ATP binding site in GyrB	ATP binding site in ParE
MD_SMO	Ser49, Asp75, Glu52, Thrl67, Gly 79, Tyrl 11, Asn48, Vall22, Gly 12L Hisl 18, Glyl 19, Val 120, HOH	Ser42, Asp68, Thrl62, GIy72, Tyrl04, Asn4l. Glyl 14, Vail 13, Glyl 12, HOH llel 15
MD_SMR	Asp75, Glu52, Thrl67, Glv79, Tyrl 11, Asn48, Val 122 Leu 117, His 118, Gly 119, Vail20, HOH	Asp68, Thr 162, Glv72, Tyrl04, Asn4L llel 15, Val 113, Gly 112, Hisl 11 HOH Glyl 14
MD_SMT	Ser49, Asp75, Glu52, Thr 167, Gly79, Tyrl 11, Asn48, Val 122, Gly 121, Gly 119, Val 120, HOH	Asp68, Thrl62.Tyrl04, Asn41, Ilell5, Ghil4, Val 113, Glyl 12, HOH
MDSMX	Ser49, Asp75, Glu52, Thr167, Gly79, Tyrl 11, Asn48, Val 122, Leu 117, HisllS.Glyl 19, Lys339, HOH	Asp68, Thrl 62, Tyrl04, Asn41, lie 115, Glyl 14, Glyl 12, Hisl IL Leul 10, Lvs333, HOH
MD_SMZ	Ser49, Asp75, Glu52, Thrl67, Gly79, Tyrl 11, Asn48, Vall22, Gly 121, Hisl 18, Glyl 19, Val 120, HOH	Asp68, Thrl62, Gly72, Tyrl04, Asn41, llel 15, Glyl 12, His 111, Leu 110, HOH
MD_SOX	Ser49, Asp75, Glu52, Thr 167, Gly79, Tyrl 11, Asn48. Val 122, Leu 117, His 118, Gly 119, Lvs339, HOH	Asp68, Tlir 162, Tyrl04, Asn41, Ik 115, Glyl 14, Glyl 12. His 111, Leul 10, HOH
MD_SPR	Ser49, Asp75, Glu52, Thr167, Gly79, Tyrl 11, Asn48, Val 122, Leu 117, His 118, Gly 119, HOH	Asp68, Thr 162, Gly72, Tyrl 04, Asn41, llel 15, Vail 13, Glyl 12, Hisl 11, HOH Gly 114
MD_STH	Ser49, Asp75, Glu52, Thr 167, Gly79, Tyrl 11, Asn48, Vall22, Glyl21, Glyl 19, Vall20, HOH	Asp68, Tlirl62, Tyrl04, Asn41, llel 15, Glyl 14, Vail 13, Glyl 12, Hisl 11, HOH
MD_SUD	Ser49, Asp75, Glu52, Thr 167, Gly79, Tyrl 11, Asn48. Val 122, Leu 117, Hisl IS. Glyll9, Val 120, HOH	Asp68, Thr 162, Glu45, Tyrl04, Asn4l, llel 15. Glyl14. ValU3, Glyl 12, Hisl 11, HOH
ADPNP	Val 120, Glyl21, Lys339, Leu 117, Hisl 18, Lysl05, Glyl 19, Asn48, Val122, Glvl04, Tyrl 11, Asp75, Thr 167, HOH	Asp68, Tyrl04, Gly97, Lys98, Asn41, llel 15, Val 113, Gly 112, His 111, Leul 10, Lys333, 1 Glyl14, HOH

Table 7.6 List of major residues involved in hydrogen bonding interactions with the designed compounds at the pterin/folate binding sites of DHPS/DHFR

Compounds	Residues involved in hydroyen bonding interactions	
	Pterin binding site in DHPS	Folate binding site in DHFR
MD_SMO	Asn 122, Asp 103, Asp 192, Arg262, Gly224, Lys228, Thr69, Ser229, Met230	Asp30, Tlc8, Tie 104, Tyrl 10, Arg55, Arg60
MD_SMR	Asn 122, Asp 103, Asp 192, Arg262, GK224, Lys228, Thr69, Ser229, Met230	Asp30, IleH. Ik 104, Tyrl 10, Arg55
MD_SMT	Asn 122, Asp 103, Aspl92, Arg262, Gly224, Lys228, Thr69, Ser229	Asp30, lies, lie 104, Tyrl 10, Arg55
MD_SMX	Asn 122, Asp 103, Aspl92, Arg262, GK224, Lys228, Thr69, Ser229, Met230, Asn204	Asp30, llc-8 Ilel04, Tvrl 10, Arg55, Arg6ll
MD_SMZ	Asn 122, Asp 103, Asp 192, Arg262, GK224, Lys228, Thr69, Ser229, Met230, Asn204	Asp30, Ue8, Ilel04, Tyrl 10, Arg55, Arg60
MD_SOX	Asn 122, Asp 103, Asp 192, Arg262, Glv224, Lys228, Thi69, Ser229, Met230, Asn204	Asp3fl. Ile8, Tie 104, Tyrl 10, Arg55, ArgSO
MD_SPR	Asn 122, Asp 103, Asp 192, Arg262, Gly224, Lys228, Thr69, Sci-229, Met23(>	Asp30, Ile8, Uel04, Tyrl 10, Arg55, Arg60
MD_STII	Asn 122, Asp 103, Asp 192, Arg262, GK 224, Lys228, Thr69, Ser229, Mel230	Asp30, Ile8, lie 104, Tyrl 10, Arg55, ArgfiO
MD_SUD	Asnl22, Asp 103, Asp 192, Arg262, Gly224, Lys228, Thr69, Ser229, Met230	Asp30, Ilc8, Ilel04, TyrllO, Arg55

interactions with Asp192 and Asn122 while the nitrogen at the 3-position also contacts the Asn122 residue, the unsubstituted amine at the 6-position interacts with Asp103, the nitrogen atom at the 5-position engages in hydrogen bond interaction with Lys228 and Arg262 at the base of the pocket (Figure 7.5A).

In addition, the N-methyl group with Arg262 and the linker group with Phe197 residue make van der Waals interactions. The linker group attached to the sulfamethoxazole extends the phenyl ring adopting the conformation similar to the *p*ABA moiety which brings it closer to interact with Thr69 residue by hydrogen bonding (Figure 7.5A). The sulfonyl group of sulfamethoxazole contacts Ser229 at the bottom of the binding pocket and the oxazole ring docked in the up position placed between loop 6 and loop 7 and also engages in hydrogen bond interactions with the residues Ser229, Met230 and Asn204.

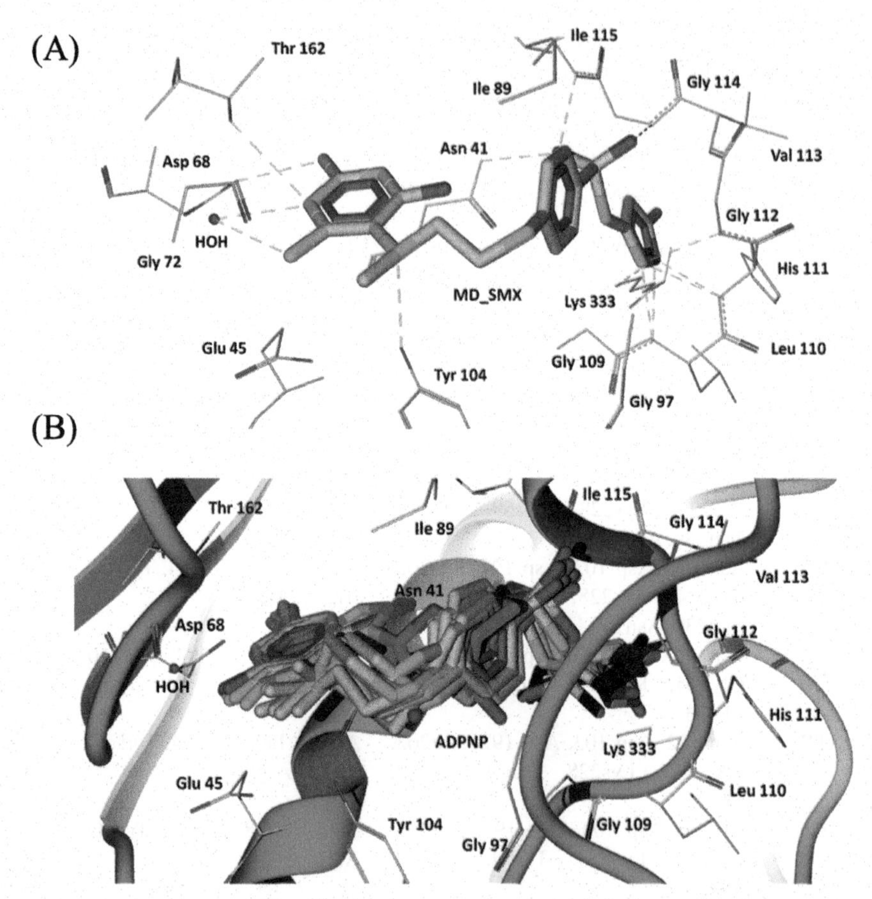

Figure 7.4 Docking of designed hybrid compounds at ATP binding site of *P. aeruginosa* ParE. (A) Binding pose of the top-scored compound MD_SMX (grey stick model); (B) Binding poses of all the designed hybrid compounds (grey stick models) superimposed with ADPNP (magenta stick model). The residues are shown as wireframe colored by their element and the hydrogen bonds are illustrated as dotted green lines.

The docking calculations also demonstrated that the pyrimidine ring and sulfonamide moieties in all the designed hybrid compounds engage in similar interactions at the pterin binding pocket of DHPS (Figure 7.5B).

7.1.1.3 Docking of proposed hybrid compounds at the folate binding site of DHFR

The docking results show that all the proposed hybrid compounds have a good fit into the folate binding site of DHFR and are positioned in a way

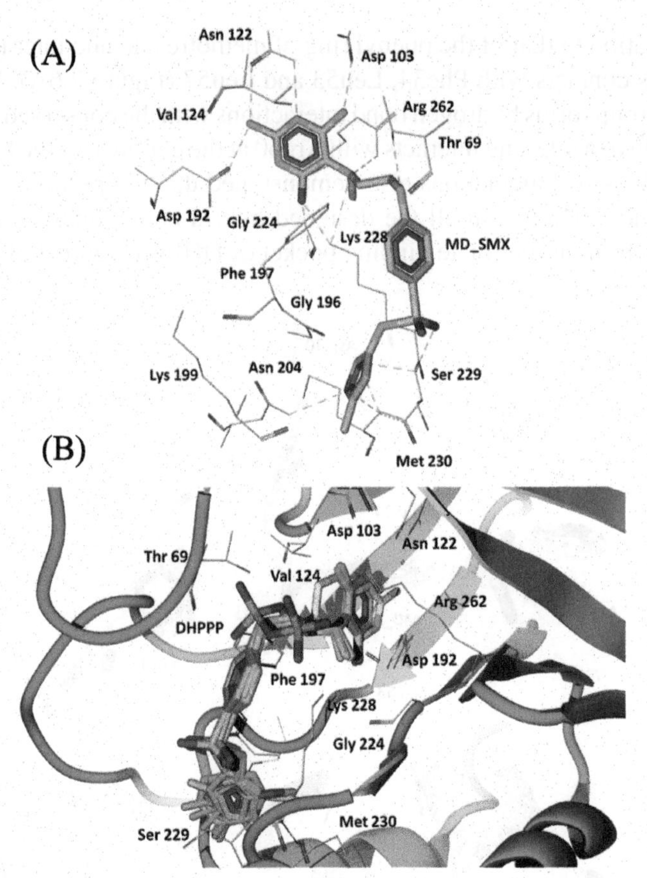

Figure 7.5 Docking of designed hybrid compounds at pterin binding site of *P. aeruginosa* DHPS. (A) Binding pose of the top-scored compound MD_SMX (grey stick model); (B) Binding poses of all the designed hybrid compounds (grey stick models) superimposed with DHPPP (yellow stick model). The residues are shown as wireframe colored by their element and the hydrogen bonds are illustrated as dotted green lines.

similar to the inhibitor (methotrexate) (Figure 7.6B). The major residues participating in hydrogen bond interactions with the designed inhibitors in the DHFR active site are tabulated in Table 7.6. The pyrimidine ring of the hybrid compound HD_EA engages the folate binding site and interacts through five major hydrogen bonds (Figure 7.6A). The 2-amino and 4-hydroxyl group engages with Asp30 while the 6-amino group interacts with Ile8, Tyr110 and Ile104 residues. The pyrimidine ring also makes van der Waals interactions with Phe34, Ile8, Ala9 and Ala10. The phenyl ring of sulfamethoxazole adopts similar orientation to that of the phenyl ring of methotrexate and interacts via van der Waals contacts with Phe34, Leu53 and Leu57 (Figure 7.6A). While, the sulfonyl group forms hydrogen bond interactions with the conserved Arg55 residue and the oxazole ring interacts with Arg60 at the entrance of the binding pocket. The docking calculations also demonstrated that the pyrimidine ring and sulfonamide moieties in all the designed hybrid compounds engage in similar interactions at the folate binding pocket of DHFR (Figure 7.6B)

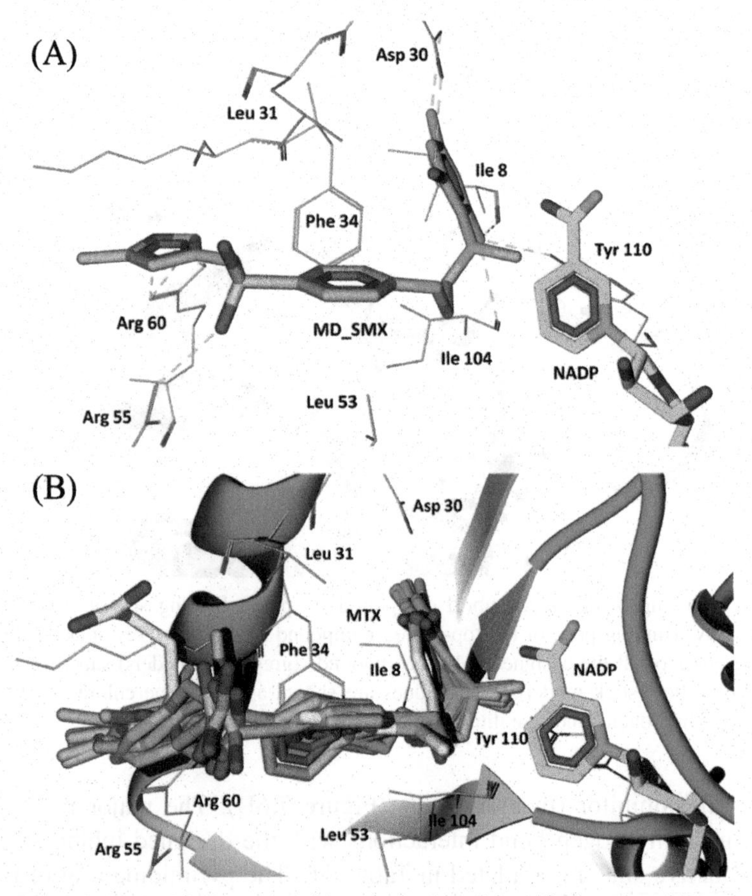

Figure 7.6 Docking of designed hybrid compounds at folate binding site of *P. aeruginosa* DHFR. (A) Binding pose of the top-scored compound MD_SMX (grey stick model); (B) Binding poses of all the designed hybrid compounds (grey stick models) superimposed with MTX (yellow stick model) and NADP shown as green stick. The residues are shown as wireframe colored by their element and the hydrogen bonds are illustrated as dotted green lines.

7.1.2 Molecular dynamics simulations

In order to show the conformational changes and binding site stability of the three systems, comparisons between the unbound enzyme complexes (I – GyrB, IV – ParE, VII – DHPS and X – DHFR), ligand-bound enzyme complexes (II – ADPNP-GyrB, V – ADPNP-ParE, VIII – DHPPP-DHPS and XI – MTX-NADP-DHFR) and top ranked hybrid compound-bound enzyme complexes (III – MD_SMX-GyrB, VI – MD_SMX-ParE, IX – MD_SMX-DHPS and XII – MD_SMX-NADP-DHFR) were analyzed by both RMSD and RMSF during 3000 ps MD simulations time (Figure 7.7A–D). The main chain average RMSD values obtained for the trajectories of the enzyme complexes (I, IV, VII & X) were about 2.060 Å, 2.001 Å, 1.44 Å and 1.972 Å, for complexes (II, V, VIII & XI) were 1.924 Å, 2.181 Å, 1.488 Å and 2.095 Å and complexes (III, VI, IX & XII) were 2.132 Å, 2.278 Å, 1.455 Å and 2.027 Å, respectively. As can be seen in the plots that after 500 ps, all the

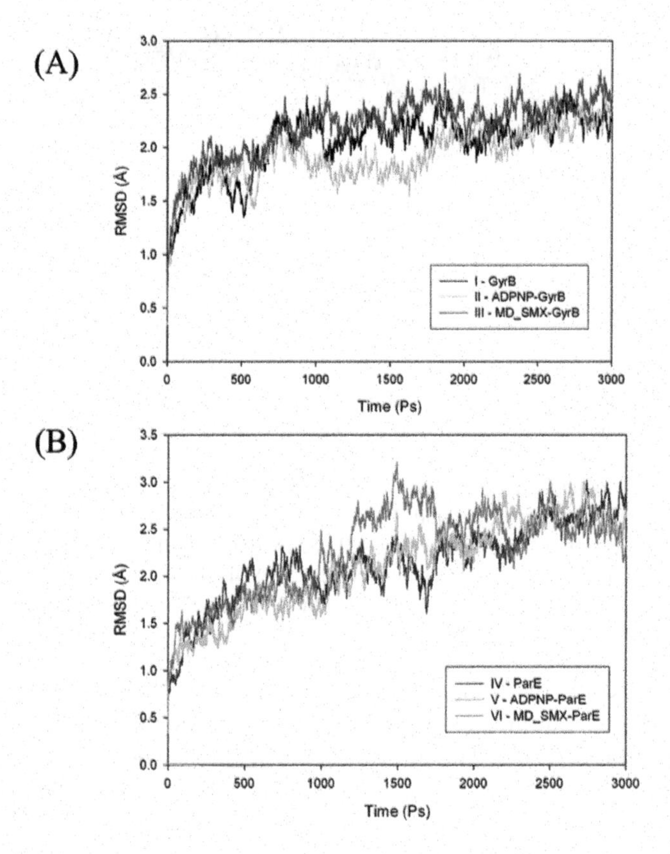

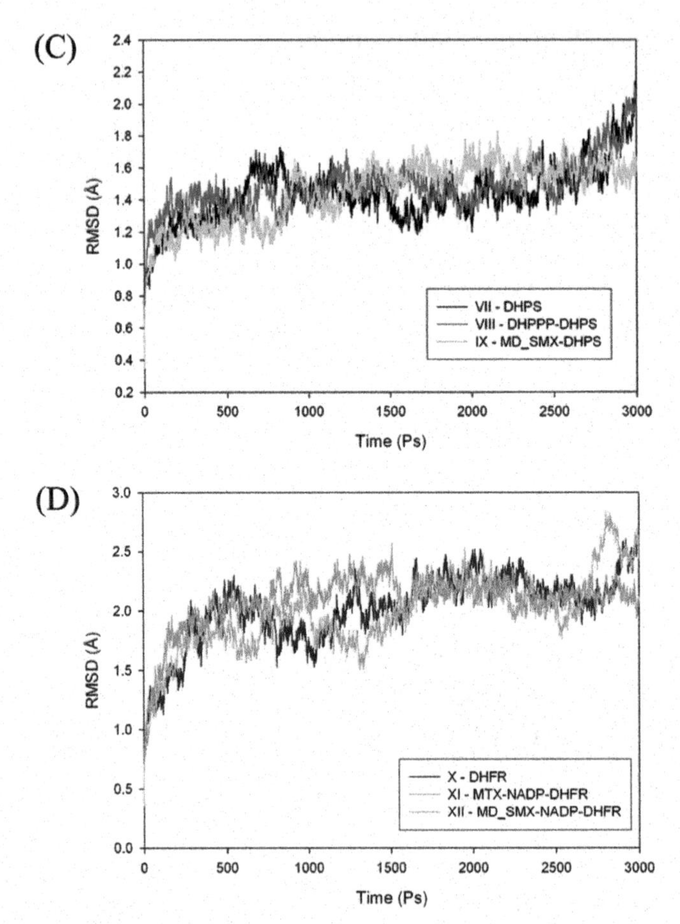

Figure 7.7 Root mean square deviation (RMSD) of backbone Cα atoms of the complexes versus time in picoseconds. (A) Unbound GyrB complex – I, ADPNP bound GyrB complex - II and top-scored hybrid compound MD_SMX bound GyrB complex – III; (B) Unbound ParE complex – IV, ADPNP bound ParE complex - V and top-scored hybrid compound MD_SMX bound ParE complex – VI; (C) Unbound DHPS complex - VII, DHPPP bound DHPS complex -VIII and top-scored hybrid compound MD_SMX bound DHPS complex - IX; (D) Unbound DHFR complex – X, methotrexate and cofactor NADP bound DHFR complex – XI and top-scored hybrid compound MD_SMX and cofactor NADP bound DHFR complex – XII.

complexes tend to converge with low values of deviation of the molecules with respect to the original atomic positions, indicating the systems are stable and equilibrated (Figure 7.7A-D). The three systems also reach equilibrium after 500 ps, however, the conformation of complex VI – MD_SMX-ParE showed a fluctuation at 1200 ps and flattens out after that. Similarly, the complexes VII – DHPS, VIII – DHPPP-DHPS, X – DHFR and XII – MD_SMX-NADP-DHFR showed fluctuation at the end of the MD simulation period about 2800 ps and further down the simulation period which was run for about 3500 ps the peaks flattens maintaining the equilibrium status (data not shown). Furthermore, analyses of the root mean square fluctuations of key active site residues versus the residue number for all the complexes are shown in Figure 7.8A–7.8H.

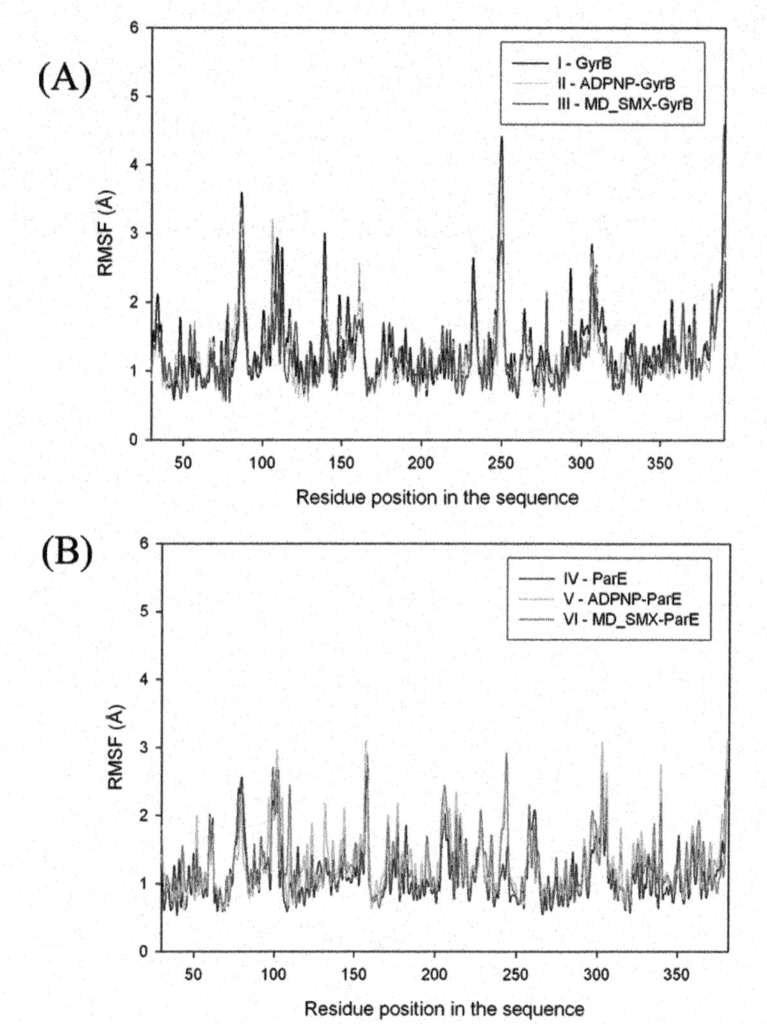

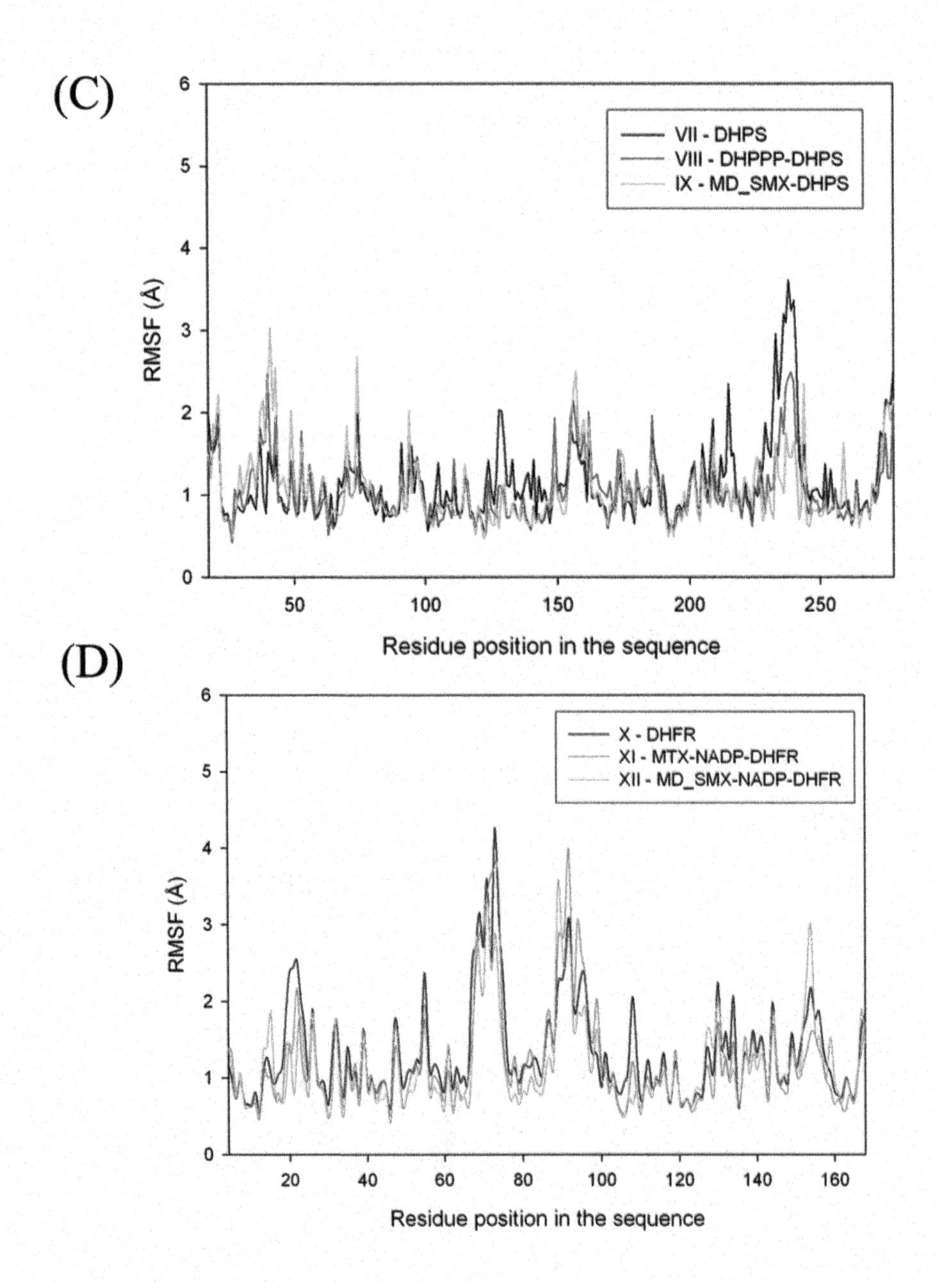
(C)
VII - DHPS
VIII - DHPPP-DHPS
IX - MD_SMX-DHPS
RMSF (Å)
0
1
2
3
4
5
6
50
100
150
200
250
Residue position in the sequence
(D)
X - DHFR
XI - MTX-NADP-DHFR
XII - MD_SMX-NADP-DHFR
RMSF (Å)
0
1
2
3
4
5
6
20
40
60
80
100
120
140
160
Residue position in the sequence

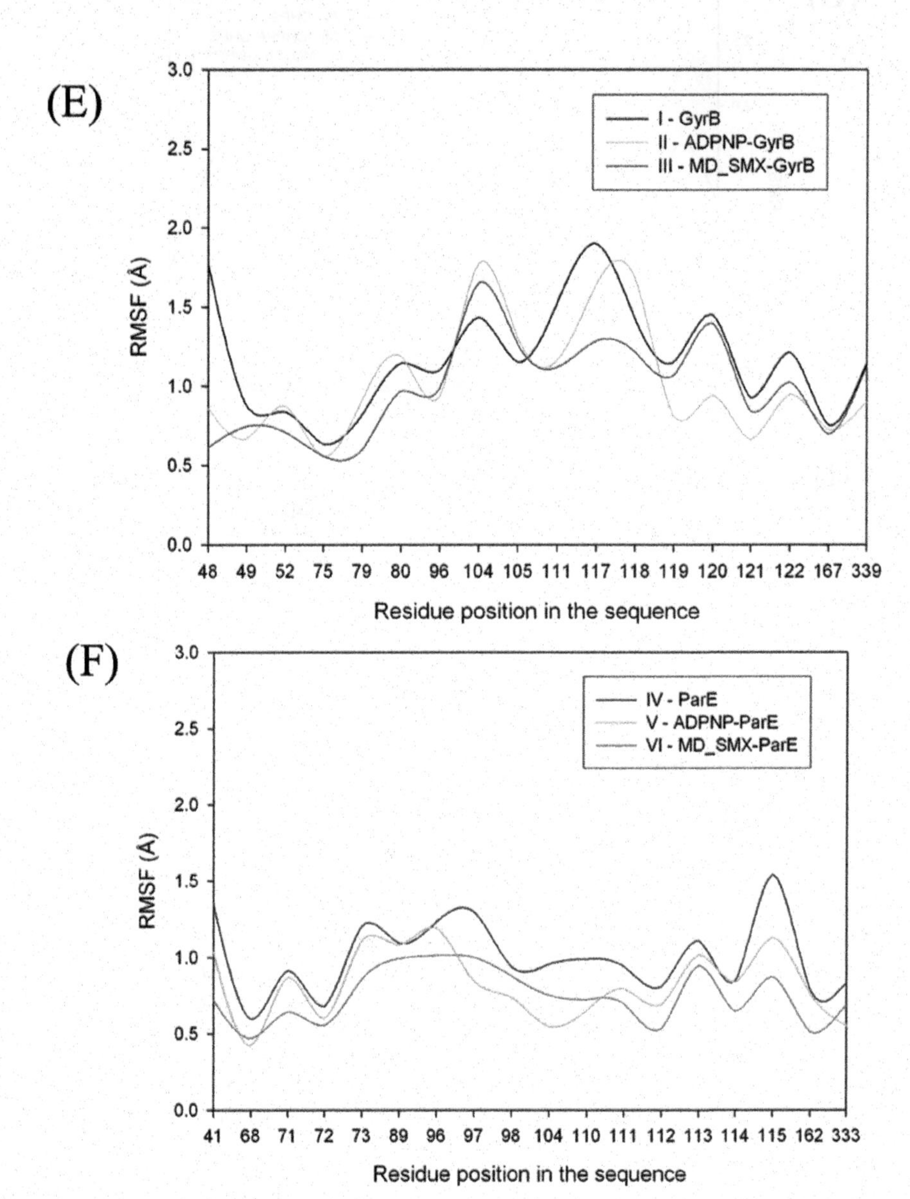
(E)
3.0
2.5
2.0
1.5
1.0
0.5
0.0
RMSF (Å)
I - GyrB
II - ADPNP-GyrB
III - MD_SMX-GyrB
48 49 52 75 79 80 96 104 105 111 117 118 119 120 121 122 167 339
Residue position in the sequence
(F)
3.0
2.5
2.0
1.5
1.0
0.5
0.0
RMSF (Å)
IV - ParE
V - ADPNP-ParE
VI - MD_SMX-ParE
41 68 71 72 73 89 96 97 98 104 110 111 112 113 114 115 162 333
Residue position in the sequence

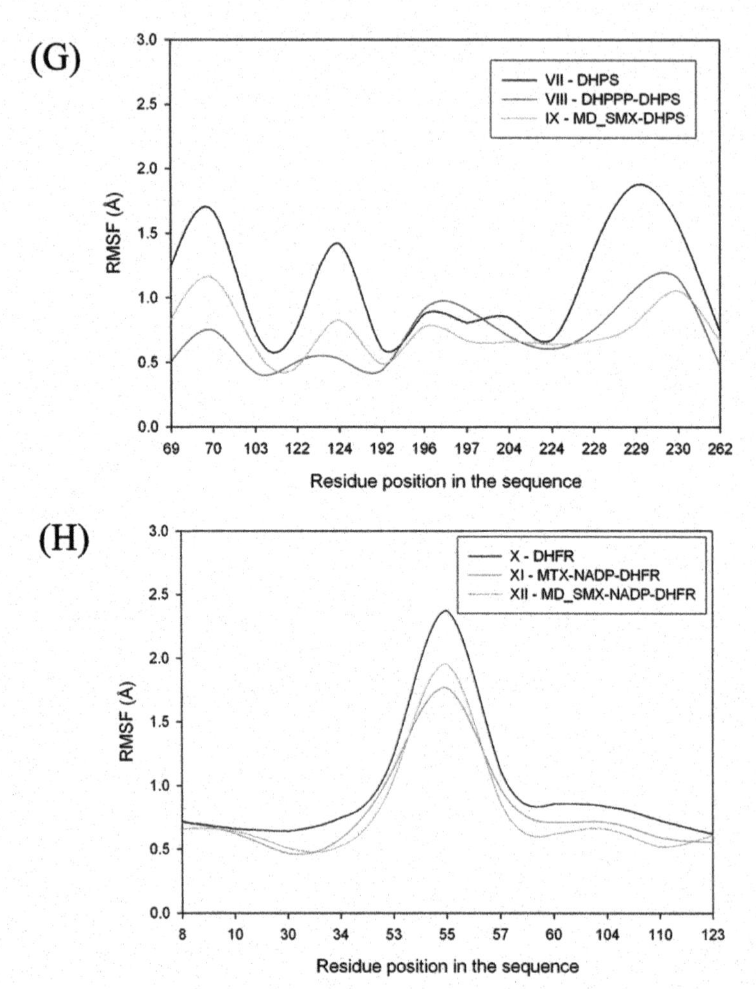

Figure 7.8 Root mean square fluctuation (RMSF) of backbone Cα atoms of the complexes versus residue number in the sequence. (A) Unbound GyrB complex – I, ADPNP bound GyrB complex - II and top-scored hybrid compound MD_SMX bound GyrB complex – III; (B) Unbound ParE complex – IV, ADPNP bound ParE complex - V and top-scored hybrid compound MD_SMX bound ParE complex – VI; (C) Unbound DHPS complex - VII, DHPPP bound DHPS complex - VIII and top-scored hybrid compound MD_SMX bound DHPS complex - IX; (D) Unbound DHFR complex – X, methotrexate and cofactor NADP bound DHFR complex – XI and top-scored hybrid compound MD_SMX and cofactor NADP bound DHFR complex – XII; (E) Closer look of GyrB active site residues RMSF during the simulation of three complexes; (F) Closer look of ParE active site residues RMSF during the simulation of three complexes; (G) Closer look of DHPS active site residues RMSF during the simulation of three complexes; (H) Closer look of DHFR active site residues RMSF during the simulation of three complexes.

The RMSF value illustrates the average displacement of each residue in relation to their average backbone structure over the whole simulation and indicates the relative flexibility that describes each of these residue elements. As expected, the RMSF plots of all the complexes in the three systems share similar RMSF distributions and dynamism other than the flexible loops in the main backbone residues that shows high fluctuations of greater than 3 Å. The enzyme complexes (I – III) which include residues in positions between (78 and 88) and the turn which includes residues in positions between (248 and 251) shows high fluctuations (Figure 7.8A). Similarly, the complexes (VII – IX) show loops that include residues between 34 and 41 and the turn which includes residues in positions between 237 and 241 (Figure 7.8C), complexes (X–XII) that includes residues between 67 and 78 and the α-helix structure which includes residues in positions between 90 and 96 which is located away from the binding site cavity, show high fluctuations (Figure 7.8D). The mobile flexibilities (RMSF) values of key active site residues in each complex fluctuates widely from 0.5 Å to 2.4 Å, indicating that the active site residue elements maintain their positions in relation to their corresponding enzyme complexes (Figure 7.8E–H). In comparison to the unbound complexes (I – GyrB, IV – ParE, VII - DHPS and X – DHFR), both ligand bound complexes (II – ADPNP-GyrB, V – ADPNP-ParE, VIII – DHPPP-DHPS and XI – MTX-NADP-DHFR) and inhibitor bound complex (III – MD_SMX-GyrB, VI – MD_SMX-ParE, IX – MD_SMX-DHPS and XII – MD_SMX-DHFR) showed smaller RMSF values indicating a tight interaction between active site residues. Thus, the low RMSD and RMSF deviations indicate that the inhibitor bound multi-enzyme complexes are stable and preserve the inhibitor interactions shown in the molecular docking experiments.

7.2 Electronic Property Analyses of all the Designed Hybrid Compounds

In order to better understand the molecular interactions and structural factors that may be involved in the activity of the designed hybrid compounds targeting multi-enzymes, stereo-electronic properties such as molecular orbital energies (the highest occupied molecular orbital (HOMO) and lowest unoccupied molecular orbital (LUMO)), reactivity index or gap (difference between HOMO and LUMO energies) and molecular electrostatic potential (MEP) were calculated using the B3LYP/6–31G** density functional theory on the geometry optimized structures. Calculated absolute energies, reactivity indexes & electronic properties of the compounds are listed in Table 7.7.

Table 7.7 Absolute energies, stereo-electronic and molecular properties calculated at B3LYP/6-31G** basis-set levels and percentage of absorption of the designed hybrid compounds from Hypothesis I

Compounds	Energy[a] (in hartress)	Energy (aq)[b] (in hartress)	Energy (S)[c] (KJ/mol)	%ABS[d]	Volume (Å^3)	PSA' (Å^2)	E_{HUMO} (eV)	E_{LUMO} (eV)	Band gap (ΔE)[f] (eV)	Dipole moment (debye)	Polarizability
MD_SMO	−1837.42	−1837.48	−164.31	53	419.5	158.1	−5.36	−1.17	−4, 19	8.44	74.41
MD_SMR	−1816.35	−1816.42	−179.89	54	416.7	156.27	−5.92	−0.99	−4.93	6.69	74.01
MD_SMT	−1855.68	−1855.75	−173.12	54	434.97	156.13	−6.01	−0.95	−5, 06	7.61	75.46
MD_JSMX	1798.05	−1798.12	−187.47	51	401.28	162.73	−6.06	−1.07	−4.99	10.29	72.75
MD_SMZ	−2137.08	−2137.14	−171.68	52	403.83	161.593	−5.96	−1.21	−4.75	11.89	73.01
MD_SOX	−1837.37	−1837.43	−169, 52	51	418.65	162.56	−5, 62	−1.21	−4, 41	10.98	74.29
MD_SPR	−1760.99	−1761.05	−157.12	56	405.1	148.72	−5.99	−0.98	−5, 01	9.26	73.05
MD_STH	−2081.74	−2081.81	−176, 64	56	392.7	149.22	−5.73	−1.12	−4, 61	9.87	72.14
MD_SLID	−1777.02	−1777.09	−183.86	54	398.51	156.26	−5.93	−1.03	−4.9	7.32	72.54

[a]Energy: Total energy in vacuo; [b]Energy(aq): Total energy in aqueous medium; [c]Energy(S): aqueous salvation energy; [d]%ABS: Absorption; [e]PSA: polar surface area; [f]ΔE: $\Delta E_{HOMO-LUMO}$

HOMO and LUMO orbital energies are related to ionization potential and electron affinity and their respective frontier orbitals are associated to the molecule's reactivity i.e., HOMO energy is susceptible to electrophilic attack (donate e^-) and LUMO energy susceptible to nucleophilic attack (accept e^-) [68]. The reactivity index (band gap) of the compounds with small difference implies high reactivity and large difference implies low reactivity in reactions [69]. The energy gap (HOMO – LUMO) of the proposed compounds was –4.19 to –5.06 and are found to be within a smaller narrow range indicating that these energies permit electron transfer and exchange making the compounds very highly reactive and similar in nature (Table 7.7).

Three dimensional HOMO and LUMO profiles over the van der Waals surface and the molecular electrostatic potential mapped onto total electron density surface of all the hybrid compounds are shown in Figure 7.9. For the three dimensional HOMO orbitals (Figure 7.9) of all the compounds, the predominant HOMO (electron-rich) orbitals were consistently observed on the amino and hydroxyl substituted aromatic pyrimidine ring of the MANIC compound derivative, indicating strong electrophilic affinity. The most significant LUMO (electron-poor) orbitals (Figure 7.9) of all the compounds were consistently observed on the sulfonamide rings, indicating the localization of strong electron acceptor elements and thereby susceptible to strong nucleophilic attack.

Additionally, the MEP is a useful descriptor for understanding electrophilic/nucleophilic sites as well as hydrogen bonding interactions. In general, the negative potential sites (red color) of MEP represent regions of electrophilic reactivity and interaction through $\pi\pi$-$\pi\pi$ bonding with the aromatic systems of interacting enzyme residues and positive potential sites (blue colour) represents regions of nucleophilic reactivity [63].

The MEP maps of all the hybrid compounds (Figure 7.9(iii)), are localized around the amino groups of the MANIC compound derivative indicating their hydrogen bond donor activity with the key active site residues in GyrB (Asp75, Ser49, Gly79 and Glu45), TopoIV (Asp68, conserved water molecule), DHPS (Asn122, Asp192 and Asp103) and DHFR (Asp30, Ile8, Tyr110 and Ile104) enzymes. The predominant red regions (negative potential sites) in the MEP profiles (Figure 7.9(iii)) of all the compounds are localized around the sulfonyl functional groups (oxygen atoms) of the sulfonamide rings, nitrogen and oxygen atom in the cyclopentane ring of sulfamethoxazole, sulfamoxole, sulfisoxazole,

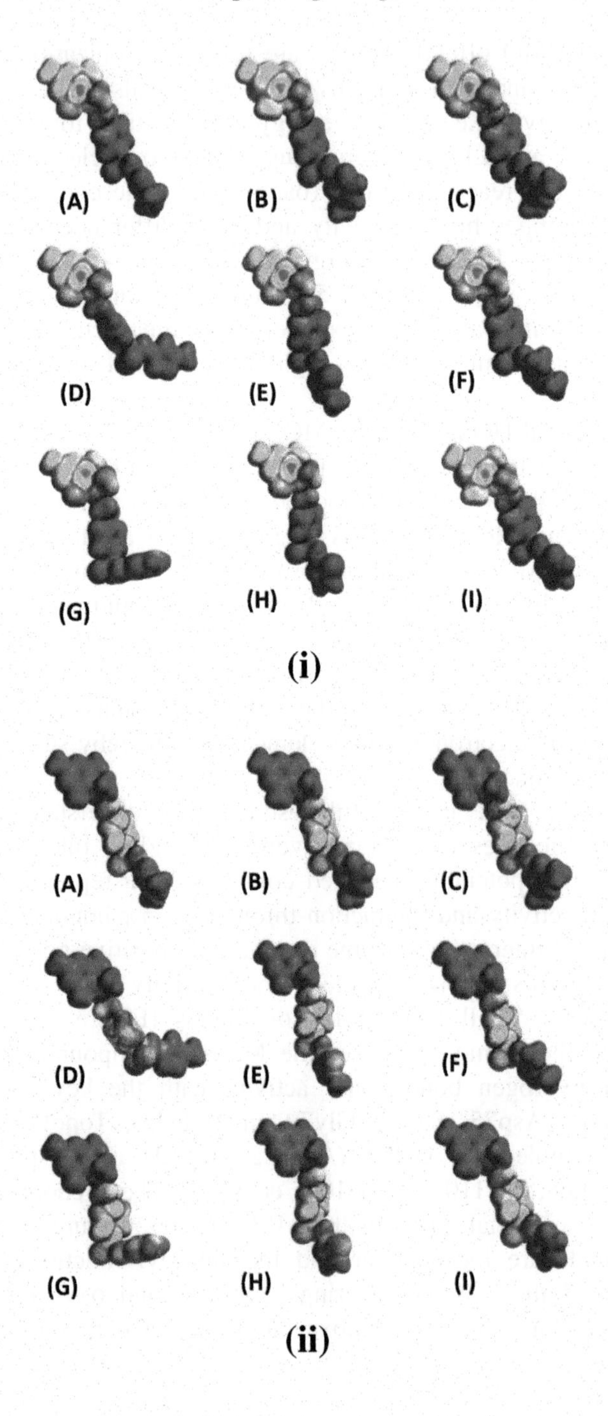
(A)
(B)
(C)
(D)
(E)
(F)
(G)
(H)
(I)
(i)
(A)
(B)
(C)
(D)
(E)
(F)
(G)
(H)
(I)
(ii)

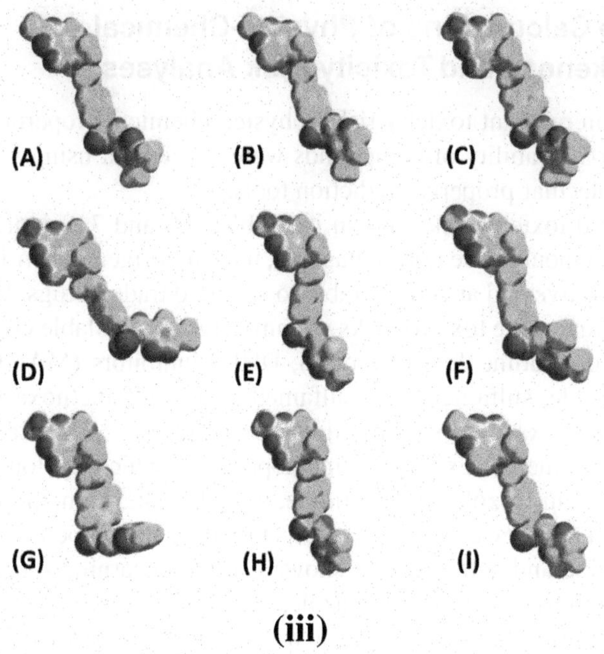

Figure 7.9 Representations of (i) HOMO; (ii) LUMO and (iii) MEP iso-density surfaces; Calculated at B3LYP/6–31G** basis-set levels on the optimized geometry of the hybrid compounds from Hypothesis III.

nitrogen atoms in the cyclopentane rings of sulfamethizole and sulfathiazole, nitrogen atoms in the pyrimidine ring of sulfadiazine, sulfapyridine, sulfamerazine, sulfamethazine and the methyl amino (nitrogen atom), suggesting that these oxygen and nitrogen atoms can act as hydrogen bond acceptors and possibly participate in hydrogen bonding interaction with donor groups of key active site residues in GyrB (Tyr111, Asn48, Val122, Lys105, Leu117, His118 and Gly119), TopoIV (Asn41, Ile115, Gly114, Gly112, His111, Leu110 and Lys333), DHPS (Lys228, Arg262, Thr69, Ser229, Met230 and Asn204) and DHFR (Arg55 and Arg60) enzymes. In addition, regions of the aromatic rings in all the compounds show red regions with decreasing intensity indicating weak positive electrostatic potential and thereby implying a hydrophobic nature of the aromatic rings. Similarly, the above mentioned molecular orbitals and MEP profile features appear to be consistent with the key pharmacophoric features required for potent multi-inhibition of quadri enzymes (GyrB/ParE, DHPS and DHFR).

7.3 *In Silico* Calculations of Physico-Chemical, Drug-Likeness and Toxicity Risk Analyses

Investigations on relevant toxicity risks, physico-chemical properties, drug-likeness and bioavailability of compounds were carried out using various *in silico*-based molecular property prediction tools.

The predicted toxicity risks shown in (Tables 7.8 and 7.9) indicate that all the hybrid compounds are non-mutagenic, non-tumorigenic, non-irritating and non-reproductive and are comparable to standard traded drugs. For comparison, we evaluated the toxicity risks of commercially available GyrB/ParE inhibitors (cyclothialidine & novobiocin), DHPS inhibitors (MANIC compound, sulfamoxole, sulfamerazine, sulfamethazine, sulfamethoxazole, sulfamethizole, sulfisoxazole, sulfapyridine, sulfathiazole & sulfadiazine), DHFR inhibitors (methotrexate & trimethoprim) of which cyclothialidine, sulfamoxole, sulfamethizole, sulfapyridine, sulfathiazole, trimethoprim and MANIC compound showed toxic free effects while, novobiocin, sulfamerazine, sulfadiazine and sulfisoxazole showed risks concerning reproductive effects, sulfamethazine, sulfamethoxazole, sulfisoxazole and methotrexate showed tumorigenic effects, sulfisoxazole was also found to be associated with mutagenic toxicity alert. Overall, the designed pyrimidine-sulfonamide hybrid scaffolds (MD_SMO, MD_SMR, MD_SMT, MD_SMX, MD_SMZ, MD_SOX, MD_SPR, MD_STH and MD_SUD) showed non-toxic behaviour with highly desirable physico-chemical parameters disclosing their potential as promising agents for antimicrobial therapy.

The Lipinski's RO5 used as a filter for drug-likeness defines four simple physico-chemical parameter ranges ($MW \leq 500$, $clog\,P \leq 5$, H-bond donors ≤ 5, H-bond acceptors ≤ 10) associated with acceptable aqueous solubility and intestinal permeability, molecules violating more than one of these rules may have problems with oral bioavailability [64]. The physico-chemical properties such as, molecular weight of the proposed hybrid compounds is below 460 (Table 7.9) and thus they are more likely to have higher absorption. The clogP value is the logarithm of its partition coefficient between n-octanol and water, measuring the compound's hydrophilicity [64]. The estimated clogPvalues for the proposed compounds is less than 5.0 indicating high lipophilicities, higher absorption and permeation. Aqueous solubility (log S) measures significantly affect the drug absorption/distribution characteristics and the estimated log S value is a unit stripped logarithm of the solubility (moles/liter). It is suggested that more than 80% of the drugs in the market have log S value greater than –4 (http://www.organic-chemistry.org/prog/peo/). The solubility measures for

Table 7.8 Calculated toxicity risks, molecular properties, drug-likeness and overall drug-score of the antibiotics investigated

Compounds	Toxicity risks prediction				Molecular properties								
	MUT[a]	TUM[b]	IRR[c]	REP[d]	MW[e]	cLP[d]	S[g]	nON h	nOH NH'	nVio latio ns	nRot b[i]	D_L[k]	D_S[l]
Cyclothialidine	●	●	●	●	625	−2.62	−2. 31	16	9	3	6	2.14	0.54
Novobiocin	●	●	●	○	598	2.81	−4.93	13	6	3	9	0.97	0.23
Methotrexate	●	○	●	●	454	−0.69	−3.77	13	7	2	9	−7.09	0.22
Trimethoprim	●	●	●	●	290	1.23	−3.32	7	4	0	5	4.95	0.87
MANIC	●	●	●	●	185	**0.03**	−2.41	8	4	0	2	−0.88	0.61
Sulfamoxole	●	●	●	●	267	1.32	−3.48	**6**	3	**0**	3	3.31	**0.86**
Sulfamerazine	●	○	●	○	**264**	**0.83**	**−2.66**	**6**	**3**	**0**	**3**	**3.31**	**0.54**
Sulfamethazine	●	○	●	○	**278**	**1.25**	**−3.03**	**6**	3	**0**	3	**3.31**	**0.32**

(Continued)

Table 7.8 Continued

Compounds	Toxicity risks prediction				Molecular properties								
	MUT[a]	TUM[b]	IRR[c]	REP[d]	MW[e]	cLP[d]	S[g]	nON h	nOH NH'	nVio latio ns	nRot b[i]	D_L[k]	D_S[l]
Sulfam ethoxazo le	●	●	●	●	**253**	**0.7**	**–3.02**	**6**	3	**0**	**3**	3.31	**0.54**
Sulfamethizole	○	○	●	○	**270**	**0.72**	**–2.58**	**6**	**3**	**0**	**3**	**4.37**	**0.92**
Sulfisoxazole	●	●	●	●	**267**	**1.02**	**–3.36**	**6**	**3**	**0**	**3**	**3.31**	**0.19**
Sulfapyridine	●	●	●	●	**249**	**1.0**	**–2.76**	5	**3**	**0**	**3**	**3.69**	**0.91**
Sulfathiazole	●	●	●	●	**255**	**1.13**	**–2.76**	**5**	**3**	**0**	3	**4.51**	**0.91**
Sulfadiazine	●	●	●	○	250	0.41	–2.29	6	3	0	3	4.09	0.56

[a]MUT: mutagenicity; [b]TUM: tumorigenicity; [c]IRR: irritating effects; [d]REP: reproductive effects; [e]MW: molecular weight; [f]cLP: cLogP; [g]S: Solubility; [h]nON: number of hydrogen bond acceptors (O and N groups); [i]nOHNH: number of hydrogen bond donors (OH and NH groups); [j]nRotb: number of rotatable bonds; [k]D_L: druglikeness; [l]D_S: drug score;

Table 7.9 Calculated toxicity risks, molecular properties, drug-likeness and overall drug-score of the designed hybrid compounds from Hypothesis III

	Toxicity risks predictii		on		Molecular properties							
Compounds	MUT[1]	TUM[1,]	IRR^	REP[d]	MW'	cLP[r]	Ss	HB A[h]	HB D'	nVio latio ns	nRot b'	D_L[fc]	D_S'
MDSMO	•	•	•	•	448	1.11	–4.91	13	5	1	8	5.73	0.62
MDSMR	•	•	•	•	445	Ł.61	–4.09	13	5	1	8	5.73	0.71
MDSMT	•	•	•	•	459	1.03	–4.45	13	5	1	8	5.73	0.65
MDSMX	•	•	•	•	434	0.49	–4.44	13	5	1	8	5.73	0.69
MD SMZ	•	•	•	•	451	0.51	–4.0	13	5	1	8	6.59	0.71
MDSOX	•	•	•	•	448	0.8	–4.78	13	5	1	8	5.73	0.64
MD SPR	•	•	•	•	430	0.79	–4.18	12	5	1	8	**6.03**	0.71
MDSTH	•	•	•	•	436	0.91	–4.18	12	5	1	8	6.71	0.71
MD SUD	•	•	•	•	431	0.19	–3.72	13	5	1	8	6.34	0.75

the proposed compounds (Table 7.9) is in the expected range of (–2.95 to –3.97), (–2.31 to –3.25), (–3.7 to –4.91) and are comparable to that of the standard commercially available drugs. The number of Lipinski's RO5 violations is one each for the hybrid compounds MD_SMO, MD_SMR, MD_SMT, MD_SMX, MD_SMZ, MD_SOX, MD_SPR, MD_STH and MD_SUD. The number of rotatable bonds and polar surface area are now widely used filter for drug-likeness and those compounds which meet only the two criteria of (1) $\leq$ 10 rotatable bond counts and (2) polar surface area (PSA) $\leq$ 140 Å^2 (or $\leq$ 12 H-bond donors and acceptors) will have increased oral bioavailability in the rat [65]. The calculated rotatable bond counts (Table 7.7) of the proposed compounds are equal to 8 and PSA are within the range for compounds. On the other hand using PSA, we calculated the percentage of absorption (%ABS) according to the equation %ABS = 109 – (0.354 x PSA), as reported by [66], the calculated absorption percentage is in the range of 55–65% for all the proposed compounds (Table 7.7). Drug-likeness scores are calculated by summing up score values, partially based on topological descriptors, fingerprints of MDL structure keys including properties such as clogP and molecular weight are estimated to be in the positive range for 80% of the traded drugs (http://www.organic-chemistry.org/prog/peo/) [67].The drug-likeness scores for the proposed compounds are positive (Table 7.7) indicating that the molecules predominantly contain fragments which are frequently present in commercial drugs and hence have the possibility of being "drug-like". Finally, the overall drug scores combining drug-likeness, cLogP, logS, molecular weight and toxicity risks describe the potential to qualify for a drug in 0 to 1 scale (http://www.organic-chemistry.org/prog/peo/) [67]. Compared to the overall drug scores of the antibiotics and MANIC compound alone, the conjugated hybrid compounds scores were in the range of (0.62 – 0.75) (Table 7.7) respectively, indicating the proposed hybrids potential to qualify as suitable drug candidates.

7.4 Conclusions

In this chapter, multi-target drug design was performed using fragment combination/knowledge-based approaches and evaluation of the design hypothesis was carried out using various *in silico*-based methods including molecular docking, dynamics simulations, physico-chemical and electronic properties predictions. This hybrid design strategy was based on the structural combination of pteridine moiety and antibiotics for simultaneous modulation of quadri-enzymes (GyrB, ParE, DHPS and DHFR) from two pharmacologically important pathways.

Designing multi-target selective hybrid compounds that can bind to more than one target is an emerging strategy with the potential to better inhibit specific multiple targets and are highly appreciated because it is much harder for the pathogen to develop resistance when an inhibitor has activity against multiple targets. This viable hybrid design framework identified in this chapter is particularly important as it combines molecular tinkering of natural products/antibiotics by optimization through structural modifications thereby restricting unwanted off-target effects with the approach to overcome the existing resistance mechanisms and delay the resistance development of multi-drug resistant pathogens due to subsequent addressing of two or more different targets. It is also significant that the design strategy implemented in this chapter has opened up an opportunity for revival of older antibiotics to bypass the resistance mechanisms through fragment combination. The approach used here for the first time integrates the experimentally determined drug-phytochemical interaction network, knowledge-based and fragment combination methods towards the identification of novel multi-action hybrid entities by covalently connecting phytochemicals with antibiotics, semi-synthetic modification of phytochemicals by coupling to substrate analogue, structural modification of antibiotics by coupling to substrate analogue. These compounds are potential drug candidates and hence identification and modification of the most biologically active promiscuous plant-derived natural products/antibiotics with the view to improve and balance their drug-like properties will offer novel agents with superior therapeutic properties against infectious diseases.

Finally, the identification of targets with common phytochemicals inhibitory profile through systemic interaction patterns and network analyses through phytochemical-antibiotic combination assays may also help to cluster common drug targets and form a base for quickening the process of antibacterial multi-target drug discovery.

References

[1] Fischbach, M. A. and C. T. Walsh (2009). "Antibiotics for emerging pathogens." Science**325**(5944): 1089–1093.

[2] Laxminarayan, R. and D. L. Heymann (2012). "Challenges of drug resistance in the developing world." BMJ**344**.

[3] Cusini, A., S. K. Rampini, V. Bansal, B. Ledergerber, S. P. Kuster, C. Ruef and R. Weber (2010). "Different patterns of inappropriate antimicrobial

use in surgical and medical units at a tertiary care hospital in Switzerland: A prevalence survey." PLoS ONE**5**(11).

[4] Ola, S. (2011). Antibiotics and Antibiotic Resistance. United States of America, John Wiley & Sons, Inc.

[5] Alvarez-Lerma, F. and S. Grau (2012). "Management of antimicrobial use in the intensive care unit." Drugs**72**(4): 447–470.

[6] Marra, A. (2011). "Antibacterial resistance: is there a way out of the woods?" Future Microbiol**6**(7): 707–709.

[7] Lambert, P. A. (2002). "Mechanisms of antibiotic resistance in Pseudomonas aeruginosa." Journal of the Royal Society of Medicine, Supplement**95**(41): 22–26.

[8] Breidenstein, E. B. M., C. de la Fuente-Núñez and R. E. W. Hancock (2011). "Pseudomonas aeruginosa: All roads lead to resistance." Trends in Microbiology**19**(8): 419–426.

[9] Schweizer, H. P. (2003). "Efflux as a mechanism of resistance to antimicrobials in Pseudomonas aeruginosa and related bacteria: Unanswered questions." Genetics and Molecular Research**2**(1): 48–62.

[10] Lister, P. D., D. J. Wolter and N. D. Hanson (2009). "Antibacterial-resistant Pseudomonas aeruginosa: Clinical impact and complex regulation of chromosomally encoded resistance mechanisms." Clinical Microbiology Reviews**22**(4): 582–610.

[11] Boucher, H. W., G. H. Talbot, J. S. Bradley, J. E. Edwards, D. Gilbert, L. B. Rice, M. Scheld, B. Spellberg and J. Bartlett (2009). "Bad Bugs, No Drugs: No ESKAPE! An Update from the Infectious Diseases Society of America." Clinical Infectious Diseases**48**(1): 1–12.

[12] Mesaros, N., P. Nordmann, P. Plésiat, M. Roussel-Delvallez, J. Van Eldere, Y. Glupczynski, Y. Van Laethem, F. Jacobs, P. Lebecque, A. Malfroot, P. M. Tulkens and F. Van Bambeke (2007). "Pseudomonas aeruginosa: Resistance and therapeutic options at the turn of the new millennium." Clinical Microbiology and Infection**13**(6): 560–578.

[13] Page, M. G. and J. Heim (2009). "Prospects for the next anti-Pseudomonas drug." Current Opinion in Pharmacology**9**(5): 558–565.

[14] Jabes, D. (2011). "The antibiotic R & D pipeline: an update." Curr Opin Microbiol**14**(5): 564–569.

[15] Moellering, R. C., Jr. (2011). "Discovering new antimicrobial agents." Int J Antimicrob Agents**37**(1): 2–9.

[16] Coates, A. R., G. Halls and Y. Hu (2011). "Novel classes of antibiotics or more of the same?" British Journal of Pharmacology**163**(1): 184–194.

[17] Kola, I. and J. Landis (2004). "Can the pharmaceutical industry reduce attrition rates?" Nat Rev Drug Discov**3**(8): 711–716.
[18] Sams-Dodd, F. (2005). "Target-based drug discovery: is something wrong?" Drug Discov Today**10**(2): 139–147.
[19] Hopkins, A. L. (2008). "Network pharmacology: The next paradigm in drug discovery." Nature Chemical Biology**4**(11): 682–690.
[20] Brötz-Oesterhelt, H. and N. A. Brunner (2008). "How many modes of action should an antibiotic have?" Current Opinion in Pharmacology**8**(5): 564–573.
[21] Korcsmáros, T., M. S. Szalay, C. Böde, I. A. Kovács and P. Csermelyt (2007). "How to design multi-target drugs: Target search options in cellular networks." Expert Opinion on Drug Discovery**2**(6): 799–808.
[22] Xie, L., L. Xie, S. L. Kinnings and P. E. Bourne (2012). "Novel Computational Approaches to Polypharmacology as a Means to Define Responses to Individual Drugs." Annual Review of Pharmacology and Toxicology**52**(1): 361–379.
[23] Zimmermann, G. R., J. Lehár and C. T. Keith (2007). "Multi-target therapeutics: when the whole is greater than the sum of the parts." Drug Discovery Today**12**(1–2): 34–42.
[24] Morphy, R. and Z. Rankovic (2009). "Designing multiple ligands - Medicinal chemistry strategies and challenges." Current Pharmaceutical Design**15**(6): 587–600.
[25] Morphy, R. and Z. Rankovic (2007). "Fragments, network biology and designing multiple ligands." Drug Discovery Today**12**(3–4): 156–160.
[26] Kaufmann, S. H. E. (2008). "Paul Ehrlich: founder of chemotherapy." Nat Rev Drug Discov**7**(5): 373–373.
[27] Chamot, E., E. B. El Amari, P. Rohner and C. Van Delden (2003). "Effectiveness of combination antimicrobial therapy for Pseudomonas aeruginosa bacteremia." Antimicrobial Agents and Chemotherapy**47**(9): 2756–2764.
[28] Bassetti, M., E. Righi and C. Viscoli (2008). "Pseudomonas aeruginosa serious infections: Mono or combination antimicrobial therapy?" Current Medicinal Chemistry**15**(5): 517–522.
[29] Johnson, S. J., E. J. Ernst and K. G. Moores (2011). "Is double coverage of gram-negative organisms necessary?" American Journal of Health-System Pharmacy**68**(2): 119–124.
[30] Traugott, K. A., K. Echevarria, P. Maxwell, K. Green and J. S. Lewis Ii (2011). "Monotherapy or combination therapy? The Pseudomonas aeruginosa conundrum." Pharmacotherapy**31**(6): 598–608.

[31] Pokrovskaya, V. and T. Baasov (2010). "Dual-acting hybrid antibiotics: A promising strategy to combat bacterial resistance." Expert Opinion on Drug Discovery**5**(9): 883–902.
[32] Labischinski, H., J. Cherian, C. Calanasan and R. S. Boyce (2010). Hybrid Antimicrobial Compounds and Their Use.
[33] Pokrovskaya, V., V. Belakhov, M. Hainrichson, S. Yaron and T. Baasov (2009). "Design, synthesis, and evaluation of novel fluoroquinolone - Aminoglycoside hybrid antibiotics." Journal of Medicinal Chemistry**52**(8): 2243–2254.
[34] Bera, S., G. G. Zhanel and F. Schweizer (2010). "Evaluation of amphiphilic aminoglycoside-peptide triazole conjugates as antibacterial agents." Bioorganic and Medicinal Chemistry Letters**20**(10): 3031–3035.
[35] Agarwal, P. and D. B. Searls (2008). "Literature mining in support of drug discovery." Briefings in Bioinformatics**9**(6): 479–492.
[36] Yang, Y., S. J. Adelstein and A. I. Kassis (2009). "Target discovery from data mining approaches." Drug Discovery Today**14**(3–4): 147–154.
[37] Balamurugan, R., F. J. Dekker and H. Waldmann (2005). "Design of compound libraries based on natural product scaffolds and protein structure similarity clustering (PSSC)." Molecular BioSystems**1**(1): 36–45.
[38] Gertsch, J. (2011). "Botanical drugs, synergy, and network pharmacology: Forth and back to intelligent mixtures." Planta Medica**77**(11): 1086–1098.
[39] Gradišar, H., P. Pristovšek, A. Plaper and R. Jerala (2007). "Green tea catechins inhibit bacterial DNA gyrase by interaction with its ATP binding site." Journal of Medicinal Chemistry**50**(2): 264–271.
[40] Bradbury, B. J. and M. J. Pucci (2008). "Recent advances in bacterial topoisomerase inhibitors." Current Opinion in Pharmacology**8**(5): 574–581.
[41] Spina, M., M. Cuccioloni, M. Mozzicafreddo, F. Montecchia, S. Pucciarelli, A. M. Eleuteri, E. Fioretti and M. Angeletti (2008). "Mechanism of inhibition of wt-dihydrofolate reductase from E. coli by tea epigallocatechin-gallate." Proteins: Structure, Function and Genetics**72**(1): 240–251.
[42] Park, K. and D. Kim (2008). "Binding similarity network of ligand." Proteins: Structure, Function, and Bioinformatics**71**(2): 960–971.
[43] Durrant, J. D., R. E. Amaro, L. Xie, M. D. Urbaniak, M. A. J. Ferguson, A. Haapalainen, Z. Chen, A. M. Di Guilmi, F. Wunder, P. E. Bourne and J. A. McCammon (2010). "A multidimensional strategy to detect

polypharmacological targets in the absence of structural and sequence homology." PLoS Computational Biology**6**(1).

[44] Oblak, M., M. Kotnik and T. Solmajer (2007). "Discovery and development of ATPase inhibitors of DNA gyrase as antibacterial agents." Current Medicinal Chemistry**14**(19): 2033–2047.

[45] James, S. P., I. Jamie, S. S. Ian, M. (2008). "Folate Biosynthesis - Reappraisal of Old and Novel Targets in the Search for New Antimicrobials." The Open Enzyme Inhibition Journal**1**: 12–33.

[46] Wiese, M., D. Schmalz and J. K. Seydel (1996). "New Antifolate 4, 4'-Diaminodiphenyl Sulfone Substituted 2, 4-Diamino-5-benzylpyrim idines. Proof of Their Dual Mode of Action and Autosynergism." Archiv der Pharmazie**329**(3): 161–168.

[47] Bennett, B. C., H. Xu, R. F. Simmerman, R. E. Lee and C. G. Dealwis (2007). "Crystal structure of the anthrax drug target, Bacillus anthracis dihydrofolate reductase." Journal of Medicinal Chemistry**50**(18): 4374–4381.

[48] Patel, O. G., E. K. Mberu, A. M. Nzila and I. G. Macreadie (2004). "Sulfa drugs strike more than once." Trends in Parasitology**20**(1): 1–3.

[49] Brvar, M., A. Perdih, M. Oblak, L. P. Mašiè and T. Solmajer (2010). "In silico discovery of 2-amino-4-(2, 4-dihydroxyphenyl)thiazoles as novel inhibitors of DNA gyrase B." Bioorganic and Medicinal Chemistry Letters**20**(3): 958–962.

[50] Frijters, R., M. van Vugt, R. Smeets, R. van Schaik, J. de Vlieg and W. Alkema (2010). "Literature mining for the discovery of hidden connections between drugs, genes and diseases." PLoS Computational Biology**6**(9).

[51] Sakharkar, M. K., P. Jayaraman, W. M. Soe, V. T. K. Chow, L. C. Sing and K. R. Sakharkar (2009). "In vitro combinations of antibiotics and phytochemicals against Pseudomonas aeruginosa." Journal of Microbiology, Immunology and Infection**42**(5): 364–370.

[52] Jayaraman P, S. M., Lim CS, Tang TH, Sakharkar KR (2010). "Activity and interactions of antibiotic and phytochemical combinations against Pseudomonas aeruginosa in vitro." Int J Biol Sci**6**: 556–558.

[53] Jayaraman P, S. M. (2011). "Insights into antifolate activity of phytochemicals against Pseudomonas aeruginosa." Journal of Drug Targeting**19**: 179–188.

[54] Gibbons, S., E. Moser and G. W. Kaatz (2004). "Catechin gallates inhibit multidrug resistance (MDR) in Staphylococcus aureus." Planta Medica**70**(12): 1240–1242.

[55] Hemaiswarya, S., A. K. Kruthiventi and M. Doble (2008). "Synergism between natural products and antibiotics against infectious diseases." Phytomedicine**15**(8): 639–652.

[56] Cushnie, T. P. T. and A. J. Lamb (2005). "Antimicrobial activity of flavonoids." International Journal of Antimicrobial Agents**26**(5): 343–356.

[57] Schechner, M., F. Sirockin, R. H. Stote and A. P. Dejaegere (2004). "Functionality maps of the ATP binding site of DNA gyrase B: Generation of a consensus model of ligand binding." Journal of Medicinal Chemistry**47**(18): 4373–4390.

[58] Lerner, M. G., A. L. Bowman and H. A. Carlson (2007). "Incorporating dynamics in E. coli dihydrofolate reductase enhances structure-based drug discovery." Journal of Chemical Information and Modeling**47**(6): 2358–2365.

[59] Hevener, K. E., M. K. Yun, J. Qi, I. D. Kerr, K. Babaoglu, J. G. Hurdle, K. Balakrishna, S. W. White and R. E. Lee (2010). "Structural studies of pterin-based inhibitors of dihydropteroate synthase." Journal of Medicinal Chemistry**53**(1): 166–177.

[60] Rogers, G. B., M. P. Carroll and K. D. Bruce (2012). "Enhancing the utility of existing antibiotics by targeting bacterial behaviour?" British Journal of Pharmacology**165**(4): 845–857.

[61] Fabbretti, A., C. O. Gualerzi and L. Brandi (2011). "How to cope with the quest for new antibiotics." FEBS Letters**585**(11): 1673–1681.

[62] Ma, X. H., Z. Shi, C. Tan, Y. Jiang, M. L. Go, B. C. Low and Y. Z. Chen (2010). "In-silico approaches to multi-target drug discovery computer aided multi-target drug design, multi-target virtual screening." Pharmaceutical Research**27**(5): 739–749.

[63] Nascimento, E. C. M. and J. B. L. Martins (2011). "Electronic structure and PCA analysis of covalent and non-covalent acetylcholinesterase inhibitors." Journal of Molecular Modeling**17**(6): 1371–1379.

[64] Lipinski, C. A. (2004). "Lead- and drug-like compounds: The rule-of-five revolution." Drug Discovery Today: Technologies**1**(4): 337–341.

[65] Veber, D. F., S. R. Johnson, H. Y. Cheng, B. R. Smith, K. W. Ward and K. D. Kopple (2002). "Molecular properties that influence the oral bioavailability of drug candidates." Journal of Medicinal Chemistry**45**(12): 2615–2623.

[66] Zhao, Y. H., M. H. Abraham, J. Le, A. Hersey, C. N. Luscombe, G. Beck, B. Sherborne and I. Cooper (2002). "Rate-limited steps of human

oral absorption and QSAR studies." Pharmaceutical Research**19**(10): 1446–1457.

[67] Maiti, R., G. H. Van Domselaar, H. Zhang and D. S. Wishart (2004). "SuperPose: A simple server for sophisticated structural superposition." Nucleic Acids Research**32**(WEB SERVER ISS.): W590–W594.

[68] Lessa, J. A., I. C. Mendes, P. R. O. da Silva, M. A. Soares, R. G. dos Santos, N. L. Speziali, N. C. Romeiro, E. J. Barreiro and H. Beraldo (2010). "2-Acetylpyridine thiosemicarbazones: Cytotoxic activity in nanomolar doses against malignant gliomas." European Journal of Medicinal Chemistry**45**(12): 5671–5677.

[69] Liu, X.-H., P.-Q. Chen, B.-L. Wang, Y.-H. Li, S.-H. Wang and Z.-M. Li (2007). "Synthesis, bioactivity, theoretical and molecular docking study of 1-cyano-N-substituted-cyclopropanecarboxamide as ketol-acid reductoisomerase inhibitor." Bioorganic and Medicinal Chemistry Letters**17**(13): 3784–3788.

[70] discovery." Briefings in Bioinformatics**9**(6): 479–492.

8

Genomics in Vaccine Development

Lars Rønn Olsen[1,2], Jing Sun[3], Christian Simon[4], Guang Lan Zhang[3,5] and Vladimir Brusic[3,5]

[1]Bioinformatics Centre, Department of Biology,
University of Copenhagen, Copenhagen, Denmark
[2]Biotech Research and Innovation Center (BRIC),
University of Copenhagen, Copenhagen, Denmark
[3]Cancer Vaccine Center, Dana-Farber Cancer Institute,
Boston, MA, USA
[4]Center for Protein Research, Department of Health Sciences,
University of Copenhagen, Copenhagen, Denmark
[5]Department of Computer Science, Metropolitan College,
Boston University, Boston, MA, USA

8.1 Introduction

Almost a century after Edward Jenner's groundbreaking work with the smallpox vaccine, the process of vaccine design was formalized by Louis Pasteur in 1880. His basic paradigm for vaccine development included the isolation, inactivation and injection of the causative microorganism. These are the basic principles which have guided vaccine development throughout the twentieth century [1]. Modern vaccinology began in mid-twentieth century based largely on breakthroughs in cell culture, bacterial polysaccharide chemistry, molecular biology and immunology [2]. Most of the successful vaccines of today were developed using this traditional, empiric approach. The eradication of smallpox (worldwide cases reduced from 2 million per year in 1959 to zero in 1978) [3], as well as an estimated two to three million deaths averted every year from diphtheria, tetanus, pertussis (whooping cough), and measles

Post-genomic Approaches in Drug and Vaccine Development, 179–204.

(http://www.who.int/hpvcentre/Global_Immunization_Data.pdf), all exemplify the significance of vaccinology for the global health.

Vaccine immunization offers the protection against a range of infectious diseases. The vast majority of successful vaccines were developed according to Pasteur's empiric principles. However, traditional vaccinology fails, for the time being, to address protection against hyper-variable and highly complex pathogens. The Pasteurian type vaccines are designed and administered to the population in a "one-size fits all" manner. These vaccines are successful despite the differences in individual vaccine responses, individual susceptibilities to infection, and individual variations in their TCR genes, antibody genes, and human leukocyte antigen (HLA) loci [4]. Considering the diversity and complexity of both pathogens and host immune system, it is not surprising that the empirical vaccine designs are of limited efficacy against some pathogens, such as emerging influenza strains and tuberculosis. There is no effective vaccine against somehypervariable pathogens such as HIV and hepatitis C and against complex pathogens such as malaria. Furthermore, vaccine administration is largely aimed at infants and children, although the average age of populations is increasing in all developed countries.

These limitations are the product of unmet challenges pertaining to three main areas of vaccine research: (i) clinical characterization of host immune system, detailed understanding of the pathogens and their variability, and the interplay between these two concepts; (ii) discovery of suitable vaccine targets and high false discovery rate, and; (iii) mode of vaccine delivery, including both technical and socioeconomic issues, i.e. distribution and manufacturing costs, particularly in third world countries. Experimental and computational genomics methods can be applied to ameliorate many of the issues in these three areas and aid the progress towards the rational vaccine design [5].

With our knowledge of the immune system and host-pathogen interaction increases, we are gradually moving away from empirical approach towards a more rational approach to vaccine design, in which genomics play a large role. Rino Rappuoli formalized the rational design approach in the "reverse vaccinology" framework–a vaccine design process anchored in the combination of computational analysis of genomic data and experimental validation of carefully selected vaccine designs [6]. Vaccine efforts are progressing in several fields including cancer [7], allergy [8] and autoimmunity [9]. In this chapter, we will focus on vaccines agains infectious diseases.

8.2 Genomics in Vaccine Development

A genome is a complete set of genes within an organism, and genomics is the structural and functional study of the genome and its variation within a population of organisms [10]. Until recently, sequencing an entire genome was expensive and experimentally laborios. However, the advent of high-throughput sequencing methods allows for rapid and increasingly affordable sequencing of entire genomes, thus significantly increasing the body of biological data [11]. For example, sequence variation data analysis leads to the discovery of risk-associated SNPs that impact cellular function and lead to specific diseases [12], epigenomic data enable us to explain the mechanisms of developmental pathways regulation [13], interactome data play an important role in annotation of functional genomics, providing genome-scale information on DNA, RNA and protein interactions [14–16]. However, while genomics play a big role in many fields of biological research, high-throughput genomics has been largely neglected in the vaccinology field.

8.2.1 Characterization of Pathogens

The most powerful genomics-based addition to the vaccinology toolbox is the ability to sequence the genomes of pathogens. The completion of the first microbial genome sequencing of *Haemophilusinfluenzae* in 1995 [17] unfolded a new and significant era in the history of biological research, and it opened prospects for new paradigm for vaccine development. This breakthrough enabled us to understand the details of pathogen evolution and the development of vaccine resistance of certain pathogen variants over time. Comparing the genomic sequences of multiple variants over time and from different geographical locations, allows for thorough intraspecies analysis of conservation and variability of antigens. From these analyses, stable antigens in viral genomes of small pathogens, such as small viruses (*e.g.* influenza, HIV, or flaviviruses), and the core proteins of the pan-genome in more complex prokaryote and eukaryote pathogen species can be determined [18]. This, in turn, enables a rational selection of vaccine targets that are more likely to confer broad and lasting immunity. Since the *Haemophilusinfluenzae* genome was sequenced, whole-genome sequencing has become a standard method for characterizing emerging pathogens, and hundreds of thousands of full genome sequences of various pathogens have been stored in public repositories such as GenBank [19].

Detailed understanding of the structure and function of the pathogen genome is of great value for rational vaccine design, most obviously because it enables functional annotation of antigenic sites in the translated proteome. Large-scale analysis of pathogen genome variants can help define immune epitopes – both antibody [20] and T-cell [21] epitopes–and offer insights into cross-reactivity and cross-protection. For example, elucidating peptide binding affinity to the HLA – one of the rate limiting steps in T-cell-based immune response [22] is of high value for identification of vaccine targets. This annotation can be performed either *in silico* prediction [23–25], *in vitro* by peptide-HLA binding assays [26–28], by functional assays [29,30] or, ideally, a combination of these methods. Large-scale analysis of pathogens requires screening of tens or hundreds of thousands of genome variants and integration of results on an unprecedented scale and it requires advanced statistical, bioinformatics, and modeling tools.

Combining the knowledge of pathogen evolution and HLA binding provides a powerful tool to elucidate vaccine targets that are well-conserved and stable. Additional analyses of pathogen life cycle may provide essential insights for identifying suitable vaccine targets. For example, gene expression profiling reveals antigen expression dynamics representing processes of clearing pathogens in the early stages of infection [31]. A detailed knowledge of pathogen gene function and relationship between variability, transmissibility, and pathogenicity supports research about directing immune response towards functionally essential proteins with low mutation rates. Comparative genomics enables modeling of patterns of epidemiological spread. Observational epidemiology is considered to be the scientific foundation of public health [32]. Advances in genomic methods enhance our understanding of epidemiology and its role in vaccine discovery. Genomics data and tools have greatly influenced epidemiological study designs, analyses, and causal inferences on 'environmental' causes of disease [33].

8.2.2 Characterization of the Host Immune System

The human immune system comprises a complex network of cells and molecules. It is the interplay between these components and invading pathogens that determines immune response. Immune response occurs in different stages and with different modes of action: from the innate immune system providing the first line of defense by removal of some foreign substances and inducing inflammation, to the adaptive immune system recognizing pathogen or non-pathogen antigens (e.g. from cancer). The adaptive immune system is the main target of prophylactic vaccines. There are two primary branches of adaptive immunity: humoral immunity, which is based on pathogen recognition by antibodies; and cellular immunity based on recognition by cellular receptors – T-cell receptor binding to an HLA-epitope complex.

Characterization of host HLA profile has traditionally been utilized in conjunction with organ transplantation in order to avoid transplant rejection, in which immune cells of the host reject mismatched transplant. HLA mismatch is the main reason for graft vs host disease (GVHD) where leukocytes from the graft recognize host tissues as foreign and cause selective damage to the liver, skin, or mucosal surfaces. The severity of GVHD is often correlated with the disparity between donor and host HLA profiles [34]. Additionally, variability of the host immune system contributes to shaping the immune response against invading pathogens. Genetic variation of the T-cell receptor genes, the antibody genes, and the HLA loci play major roles in the recognition of pathogen antigens. Somatic hypermutation of TCR and paratope genes allow for an extremely large diversity. The recognition of antigens, both B- and T-cell epitopes, are dependent on genetic factors, as well as previous immunological history and are, therefore individual.

The host HLA profiles must be taken into consideration in the identification of potential vaccine targets. HLAs are amongst the most polymorphic molecules in the human genome, and represent the most variable factor of human immune recognition. HLA genes belong to one of two classes: class I and class II. Comprised of more than 200 genes, the key products of HLA are encoded by three loci that produce three proteins (HLA-A, -B, and -C) for class I, and by six loci that produce three proteins for class II (HLA DR composed of HLA-DRA and DRB1; HLA-DQ composed of HLA-DQA1 and DQB1, and HLA-DP composed of HLA-DPA1 and HLADPB1) [35]. There are 9310 HLA allelic variants reported in IMGT/HLA database Release 3.12.0 as of April 17, 2013 [36]. HLA alleles that bind overlapping sets of peptides belong to the same HLA supergroups that are collectively referred to as a HLA supertype [37]. Instead of classifying HLA polymorphisms on the basis of serological reactivity, sequence or evolutionary relationships, the classification of HLA molecules into supertypes on the basis of shared structural features and peptide binding motifs could have important practical and theoretical consequences in the development of epitope-based vaccines [38]. It was shown that relatively small collections of supertypes cover a large proportion of the population, for example, approximately 88% for A2, A3, and B7 supertypes or 99% for A1, A2, A3, B7, A24, and B44 supertypes in different ethnic groups [39].

HLA molecules exhibit specific binding to pathogen peptides and tolerance to self-peptides and play an important role in determining control and susceptibility to invading pathogens and cancers. Individuals' HLA profiles have strong association with susceptibility or resistance to infections. For example, HIV-positive individuals with HLA-B57 and HLA-B27 supertypes (known as HIV elite controllers), show much slower progression of AIDS than those with other HLA supertypes[40]. As HLA loci are hereditary, alleles are often geographically clustered, meaning that some populations are more susceptible to some diseases, for example, Epstein-Barr virus related cancers [41]. The distribution of HLA loci in at-risk populations is highly valuable information for vaccine design. High-throughput and high resolution HLA genotyping methods, such as next-generation sequencing and microarrays, enable fast and cost effective immune profiling, and thus facilitate the development of effective personalized vaccines targeting subpopulations [42–44]. The main challenge in genomic identification of HLA alleles is combinatorial complexity of HLA genes where sophisticated algorithms are needed for the assembly of reads. This is a non-trivial problem that requires a validatory sequencing or sophisticated knowledge-based systems.

In addition to HLA and antibody related genes and their accessory molecules, other classes of genes may influence immune response. These include viral receptors and other pattern recognition receptors, signaling molecules, cytokine and cytokine receptor genes, perforin and granzymes, death receptors, and many others [4]. Immune responses to an invading pathogen or to an administered vaccine is the cumulative effect of a large number of products, which, when fully characterized, should theoretically be predictable in an approach termed "immune response network theory" [45, 46].

8.2.2.1 Personalized vaccines

Vaccinomics is the branch of vaccinology that deals with the vaccine analysis and development. It studies variations in individual's immune responses to vaccines by combining the strengths of immunogenetics and immunogenomics including both single nucleotide polymorphism (SNP) discovery and functional validation with immune profiling and systems biology approaches [45]. Analyzing infection and response in a vaccinomics setting was used to uncover complex interactions between CD4, CD8 and humoral response against vaccinia virus [47], and may also be used to predict adverse effects of vaccination [48].

Integrated analysis of the host and pathogen can also be used to design highly efficient and specific vaccines. The variations in individual immune-response genes are thought to be predictive of vaccination outcomes. In the US measles epidemics of 1989–1991, 10% of previously vaccinated children were not protected [49]. Genetic polymorphisms of the immune system significantly influence the variation in immune responses to viral vaccines, causing differential responses to both vaccines and pathogens. Selecting antigens and designing vaccines to cater for individual immune profiles discovered through vaccinomics methods, is promising to increase vaccine efficiency, minimize adverse effects, and increase the potential of vaccines as therapeutic treatments.

8.3 Application

Making sense of genomics data both human and pathogens, require sophisticated and comprehensive computational analysis and specialized algorithms. Applications in vaccine development are more complicated because of combinatorial complexity of the human immune system as well as combinatorial complexity of pathogens that make vaccine genomics resource consuming both in terms of quantity and complexity of data. In the following example, we will focus on computational methods for identification and selection of potential vaccine targets for inclusion in vaccine constructs against viral pathogens.

8.3.1 Host Diversity Characterization

Host HLA profiles were traditionally characterized in a number of ways, including serological methods, cellular methods, and molecular techniques. Presently, the practice relies more and more on high-throughput genomics methods, such as HLA microarrays and next generation sequencing [43,50,51]. These experimental methods in turn relies on bioinformatics methods for sequence mapping and predictions algorithms for microarray data interpretation [44]. For personalized vaccines, the HLA profile of the subject is by far the most important, but population coverage of the included epitopes can be calculated using tools provided by the Immune Epitope Database and Analysis Resources (IEDB) [52,53].

8.3.2 Pathogen Diversity Characterization

For higher order pathogens, such as bacteria, fungi, and parasites, characterization of the pan-genome can be the first step in cataloguing potential antigens for vaccine targeting [18]. In lower order pathogens, such as viral pathogens, genomes are typically of lower variability on the macro-scale, i.e. they often express the same set of proteins (some exceptions do exist such as the alternative reading frame product PB1-F2 of influenza [54]), an issue which has also been addressed by bioinformatics methods [55]. Conservation and variability analysis of viral pathogens is therefore often done by calculating the frequency of the consensus nucleotide or amino acid on each position of the genome or translated proteome [56, 57]. When searching for peptide antigens it may, however, be advantageous to calculate conservation and variability of peptides rather than single amino acids [21].

8.3.3 Vaccine Target Prediction

8.3.3.1 T-cell epitopes

T-cell epitopes are short peptide fragments of 8–12 and 13–25 amino acids in length for HLA class I and II, respectively [58, 59]. Before proteins are transported to the ER where they bind HLA, they are intracellularly processed by proteasomal cleavage in the cytosol and transported to the ER by TAP proteins. The HLA-peptide complex is then transported to the cell surface, where it is recognized by circulating T-cells through TCR. Each cellular process involved in peptide processing, transport, binding to HLA and presentation of the cell surface is a rate-limiting step in the classical T-cell-mediated immunity, rendering prediction of immunogenicity a highly challenging task.

The prediction of peptide processing events, such as proteasomal cleavage [60–62] and TAP transport [63–65] has also been explored. However, the performance of these methods is still not optimal [66]. Algorithms for predicting HLA binding affinity are reasonable accurate for most HLA alleles, and highly accurate for a some HLA alleles [23–25]. Several algorithms for prediction of HLA class I and class II peptide binding affinity exists (reviewed in [24, 25]). Currently, the best overall performing predictor for HLA class I binding is the artificial neural network (ANN) and weight matrix-based prediction tool, netMHC 3.2 [67] and for class II, the best performing algorithm is the ANN predictor netMHCII 2.2 [68]. Other popular classification algorithms include BIMAS [69] and SYFPEITHI [70].

For a peptide to be immunogenic, it must be adequately pre-processed and bind to the HLA. However, this is not always sufficient, as it has been shown that some peptides were unable to elicit immune response in spite of strong HLA binding. This phenomenon is known as "holes in the T-cell repertoire" [71]. A better predictor for immunogenicity may therefore be the stability of the peptide-HLA complex, as highly immunogenic peptides typically bind HLA more stably [72]. Similarly, there are indications that stronger binding of the peptide-HLA complex to the TCR is a good predictor of the immunogenicity of a peptide [73], however, only 21 crystal structures of the peptide-HLA-TCR complex are solved at present, which is insufficient data to train a functional classifier. Some more advanced prediction methods consider a large number of physiochemical properties for predictions [74, 75], but these fail to outperform the simple HLA binding prediction algorithms.

Although both computational and experimental methods highly accurate, annotating genomes with T-cell epitopes is only one of many steps in selecting appropriate vaccine targets. Many parameters of a good vaccine target conferring efficient, lasting immunity, still remains to be considered after determining HLA binding: multiple rate-limiting steps of peptide pre-processing [22], confirming *in vivo* expression, conservation across pathogen population [56] response across host population, and stability over time, among others. Genomics, therefore, helps identify the diversity of T-cell epitope candidates and diversity of HLA molecules along with possible mutations in human individuals of antigen processing and presentation machinery.

8.3.3.2 B-cell epitopes

B-cells recognize discontinuous epitopes on antigen proteins through B-cell receptors. This antigen binding site, or B-cell receptor, is a membrane-bound antibody molecule [76]. Each distinct antibody has a unique antigen binding site, called the paratope, which is produced through the processes of recombination, hyper-mutation and affinity maturation. The binding between antigen and antibody is specific, but in some cases, antibodies display cross-reactivity against other antigens.

Dengue virus is a mosquito-borne virus of the *Flavivirus* genus [77]. Over 2.5 billion people – over 40% of the world's population – are now at risk from dengue. WHO currently estimates there may be 50–100 million dengue infections worldwide every year (WHO: Dengue and severe dengue. Available: http://www.who.int/ mediacentre/ factsheets/fs117/en/index .html.

Accessed September 2013). However, no effective vaccine has been licensed for human use yet. Vaccine development has been hampered by potential complications following secondary DENV infections, known as antibody dependent enhancement (ADE). Recent studies have further shown that natural variation within one serotype has a strong influence on antibody binding and neutralization [78]. Vaccines with broadly neutralizing ability across and within serotypes are therefore required for effective dengue vaccine.

The cross-reactivity of antibodies is essential in vaccine development for pathogens with variations among strains, which include, but not limited to, flaviviruses such as dengue, or West Nile [79], but also orthomyxoviruses (influenza) [80], arenaviruses such as lymphocytic choriomeningitis virus [81], human immunodeficiency virus [82], and human papillomavirus (HPV) [83], among others. Generally, the cross-reactivity of neutralizing antibodies results from common B-cell epitopes shared by different strains. The description of B-cell epitopes, measurement of B-cell epitope similarity among different strains. B-cell epitope variability among specific pathogen populations represents important issues for vaccine design. Systematic bioinformatics analysis of antibody-based cross-reactivity is enabled by the development of genomics: characterization of an increasing number of neutralizing antibodies specific for pathogens, increasing number of known 3D structures of antigen-antibody interactions, rapid growth of the number of sequences of viral variants, and the emergence of new viral strains. Genomics enables sequencing of antibody repertoires [84]. Next Generation Sequencing may enable us to understand detailed workings of the immune system, for example, shed light on antibody cross-reactivity and cross-protection. It will improve our capacity for selection, design, and improvement of therapeutic candidates, improve drug discovery and design, and generate better vaccines.

8.3.4 Selection of Vaccine Targets

Effective vaccine targets elicit strong immune responses in patients. The design of personalized vaccines involves identifying targets that the host immune system responds to, i.e. matching host HLA profile with peptide targets. The targets should ideally be present in all variants of the pathogen targeted by the vaccine. If no single target holds the capacity to do so, several targets should be combined to provide broad coverage–multi–epitope vaccine strategies. For some pathogens, such as flaviviruses, such strategies are required to avoid the complications associated with secondary infection by a different subtype than the primary infection. The design of polyvalent vaccines – those targeting several subtypes of the same virus – requires careful selection of targets. For peptide vaccines, a number of methods for compressing pathogen diversity into artificial constructs are available, for example, the consensus method [85,86], the ancestor method [87], and center of the tree method [88] are based on assembling individual amino acids into "centralized" consensus peptides designed to compress population diversity into a small set of artificial immunogens. Other methods, such as the mosaic vaccine design [89], attempts to achieve polyvalent coverage of the viral diversity by assembling naturally occurring peptides into longer artificial polypeptides. These methods find common ground in a systematic exclusion of rare peptides, regardless of whether these are predicted to be HLA binders, in spite of the fact that it has been shown that HLA class I epitopes in highly variable viruses, such as dengue virus, have low targeting efficiency, i.e. low correlation between HLA binding affinity and conservation of the epitopes [90]. Furthermore, some low frequency peptides have been shown to be highly efficient T-cell epitopes in HIV [91]. We previously proposed an approach based on analysis of peptide conservation rather than residue conservation [21]. This is achieved by the alignment of multiple sequences of pathogen proteins and iterative analysis of peptides in a sliding window (Figure 8.1) of a size corresponding to the length of T-cell epitopes (8–22 amino acids in length). This approach facilitates the selection of multiple targets of varying frequency that collectively cover the diversity of the pathogen. The potential binding affinity of each peptide in each block is predicted so that the block as a whole can be assessed for immunogenic potential. If all peptides in a block are potential binders, the pool of peptides in the block can be included in a multi-epitope vaccines.

8.4 Reverse Vaccinology Pipeline for Viral Pathogens

To demonstrate the principles presented in this chapter, we analyzed 82 sequences of H7N9 influenza A hemagglutinin proteins. The sequences were extracted from the influenza knowledgebase, FluKB (http://research4.dfci.harvard.edu/cvc/flukb/). We formalized the approach into a five-step workflow (Figure 8.2) adapted from the flavivirus specific workflow in the FLAVIdB database [92]. Each of the five steps is discussed in detail below and the results of each analysis are presented.

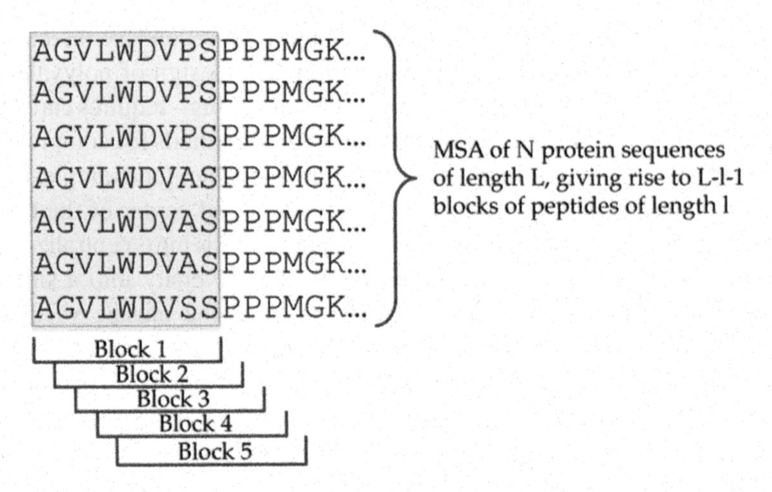

Figure 8.1 Analysis of peptides from multiple sequence alignments. The window of length 8–22 (corresponding to the length of HLA class I and II binding peptides) is moved across the entire protein of interest. Each block contains a number of unique peptides, for which HLA binding potential can be predicted.

8.4.1 Step 1: Pathogen and Host Characterization

The preliminary step in pathogen characterization involves whole genome sequencing of as many pathogen variants as possible, and preferably isolates distributed equally over time and geographical location. The more isolates that are sequenced, the more accurate the predictions of virus evolution will be. Currently, more than 67,000 complete genomes are sequenced (http://www.fludb.org/), providing a solid basis for genomic characterization. However, owing to large amount of variation between influenza subtypes, broadly neutralizing vaccines require seasonal reformulations to counteract the evolution of the virus. The genes of each new isolate are mapped onto reference sequences and annotated for protein products.

8.4.2 Step 2: Cataloguing Potential Antigen

In simple pathogens, such as small viruses, all gene products theoretically hold antigenic potential, whereas in more complex diseases, such as complex pathogens or cancer, not all proteins are useful vaccine targets. Thus, cataloguing potential antigens in viral pathogens is done by sequencing the genomes of a large number of isolates, performing comparative genomics analysis, and cataloguing gene products expressed in all isolates. In influenza, the genome consists of eight genes coding for at least ten products (Figure 8.3), since six

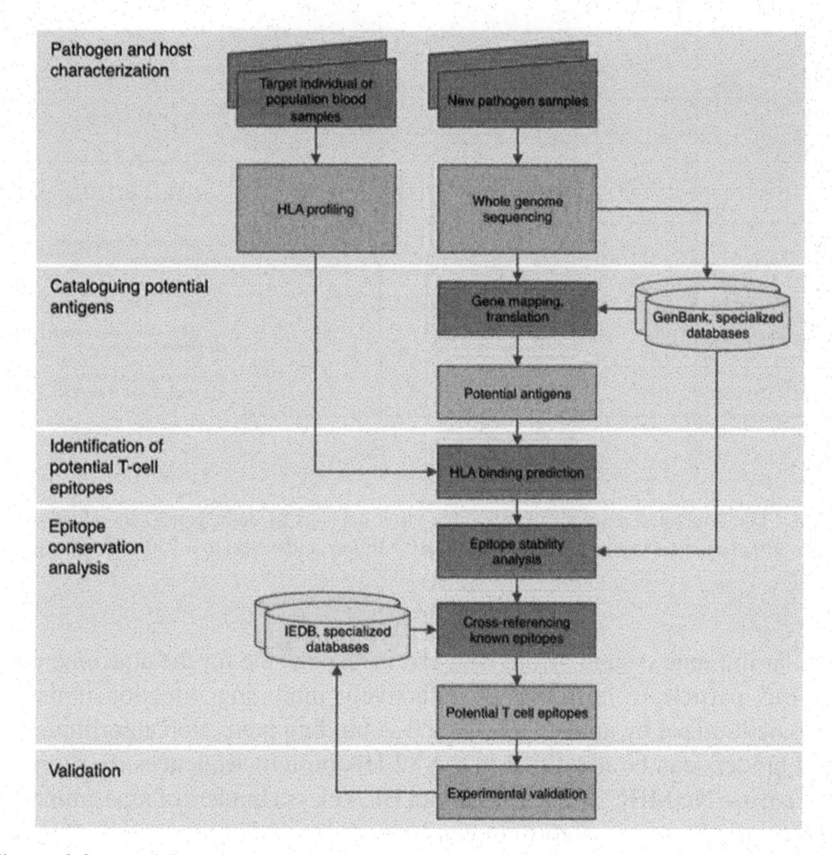

Figure 8.2 Workflow for reverse vaccinology discovery of vaccine targets against viral pathogens, utilizing genomics methods and computational analysis.

of the eight segments are monocistronic (coding for one protein) while the remaining two are dicistronic (coding for two proteins). Only surface proteins (HA and NA) can be targeted by antibodies, whereas all proteins are potential targets for T-cells. Another important function of thorough antigen cataloguing is to avoid targeting pseudogenes and gene products with high homology to human genes, as this may render the vaccine inefficient, or induce adverse effects related to autoimmunity.

8.4.3 Step 3: Identification of Potential T-cell Epitopes

In the following example, we focus on predicting potential T-cell epitopes in the HA protein. HA is a natural vaccine target, as HA is a glycoprotein on the surface of the influenza virus and thus visible to both the humoral

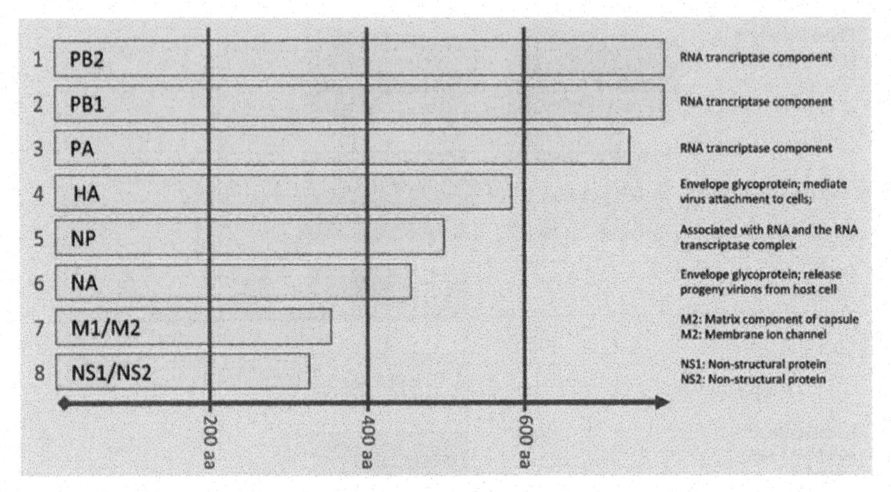

Figure 8.3 The annotated genome of influenza virus consists of eight genes, encoding at least 10 protein products, ranging in length from ~100 amino acids (M1) to ~750 amino acids (PB1/PB2).

and cellular immune system. Moreover, HA is responsible for the attachment of the virus particle to host, hereby effectively inhibiting infection if the protein is neutralized by antigens. Using HLA binding prediction algorithms, potential binders can be identified in the 82 HA protein sequences. For this example we use NetMHC 3.0 [93] to predict HLA class I binders of nine amino acids in length. The same algorithm can be used to predict binders of eight, ten, eleven and twelve amino acids in length. Predictions reveal 18 peptides that potentially bind HLA A*02:01. However, before these can be used in vaccine constructs, they should be assembled into constructs that encompass the diversity of all 82 isolates of H7N9. This can, for example, be done using the aforementioned block conservation method.

8.4.4 Step 4: Epitope Conservation Analysis

Block conservation can be visualized using a bar plot displaying the number of different peptides in each. The number of predicted binders and their accumulated frequency in a given block determines the suitability of the block for polyvalent vaccine designs: if all peptides are predicted binders to the same HLA with similar affinity, the block may be valuable to examine further for potential use in multi-epitope vaccine constructs. The predicted binding affinities of peptide blocks to the selected HLA alleles can be visualized using a heat map displayed below the conservation graph. In this heat map, each column corresponds to the binding affinity of the given position in the MSA, and each row corresponds to an HLA allele. This approach to visualization allows simultaneous display of peptide conservation and overview of predicted binding to multiple HLA alleles (Figure 8.4). Block conservation analysis can be performed at http://met-hilab.bu.edu/blockcons.

Detailed visual examination of block 244 can be performed using sequence logos [94] such as WebLogo [95] and BlockLogo [96] in combination with prediction results. Figure 8.5 shows the variation of residues (top panel) and peptides (bottom panel). Although a large number of peptides could theoretically be derived from the variation seen in the top panel, the bottom panel shows that four peptides encompass most of the variation in the block. These four peptides cover 98.75% of all sequenced variants of the influenza A H7N9, and all are predicted to bind HLA A*02:01 (Table 8.1). If preprocessing and *in vivo* binding can be experimentally confirmed, these peptides could be valuable candidates for a multi-epitope vaccine construct.

8.4.5 Step 5: Validation

HLA binding is a central, but only one of a number of rate limiting steps in recognition of peptides, that is necessary for the destruction of infected cells by T-cells of the immune system. Therefore, experimental validation of the identified peptides is necessary to ensure that these are indeed processed, presented, and recognized by circulating T-cells. Validation of HLA binding can be performed using basic immuno assays [26] flow cytometry-based methods [27] or mass-spectrometry techniques [28]. Once peptides have been *in vitro* validated they are subject to preclinical trials to validate *in vivo* response.

Table 8.1 HLA binding predictions of each peptide present in the block of 9-mer peptides starting at position 244 of 82 isolates of influenza A H7N9 hemagglutinin. Prediction for HLA A*02:01 allele is listed

#	Peptide	Frequency	Accumulated frequency	Number of sequences	Binding affinity to HLA A*0201
1	LMLNPNDTV	71.25%	71.25%	57	53 nM
2	LILNPNDTV	10.00%	81.25%	8	458 nM
3	LLLDPNDTV	10.00%	91.25%	8	20 nM
4	LMLNPNDTI	7.50%	98.75%	6	312 nM
5	MLLDPNDTV	1.25%	100.00%	1	10 nM

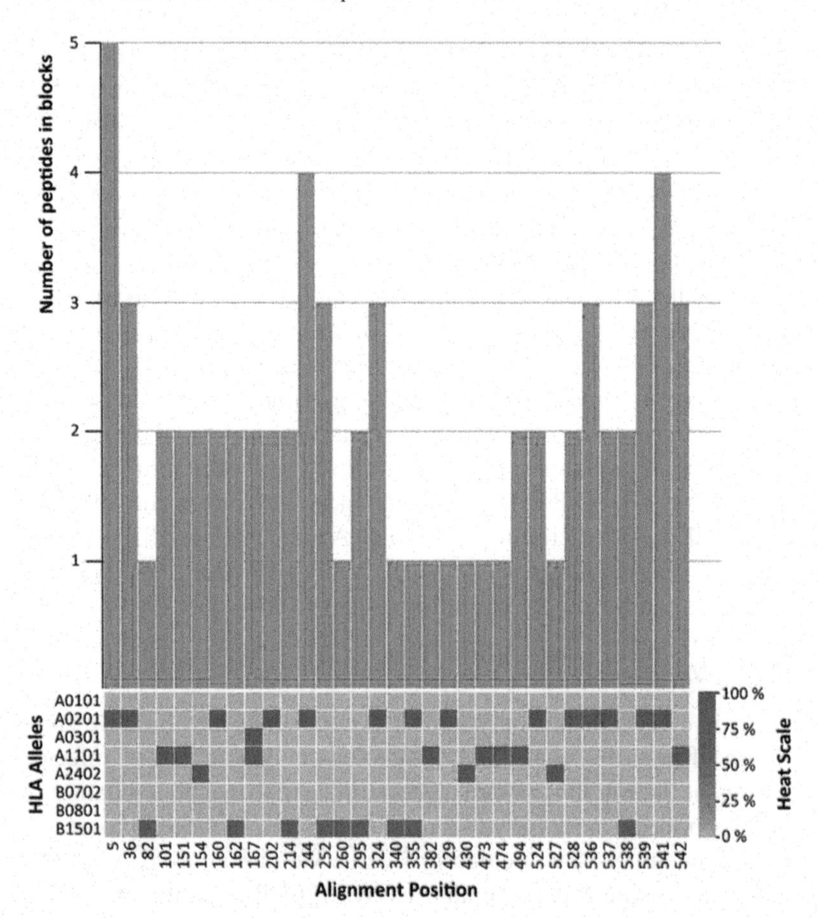

Figure 8.4 Visualization of the block conservation analysis and HLA class I binding affinity predictions in 82 isolates of influenza A H7N9 hemagglutinin, in which at least 95% of the peptides were predicted to bind at least one of eight HLA alleles used in this example (A*01:01, A*02:01, A*03:01, A*11:01, A*24:02, B*07:02, B*08:01, B*15:01). The bars show the minimum number of peptides in a block (Y axis) at a given starting position in the MSA (X axis) required for fulfilling the user defined coverage threshold (in this case 95%). The heat map below the bar show the percentage of peptides in the block predicted to bind to each of the HLA alleles predicted for in these examples. The color of each position in the heat map matrix ranges from gray (0% accumulated conservation by predicted binders in the block for the given allele) to red (blocks predicted to bind to the given allele with a minimum binding affinity of 500 nM represents 95% conservation in the block). In this figure, only blocks in which 95% of the peptides are predicted binders are shown. The starting positions of the blocks are shown below the heat map. Figure generated using Block conservation webserver (http://met-hilab.bu.edu/blockcons).

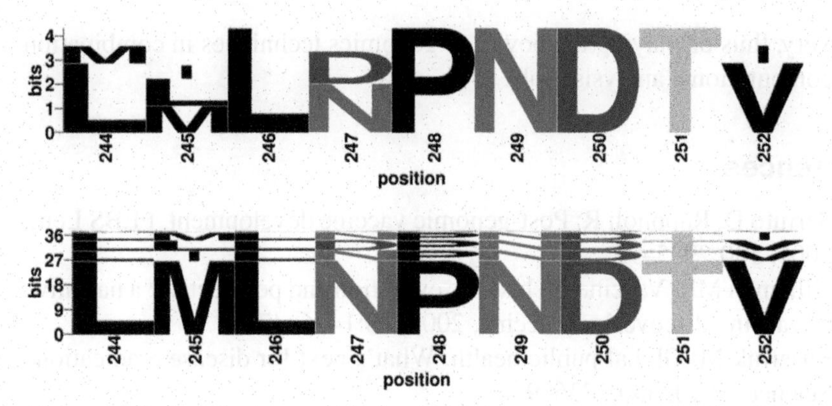

Figure 8.5 (Top panel) Sequence logo plot of the residues in the 9-residue block starting at position 244 of 82 isolates of influenza A H7N9 hemagglutinin generated using WebLogo. (Bottom panel) BlockLogo of the peptides in the 244–252 block. The residue position in the MSA is shown on the X-axis, and the information content is shown on the Y-axis. See Table 8.1 for peptide frequencies and HLA binding affinity predictions.

8.5 Conclusions

Genomics methods have made an enormous impact on vaccine research and development. The greatest contribution is arguably the advent of high–throughput sequencing techniques and computational methods for analysis of the large amounts of pathogen sequence data and the ability to perform large-scale profiling of human genes involved in immune responses. These methods have facilitated thorough analysis of host immune system diversity, thus enabling extensive research on the modalities of immune responses. Rapid characterization of individual and population HLA profiles and accessory molecules have greatly increased our understanding of cellular immune response against invading pathogens against which this mode of response is utilized. Combing information about human immune system diversity with the analysis of pathogen antigen diversity, is likewise facilitated by high–throughput genomics methods. They provide the technological infrastructure for reverse vaccinology – rational large-scale selection of vaccine targets for both general or personalized vaccine designs.

In this chapter, we have provided an example of a reverse vaccinology workflow for vaccine target selection in viral pathogens and showed the analyzis of 82 isolates of the emerging influenza A H7N9. We found 33 potential target regions in the hemagglutinin proteins of the analyzed isolates using the block conservation method for multi epitope vaccine target discovery, thus displaying the power of genomics techniques in combination with computational analysis tools.

References

[1] Serruto D, Rappuoli R: Post-genomic vaccine development. FEBS Lett., 2006, 580:2985–92.
[2] Hilleman MR: Vaccines in historic evolution and perspective: a narrative of vaccine discoveries. Vaccine, 2000, 18:1436–47.
[3] Enserink M: Global public health. What's next for disease eradication? Science, 2010, 330:1736–9.
[4] Poland G a, Kennedy RB, Ovsyannikova IG: Vaccinomics and personalized vaccinology: is science leading us toward a new path of directed vaccine development and discovery? PLoSPathog., 2011, 7:e1002344.
[5] Mooney M, McWeeney S, Sékaly R-P: Systems immunogenetics of vaccines. Semin. Immunol., 2013, 25: 124–9.
[6] Rappuoli R: Reverse vaccinology. Curr. Opin. Microbiol., 2000, 3: 445–50.
[7] Palucka K, Ueno H, Banchereau J. Recent developments in cancer vaccines. J Immunol., 2011, 186: 1325–31.
[8] Linhart B, Valenta R. Vaccines for allergy. CurrOpinImmunol., 2012, 24: 354–60.
[9] Zaccone P, Cooke A. Vaccine against autoimmune disease: can helminths or their products provide a therapy? CurrOpinImmunol., 2013, 25: 418–23.
[10] McKusick VA: Genomics: structural and functional studies of genomes. Genomics, 1997, 45: 244–9.
[11] Joyce AR, Palsson BØ: The model organism as a system: integrating "omics" data sets. Nat. Rev. Mol. Cell Biol., 2006, 7: 198–210.
[12] Maynard ND, Chen J, Stuart RK, Fan J-B, Ren B: Genome-wide mapping of allele-specific protein-DNA interactions in human cells. Nat. Methods, 2008, 5: 307–9.
[13] Hellman A, Chess A: Gene body-specific methylation on the active X chromosome. Science, 2007, 315: 1141–3.
[14] Chi SW, Zang JB, Mele A, Darnell RB: Argonaute HITS-CLIP decodes microRNA-mRNA interaction maps. Nature, 2009, 460: 479–86.
[15] Hutchins JRA, Toyoda Y, Hegemann B, Poser I, Hériché J-K, Sykora MM, Augsburg M, Hudecz O, Buschhorn BA, Bulkescher J, Conrad C, Comartin D, Schleiffer A, Sarov M, Pozniakovsky A, Slabicki MM,

Schloissnig S, Steinmacher I, Leuschner M, Ssykor A, Lawo S, Pelletier L, Stark H, Nasmyth K, Ellenberg J, Durbin R, Buchholz F, Mechtler K, Hyman AA, Peters J-M: Systematic analysis of human protein complexes identifies chromosome segregation proteins. Science, (80-.). 2010, 328: 593–9.

[16] Walhout AJ, Vidal M: Protein interaction maps for model organisms. Nat. Rev. Mol. Cell Biol., 2001, 2: 55–62.

[17] Fleischmann RD, Adams MD, White O, Clayton RA, Kirkness EF, Kerlavage AR, Bult CJ, Tomb JF, Dougherty BA, Merrick JM: Whole-genome random sequencing and assembly of Haemophilusinfluenzae Rd. Science, (80-.). 1995, 269: 496–512.

[18] Muzzi A, Masignani V, Rappuoli R: The pan-genome: towards a knowledge-based discovery of novel targets for vaccines and antibacterials. Drug Discov. Today, 2007, 12: 429–39.

[19] Benson D a, Karsch-Mizrachi I, Clark K, Lipman DJ, Ostell J, Sayers EW: GenBank. Nucleic Acids Res., 2012, 40:D48–53.

[20] Sun J, Kudahl UJ, Simon C, Zhiwei C, Reinherz EL, Brusic V. Large-scale analysis of B-cell epitopes on influenza virus hemagglutinin - implications for cross-reactivity of neutralizing antibodies. Front. Immunol., 2013.

[21] Olsen LR, Zhang GL, Keskin DB, Reinherz EL, Brusic V: Conservation analysis of dengue virus T-cell epitope-based vaccine candidates using peptide block entropy. Front. Immunol., 2011, 2: 1–15.

[22] Montoya M, Del Val M: Intracellular rate-limiting steps in MHC class I antigen processing. J. Immunol., 1999, 163: 1914–22.

[23] Zhang GL, Ansari HR, Bradley P, Cawley GC, Hertz T, Hu X, Jojic N, Kim Y, Kohlbacher O, Lund O, Lundegaard C, Magaret C a, Nielsen M, Papadopoulos H, Raghava GPS, Tal V-S, Xue LC, Yanover C, Zhu S, Rock MT, Crowe JE, Panayiotou C, Polycarpou MM, Duch W, Brusic V: Machine learning competition in immunology - Prediction of HLA class I binding peptides. J. Immunol. Methods, 2011, 374: 1–4.

[24] Lin HH, Ray S, Tongchusak S, Reinherz EL, Brusic V: Evaluation of MHC class I peptide binding prediction servers: applications for vaccine research. BMC Immunol., 2008, 9: 8.

[25] Lin HH, Zhang GL, Tongchusak S, Reinherz EL, Brusic V: Evaluation of MHC-II peptide binding prediction servers: applications for vaccine research. BMC Bioinformatics, 2008, 9 Suppl 12:S22.

[26] Sylvester-Hvid C, Kristensen N, Blicher T, Ferré H, Lauemøller SL, Wolf XA, Lamberth K, Nissen MH, Pedersen LØ, Buus S: Establishment

of a quantitative ELISA capable of determining peptide - MHC class I interaction. Tissue Antigens, 2002, 59: 251–8.

[27] Andersen RS, Kvistborg P, Frøsig TM, Pedersen NW, Lyngaa R, Bakker AH, Shu CJ, Straten PT, Schumacher TN, Hadrup SR: Parallel detection of antigen-specific T-cell responses by combinatorial encoding of MHC multimers. Nat. Protoc., 2012, 7: 891–902.

[28] Reinhold B, Keskin DB, Reinherz EL: Molecular Detection of Targeted Major Histocompatibility Complex I-Bound Peptides Using a Probabilistic Measure and Nanospray MS(3) on a Hybrid Quadrupole-Linear Ion Trap. Anal. Chem., 2010, 82: 9090–9099.

[29] Miyahira Y, Murata K, Rodriguez D, Rodriguez JR, Esteban M, Rodrigues MM, Zavala F: Quantification of antigen specific CD8$^+$T-cells using an ELISPOT assay. J Immunol Methods, 1995, 181: 45–54.

[30] Sette A, Vitiello A, Reherman B, Fowler P, Nayersina R, Kast WM, Melief CJ, Oseroff C, Yuan L, Ruppert J, Sidney J, del Guercio MF, Southwood S, Kubo RT, Chesnut RW, Grey HM, Chisari FV. The relationship between class I binding affinity and immunogenicity of potential cytotoxic T-cell epitopes. J Immunol., 1994, 153: 5586–92.

[31] Assarsson E, Greenbaum JA, Sundström M, Schaffer L, Hammond JA, Pasquetto V, Oseroff C, Hendrickson RC, Lefkowitz EJ, Tscharke DC, Sidney J, Grey HM, Head SR, Peters B, Sette A: Kinetic analysis of a complete poxvirus transcriptome reveals an immediate-early class of genes. Proc. Natl. Acad. Sci. U. S. A., 2008, 105: 2140–5.

[32] Weed DL, Mink PJ: Roles and responsibilities of epidemiologists. Ann. Epidemiol., 2002, 12: 67–72.

[33] Khoury MJ, Millikan R, Little J, Gwinn M: The emergence of epidemiology in the genomics age. Int. J. Epidemiol., 2004, 33: 936–44.

[34] Riddell SR, Appelbaum FR: Graft-versus-host disease: a surge of developments. PLoS Med., 2007, 4:e198.

[35] Erlich H: HLA DNA typing: past, present, and future. Tissue Antigens, 2012, 80: 1–11.

[36] Robinson J, Halliwell J a, McWilliam H, Lopez R, Parham P, Marsh SGE: The IMGT/HLA database. Nucleic Acids Res., 2013, 41:D1222–7.

[37] Sidney J, Peters B, Frahm N, Brander C, Sette A. HLA class I supertypes: a revised and updated classification. BMC Immunol., 2008;9: 1.

[38] Sette A, Newman M, Livingston B, McKinney D, Sidney J, Ishioka G, Tangri S, Alexander J, Fikes J, Chesnut R: Optimizing vaccine design for cellular processing, MHC binding and TCR recognition. Tissue Antigens, 2002, 59: 443–51.

[39] Sette A, Sidney J: Nine major HLA class I supertypes account for the vast preponderance of HLA-A and -B polymorphism. Immunogenetics, 1999, 50: 201–12.

[40] Kosmrlj A, Read EL, Qi Y, Allen TM, Altfeld M, Deeks SG, Pereyra F, Carrington M, Walker BD, Chakraborty AK: Effects of thymic selection of the T-cell repertoire on HLA class I-associated control of HIV infection. Nature, 2010, 465: 350–4.

[41] Hildesheim A, Apple RJ, Chen C-J, Wang SS, Cheng Y-J, Klitz W, Mack SJ, Chen I-H, Hsu M-M, Yang C-S, Brinton LA, Levine PH, Erlich HA: Association of HLA class I and II alleles and extended haplotypes with nasopharyngeal carcinoma in Taiwan. J. Natl. Cancer Inst., 2002, 94: 1780–9.

[42] Lank SM, Golbach BA, Creager HM, Wiseman RW, Keskin DB, Reinherz EL, Brusic V, O'Connor DH: Ultra-high resolution HLA genotyping and allele discovery by highly multiplexed cDNA amplicon pyrosequencing. BMC Genomics, 2012, 13: 378.

[43] Holcomb CL, Höglund B, Anderson MW, Blake LA, Böhme I, Egholm M, Ferriola D, Gabriel C, Gelber SE, Goodridge D, Hawbecker S, Klein R, Ladner M, Lind C, Monos D, Pando MJ, Pröll J, Sayer DC, Schmitz-Agheguian G, Simen BB, Thiele B, Trachtenberg EA, Tyan DB, Wassmuth R, White S, Erlich HA: A multi-site study using high-resolution HLA genotyping by next generation sequencing. Tissue Antigens, 2011, 77: 206–17.

[44] Zhang GL, Keskin DB, Reinherz EL, Brusic V: A cDNA microarray for rapid and economical identification of HLA profiles of individuals. In 2011 IEEE Int. Conf. Bioinforma. Biomed. Work. IEEE; 2011: 677–679.

[45] Poland GA, Ovsyannikova IG, Jacobson RM, Smith DI: Heterogeneity in vaccine immune response: the role of immunogenetics and the emerging field of vaccinomics. Clin. Pharmacol. Ther., 2007, 82: 653–64.

[46] Poland GA, Ovsyannikova IG, Jacobson RM: Application of pharmacogenomics to vaccines. Pharmacogenomics, 2009, 10: 837–52.

[47] Moutaftsi M, Tscharke DC, Vaughan K, Koelle DM, Stern L, Calvo-Calle M, Ennis F, Terajima M, Sutter G, Crotty S, Drexler I, Franchini G, Yewdell JW, Head SR, Blum J, Peters B, Sette A: Uncovering the interplay between CD8, CD4 and antibody responses to complex pathogens. Future Microbiol., 2010, 5: 221–39.

[48] Reif DM, Motsinger-Reif AA, McKinney BA, Rock MT, Crowe JE, Moore JH: Integrated analysis of genetic and proteomic data identifies biomarkers associated with adverse events following smallpox vaccination. Genes Immun., 2009, 10: 112–9.
[49] Jacobson RM, Poland GA: The genetic basis for measles vaccine failure. ActaPaediatr. Suppl., 2004, 93: 43–6; discussion 46–7.
[50] Feng C, Putonti C, Zhang M, Eggers R, Mitra R, Hogan M, Jayaraman K, Fofanov Y: Ultraspecific probes for high throughput HLA typing. BMC Genomics, 2009, 10: 85.
[51] Lank SM, Wiseman RW, Dudley DM, O'Connor DH: A novel single cDNA amplicon pyrosequencing method for high-throughput, cost-effective sequence-based HLA class I genotyping. Hum. Immunol., 2010, 71: 1011–7.
[52] Bui H-H, Sidney J, Dinh K, Southwood S, Newman MJ, Sette A: Predicting population coverage of T-cell epitope-based diagnostics and vaccines. BMC Bioinformatics, 2006, 7: 153.
[53] Vita R, Zarebski L, Greenbaum JA, Emami H, Hoof I, Salimi N, Damle R, Sette A, Peters B: The immune epitope database 2.0. Nucleic Acids Res., 2010, 38:D854–62.
[54] Coleman JR: The PB1-F2 protein of Influenza A virus: increasing pathogenicity by disrupting alveolar macrophages. Virol. J., 2007, 4: 9.
[55] DeLuca DS, Keskin DB, Zhang GL, Reinherz EL, Brusic V: PB1-F2 Finder: scanning influenza sequences for PB1-F2 encoding RNA segments. BMC Bioinformatics, 2011, 12 Suppl 1:S6.
[56] Khan AM, Miotto O, Nascimento EJM, Srinivasan KN, Heiny AT, Zhang GL, Marques ET, Tan TW, Brusic V, Salmon J, August JT: Conservation and variability of dengue virus proteins: implications for vaccine design. PLoSNegl. Trop. Dis., 2008, 2:e272.
[57] Tan PT, Khan AM, August JT: Highly conserved influenza A sequences as T-cell epitopes-based vaccine targets to address the viral variability. Hum. Vaccin., 2011, 7: 402–9.
[58] Falk K, Rötzschke O, Stevanoviæ S, Jung G, Rammensee HG: Allele-specific motifs revealed by sequencing of self-peptides eluted from MHC molecules. Nature, 1991, 351: 290–6.
[59] Reinherz EL, Tan K, Tang L, Kern P, Liu J, Xiong Y, Hussey RE, Smolyar A, Hare B, Zhang R, Joachimiak A, Chang HC, Wagner G, Wang J: The crystal structure of a T-cell receptor in complex with peptide and MHC class II. Science, 1999, 286: 1913–21.

[60] Holzhütter HG, Kloetzel PM: A kinetic model of vertebrate 20S proteasome accounting for the generation of major proteolytic fragments from oligomeric peptide substrates. Biophys. J., 2000, 79: 1196–205.

[61] Kuttler C, Nussbaum AK, Dick TP, Rammensee HG, Schild H, Hadeler KP: An algorithm for the prediction of proteasomal cleavages. J. Mol. Biol., 2000, 298: 417–29.

[62] Keşmir C, Nussbaum AK, Schild H, Detours V, Brunak S: Therapeutic cancer vaccines in combination with conventional therapy. Protein Eng., 2002, 15: 287–96.

[63] Peters B, Bulik S, Tampe R, Van Endert PM, Holzhütter H-G: Identifying MHC class I epitopes by predicting the TAP transport efficiency of epitope precursors. J. Immunol., 2003, 171: 1741–9.

[64] Zhang GL, Petrovsky N, Kwoh CK, August JT, Brusic V: PRED(TAP): a system for prediction of peptide binding to the human transporter associated with antigen processing. Immunome Res., 2006, 2: 3.

[65] Bhasin M, Lata S, Raghava GPS: Therapeutic cancer vaccines in combination with conventional therapy. Methods Mol. Biol., 2007, 409: 381–6.

[66] Saxová P, Buus S, Brunak S, Ke°mir C: Predicting proteasomal cleavage sites: a comparison of available methods. Int. Immunol., 2003, 15: 781–7.

[67] Lundegaard C, Lund O, Nielsen M: Accurate approximation method for prediction of class I MHC affinities for peptides of length 8, 10 and 11 using prediction tools trained on 9mers. Bioinformatics, 2008, 24: 1397–8.

[68] Nielsen M, Lund O: NN-align. An artificial neural network-based alignment algorithm for MHC class II peptide binding prediction. BMC Bioinformatics, 2009, 10: 296.

[69] Parker KC, Bednarek M a, Coligan JE: Scheme for ranking potential HLA-A2 binding peptides based on independent binding of individual peptide side-chains. J. Immunol., 1994, 152: 163–75.

[70] Schuler MM, Nastke M-D, Stevanoviæ S: SYFPEITHI: database for searching and T-cell epitope prediction. Methods Mol. Biol., 2007, 409: 75–93.

[71] Frankild S, de Boer RJ, Lund O, Nielsen M, Kesmir C: Amino acid similarity accounts for T-cell cross-reactivity and for "holes" in the T-cell repertoire. PLoS One, 2008, 3:e1831.

[72] Harndahl M, Rasmussen M, Roder G, Dalgaard Pedersen I, Sørensen M, Nielsen M, Buus S: Peptide-MHC class I stability is a better predictor

than peptide affinity of CTL immunogenicity. Eur. J. Immunol., 2012, 42: 1405–16.

[73] Pierce BG, Weng Z: A flexible docking approach for prediction of T-cell receptor-peptide-MHC complexes. Protein Sci., 2013, 22: 35–46.

[74] Tung C-W, Ziehm M, Kämper A, Kohlbacher O, Ho S-Y: POPISK: T-cell reactivity prediction using support vector machines and string kernels. BMC Bioinformatics, 2011, 12: 446.

[75] Saethang T, Hirose O, Kimkong I, Tran VA, Dang XT, Nguyen LAT, Le TKT, Kubo M, Yamada Y, Satou K: Therapeutic cancer vaccines in combination with conventional therapy. J. Immunol. Methods, 2012.

[76] Kindt TJ, Goldsby RA, Osborne BA, Kuby J: Kuby Immunology. W. H. Freeman; 2007: 574.

[77] Lindenbach BD, Thiel H-J, Rice CM: Flaviviridae: The Viruses and Their Replication. 5th edition. Phiadelphia: Lippincott-Raven Publishers; 2007: 931–959.

[78] Wahala WMPB, Donaldson EF, de Alwis R, Accavitti-Loper MA, Baric RS, de Silva AM: Natural strain variation and antibody neutralization of dengue serotype 3 viruses. PLoSPathog., 2010, 6:e1000821.

[79] De Madrid AT, Porterfield JS: The flaviviruses (group B arboviruses): a cross-neutralization study. J. Gen. Virol., 1974, 23: 91–6.

[80] Hancock K, Veguilla V, Lu X, Zhong W, Butler EN, Sun H, Liu F, Dong L, DeVos JR, Gargiullo PM, Brammer TL, Cox NJ, Tumpey TM, Katz JM: Cross-reactive antibody responses to the 2009 pandemic H1N1 influenza virus. N. Engl. J. Med., 2009, 361: 1945–52.

[81] Sanchez A, Pifat DY, Kenyon RH, Peters CJ, McCormick JB, Kiley MP: Junin virus monoclonal antibodies: characterization and cross-reactivity with other arenaviruses. J. Gen. Virol., 1989, 70 (Pt 5): 1125–32.

[82] Wu X, Yang Z-Y, Li Y, Hogerkorp C-M, Schief WR, Seaman MS, Zhou T, Schmidt SD, Wu L, Xu L, Longo NS, McKee K, O'Dell S, Louder MK, Wycuff DL, Feng Y, Nason M, Doria-Rose N, Connors M, Kwong PD, Roederer M, Wyatt RT, Nabel GJ, Mascola JR: Rational design of envelope identifies broadly neutralizing human monoclonal antibodies to HIV-1. Science, 2010, 329: 856–61.

[83] Kawana K, Kawana Y, Yoshikawa H, Taketani Y, Yoshiike K, Kanda T: Nasal immunization of mice with peptide having a cross-neutralization epitope on minor capsid protein L2 of human papillomavirus type 16 elicit systemic and mucosal antibodies. Vaccine, 2001, 19: 1496–502.

[84] Fischer N. Sequencing antibody repertoires: the next generation. MAbs., 2011, 3(1): 17–20.

[85] Gaschen B, Taylor J, Yusim K, Foley B, Gao F, Lang D, Novitsky V, Haynes B, Hahn BH, Bhattacharya T, Korber B: Diversity considerations in HIV-1 vaccine selection. Science, 2002, 296: 2354–60.

[86] De Groot AS, Marcon L, Bishop E a, Rivera D, Kutzler M, Weiner DB, Martin W: HIV vaccine development by computer assisted design: the GAIA vaccine. Vaccine, 2005, 23: 2136–48.

[87] Gao F, Weaver EA, Lu Z, Li Y, Liao H, Ma B, Alam SM, Scearce RM, Sutherland LL, Yu J, Decker JM, Shaw GM, Montefiori DC, Korber BT, Hahn BH, Haynes BF: Antigenicity and immunogenicity of a synthetic human immunodeficiency virus type 1 group m consensus envelope glycoprotein. J. Virol., 2005, 79: 1154–63.

[88] Nickle DC, Rolland M, Jensen M a, Pond SLK, Deng W, Seligman M, Heckerman D, Mullins JI, Jojic N: Coping with viral diversity in HIV vaccine design. PLoSComput. Biol., 2007, 3:e75.

[89] Fischer W, Perkins S, Theiler J, Bhattacharya T, Yusim K, Funkhouser R, Kuiken C, Haynes B, Letvin NL, Walker BD, Hahn BH, Korber BT: Polyvalent vaccines for optimal coverage of potential T-cell epitopes in global HIV-1 variants. Nat. Med., 2007, 13: 100–6.

[90] Hertz T, Nolan D, James I, John M, Gaudieri S, Phillips E, Huang JC, Riadi G, Mallal S, Jojic N: Mapping the landscape of host-pathogen coevolution: HLA class I binding and its relationship with evolutionary conservation in human and viral proteins. J. Virol., 2011, 85: 1310–21.

[91] Rolland M, Frahm N, Nickle DC, Jojic N, Deng W, Allen TM, Brander C, Heckerman DE, Mullins JI: Increased breadth and depth of cytotoxic T lymphocytes responses against HIV-1-B Nef by inclusion of epitope variant sequences. PLoS One, 2011, 6:e17969.

[92] Olsen LR, Zhang GL, Reinherz EL, Brusic V: FLAVIdB: A data mining system for knowledge discovery in flaviviruses with direct applications in immunology and vaccinology. Immunome Res., 2011, 7: 1–9.

[93] Lundegaard C, Lund O, Nielsen M: Prediction of epitopes using neural network based methods. J. Immunol. Methods, 2011, 374: 26–34.

[94] Schneider TD, Stephens RM: Sequence logos: a new way to display consensus sequences. Nucleic Acids Res., 1990, 18: 6097–100.

[95] Crooks GE, Hon G, Chandonia J-M, Brenner SE: WebLogo: a sequence logo generator. Genome Res., 2004, 14: 1188–90.

[96] Olsen LR, Kudahl UJ, Simon C, Sun J, Schönbach C, Reinherz EL, Zhang GL, Brusic V: BlockLogo: Sequence logo visualization of peptide and sequence motif conservation. J. Immunol. Methods, 2013.

9

Toward the Computer-aided Discovery and Design of Epitope Ensemble Vaccines

Nilay Mahida[1], Martin Blythe[2], Matthew N Davies[3], Irini A Doytchinova[4] and Darren R Flower[1]

[1]School of Life and Health Sciences,
University of Aston, Aston Triangle,
Birmingham, United Kingdom, B4 7ET
[2]University Medical School, Queen's Medical Centre
University of Nottingham, Nottingham NG7 2UH
[3]Department of Twin Research & Genetic Epidemiology
King's College London, St Thomas' Hospital Campus
Westminster Bridge Road London SE1 7EH
[4]School of Pharmacy, Medical University of Sofia, 2 Dunav st.,
1000 Sofia, Bulgaria

9.1 Exordium

Currently, and for infectious disease at least, sanitation and vaccination seem to be by far-and-away the most efficient and the most cost-effective prophylactic treatments available [1]. While sanitation and potable water supply are matters primarily of infrastructure provision – since we know well how to provide adequate sewage and drinkable water – albeit ones prone to idiosyncratic and context-dependent geographical and political problems of all kinds, vaccination poses questions about orchestrating the delivery of extant vaccines and of developing new vaccines to combat recalcitrant diseases, such as malaria or HIV; recurrent diseases, such as tuberculosis; seasonally varying diseases, such as influenza; and a burgeoning number of newly emergent diseases, of which West Nile Fever is a prime example.

Post-genomic Approaches in Drug and Vaccine Development, *205–244.*

The deployment of existing vaccines remains problematic; typically falling foul of a plethora of confounding issues of supply and deployment, long after science has placed its work into the hands of politicians [2]. Many issues stymie the success of vaccines and vaccination: chronic under-funding; the exigencies of cold chain supply; sporadic take-up of vaccination by scattered, poorly educated populations in the developing world swayed perhaps by religious misinformation; or compromised herd–immunity through the partial take-up by the cosseted middle-class citizens of first-world nations who would rather their child, or more likely the children of others, die rather than develop autism. Vaccine-preventable diseases still kill millions. All told, contagious disease gives rise to about 1 in 4 deaths worldwide, particularly in children under five. In developed countries, diphtheria, polio, or measles kill less than 0.1%, yet elsewhere deaths from such infectious diseases is significant. Pertussis, tetanus, Influenza, Hib, and Hepatitis B are all responsible for fatalities numbering over one hundred thousand. Perhaps the most execrable situation concerns Measles, which accounts for over a half-a-million fatalities worldwide. Thus far, only one disease, smallpox, has been completely eradicated. Polio or Poliomyelitis is the next nearest to full eradication, having long been targeted by a systematic, coordinated, worldwide eradication campaign. In 1991, one such program run by the Pan American Health Organization effected a partial eradication of polio in the Western Hemisphere. Subsequently, the Global Polio Eradication Program has radically reduced polio throughout the rest of the world, so that we can count global cases in the tens or hundreds rather than ten thousand times that number.

Since the days of Pasteur, conventional vaccine development has largely depended on a somewhat inelegant and predominantly empirical approach which uses whole microbial organisms that have either been killed, typically by heat or chemical treatment, or which represent an attenuated form of the pathogen that replicates but does not cause infection [1–4]. This approach to vaccinology has proved successful, and had a significant effect on the mortality rates caused by of a number of diseases including smallpox, polio and diphtheria [5].

However, there can be considerable obstacles in applying this approach. This can include difficulty in cultivating the pathogen *in vitro*, as in the case of the Hepatitis B virus (HBV) and parasites such as malaria and schistosoma. Determining that cultured pathogens are sufficiently killed or attenuated for safe inoculation is also of major concern. Recombinant DNA technology has enabled the development of protein-based vaccines that can address these issues. A recombinant protein of the HBV surface antigen expressed by DNA-transfected cells has lead to a human vaccine against the virus [6]. Other recombinant protein vaccines are being developed and tested including those for tuberculosis [7], herpes simplex virus (HSV-1) [8, 9], the influenza virus [10], as well as cancer [11].

The development of new vaccines however can be addressed in the first instance by an increasingly rational scientific approach. Immunovaccinology is rapidly displacing empirical vaccinology as the principal tool in vaccine discovery. Just over a decade ago, Rino Rappuoli coined the term "reverse vaccinology" to describe the development of vaccines using a genome-centred approach, rather than the lumbering and laborious empirical methods favoured by most involved in vaccine discovery and development. Reverse vaccinology is beginning to deliver on its potential: In January 2013, Novartis received EU approval for the recombinant sub-unit vaccine Bexsero[12, 13]. Bexsero targets the primary cause of life-threatening bacterial meningitis in Europe: meningococcal serogroup B (MenB), which can be fatal or cause serious, life-long disabilities. Bexsero is indicated for all age groups, including infants, from 2 months onwards. Although rare, this disease can affect healthy people rapidly and without warning. Symptoms resemble those of flu, making it difficult to diagnose in a lucently accurate manner in its initial stages. Up to one in five survivors may suffer from severe disabilities including brain damage, loss of limbs, or hearing.

During the development of Bexsero, new protective protein antigens were identified using genomics: initially over 600 surface-exposed proteins were predicted from the *N. meningitidis*proteome as molecules liable to host immune surveillance, of which about 350 were then expressed in *E. coli*. This number was reduced by using these proteins to immunize 350 sets of mice, identifying 91 that could induced antibodies *in vivo*, 29 of which killed *N. meningitidis in vitro*. By comparing the genomes of 31 clinical strains, a further subset of proteins offering broad protection could be identified.

Reverse vaccinology has become perhaps the most famous and best well-developed approaches amongst many new, advanced approaches now available within the discipline of vaccinology. Indeed, a whole range of other, high-technology methods and techniques have been and are being developed to compliment and optimise Reverse vaccinology. Most utilise the plethora of genomes now available for microorganisms of all kinds. Expressing recombinant microbial proteins identified or suspected of being immunogens allows many of the difficulties of culturing microbes and identifying their immunogenic components to be circumvented, allowing direct evaluation of putative candidate subunit vaccines. Capitalizing on the high-profile success of reverse vaccinology and related techniques, these new technical approaches offer new hope in our constant struggle with infection; all we need is for this technology to be fostered, developed, and utilized.

There are several main types of successful vaccines: so-called attenuated – weakened but live – pathogen-based vaccines, such as BCG; whole pathogen-based vaccines killed typically after heat or chemical treatment, such as influenza vaccines; so-called subunit or toxoid protein-based vaccines, such as HBV; and finally vaccines based on extended carbohydrate epitopes [1, 2]. Other types of vaccine – primarily epitope-based vaccines or vaccines based upon dendritic cells – are available but have yet fully to prove their worth. Vaccines comprising one or a few epitopes are in particular seen to promise much in areas of medicine where the search for conventional vaccines have signally failed; for example, in disease areas such as malaria, cancer, and HIV. We use the term "epitope ensemble vaccine" as a broad, all-encompassing collective to describe vaccines based on epitopes, since a vaccine can *inter alia* be based on one or more epitopes delivered explicitly as a mixture of different peptides or as a single peptide comprising multiple vaccines or implicitly as naked DNA or encoded within some microbial vector. This chapter explores several of the problems facing epitope ensemble vaccines, and seeks to explore the causes of success or failure within the field.

The move away from whole-pathogen vaccines towards subunit and epitope ensemble vaccines has been prompted by several safety concerns [1, 2]. Subunit or epitope-based vaccines cannot and thus do not revert to a virulent form, a phenomenon apparent amongst certain attenuated vaccines, which can prompt a wide range of adverse effects, from severe morbidity to mortality. Secondly, subunit and epitope ensemble vaccines have the potential to obviate the problem of the immune system becoming sensitized to self-antigens, where the epitopic components of self-antigens may be presented as part of a foreign protein, engendering an auto-immune response. This second concern is addressed by the ability to design out of subunit vaccines cross-reactive epitopes shared by the host genome, or the constraint of not selecting such epitopes for inclusion in vaccines based on epitope ensemble.

The ability of peptides representing B-cell epitopes to induce neutralizing antibodies, and to a lesser extent, protection from pathogen, was documented as early as 1982 [14, 15]. Vaccines based upon peptides would represent the distillation of those parts of a pathogen that can induce host protection. Bittle and colleagues were the first to demonstrate the feasibility of peptides in vaccine design by using a linear B-cell epitope identified from the Foot and Mouth Virus (FMV) viral capsid protein that offered protection to viral challenge. The performance of this vaccine in protecting a population was weaker in comparison to vaccines developed conventionally [16]. Taboga and colleges conducted an evaluation of FMV peptide vaccines and concluded that in many of the cattle not protected from infection was a result of viral antigenic variation, and that the inclusion of helper T-cell epitopes was a determining factor in host protection [17].

The first viable epitope vaccine was reported in 1994, and gave protection to dogs from canine parvovirus [18]. The selected peptide was conjugated to a carrier protein that conveyed Helper T-cell epitopes, which clearly augmented the efficacy of the vaccine. A number of epitope ensemble vaccines containing T-cell epitopes are in clinical trials. These include vaccines for Alzheimer's disease [19, 20], and for the malaria parasite [21, 22], as well as a poly-epitope vaccine for the measles virus [23, 24].

Epitope-based vaccines have thus been in development for the last three decades, but tend to reach phase III trials and progress no further. Much of this is due to their low intrinsic immunogenicity: compared to whole pathogen vaccines, individual epitopes provide too few – and too homogeneous – a challenge to the immune system to be fully effective as vaccines. In the remainder of this chapter, we will explore definitions of the epitope, how bioinformatics can help in the identification of epitopes, the current status of epitope ensemble vaccines, and conclude with a critique of present computational approaches to the design of epitope ensemble vaccines.

9.2 The Informatics of Epitology

The word "epitope" is a term of true ubiquity within the silo language of immunologists, and of biologists more widely. Paradoxically, the OED only traces its usage to 1960 and a review in the Annual Review of Microbiology by Jerne: "The suffixes -tope and -type are used in words denoting certain units of immunological significance. An antigen particle carries several epitopes (= surface configurations, single determinants, structural themes, immunogenic elements, haptenic groups, antigenic patterns, specific areas)" [25]. This quotation highlights the many alternative and competing terms used to describe a key concept in immunology. Modern usage has extended and qualified the use of epitope, most notably by including cellular immunology, and the major histocompatibility complex (MHC), along with humoral immunology in defining the notion of the epitope. The word epitope is also used more widely in biology and other disciplines for a quantum motif of molecular recognition not involving antibody or MHC: see [26] and [27] for recent examples of such usage.

For our purposes, an epitope can be defined as the smallest fragment of a protein that can independently induce an immune response.

A B-cell or, more generally, an antibody-mediated epitope can be defined as the distinct surface region of a protein bound by an antibody's variable domain. All antibodies have two binding sites, comprising three hypervariable complementary determining (CDR) loops, which form the antigen-binding "paratope". The antigen surface that interacts with this paratope is referred to as an "epitope".

Antibody-mediated epitopes are of two types: discontinuous (or conformational) or continuous (or linear or sequential). Discontinuous epitopes are regions of an antigen that are separated in sequence but proximal in the folded protein. Discontinuous epitopes may contain residues not involved directly with the antibody, with some residues more vital than others. Continuous epitopes are short protein regions comprising a few amino acid residues that are believed to be cross-reactive with antibodies raised against the whole protein.

Various experimental approaches exist for determining epitope location within protein antigens. Methods that determine linear epitope location are mostly based on the supposed ability of antibodies to cross-react with peptide fragments of the original antigen. Early methods used chemically or proteolytically degraded, or recombinant expressed fragments of the studied protein. By identifying those regions essential for antibody binding an approximate location of linear, and sometimes discontinuous epitopes, could then be deduced [28]. Some of the more recently developed methods use phage display of random peptide libraries, or analysis of peptide fragments representing overlapping segments of the parent protein sequence.

To identify discontinuous epitopes, the 3-dimensional structure of the antigen is examined, with site-directed mutations used to confirm residues implicated in antibody binding. By examining antibody–antigen complexes using by X-ray crystallography, detailed information about the interaction can be identified. This approach has been used to identify epitopes and analyze antibody–antigen interactions for many proteins.

Arguably, PEPSCAN is the key experimental technique used to identify linear B-cell epitopes [29]. The method involves synthesizing peptides representing overlapping segments of the parent protein, and then linking these to the surface of plastic pins, with one peptide per pin. This peptide array is then analyzed using an Enzyme-Linked Absorption Assay (ELISA), with either polyclonal or monoclonal antibodies specific for the parent protein used to identify peptides representing antibody binding regions through antibody cross-reactivity [30]. By analyzing those peptides that immediately overlap each other for a common reactive sequence the location of the epitope can then be deduced [31, 32]. The resolution thus depends on the extent of overlap between one peptide and the next. Epitopes corresponding to monoclonal antibodies have more easily identifiable boundaries; the precise location of epitopes bound by polyclonal antibodies is less clear cut.

Phage display for linear epitope mapping was established by Smith [33]. The method uses libraries of randomized peptide fragments fused to the coat proteins of filamentous phage [34]. Phage display for peptide libraries are produced by shotgun cloning of random oligonucleotides onto the 5' ends of either the PIII or PVIII genes of filamentous phage. After purifying phage particles expressing different peptides, an ELISA-based antibody-binding assay is used with either affinity-purified polyclonal or monoclonal antibodies specific for the original antigen. Phage display is predominantly used to examine monoclonal binding. By identifying those phage-expressed peptides bound by an individual monoclonal antibody, and determining the amino acid sequences of those peptides, the location of a linear epitope within the parent protein maybe deduced from sequence-similarity profiling.

Phage-display libraries containing random sequence peptides are also used to identify epitope mimics, termed 'mimotopes'. These peptide sequences may show no sequence similarity to the epitope to which the binding antibody/antibodies naturally recognizes and are thought to confer electrostatic similarities to the native epitope [28]. Mimotopes of discontinuous epitopes may be identified, though it may not be possible to deduce the location of the epitope within the parent protein.

A T-cell epitope is the smallest peptide, carbohydrate, lipid, or other molecular structure bound by a major histocompatibility complex (MHC), which a T-cell can subsequently recognzise [4]. Cytotoxic T-cells recognize molecules presented by class I MHC and helper T-cells recognize molecules presented by class II MHC. MHC Class I and Class II molecules bind distinct peptides. Class I molecules bind within a single closed groove peptides 8–15 amino acids in length. The peptide is anchored by interacting with residues at both termini of the peptide, while the central part is inherent more variable. Class II peptides are 12–25 amino acids in length; however, only nine residues bind into the groove, which is open at both ends allowing peptides to potentially protrude at either end. Class II peptide side chains bind into cavities within the groove, forming hydrogen bonds to the side chains of MHC residues.

Methods abound for the measurement of T-cell responses [35–38]. At one extreme, Delayed-Type Hypersensitivity test (DTH) is a long-used clinical test that is neither particularly sensitive nor amenable to ready standardization. It places antigenic material into part of the arm: any swelling that results, is usually taken to indicate that the host immune system has a memory response to the antigen investigated.

Methods differ for class I and class II presentation. $CD8^+$ T-cells - cytotoxic T-lymphocytes (CTL) - lyse cells after activation. This can be monitored with ^{51}Cr, or radiolabelled thymidine, released during CTL lysis. The proliferative response of $CD4^+$ T-cells, acting via macrophage or B-cell activation, is measurable with tritiated thymidine incorporated into T-cell DNA. ELISpot assays measure the production of cytokines by class I and/or class II T-cells after antigen exposure. More recently, RT-PCR and tetramers have become popular as approaches to T-cell response detection.

Although numerous approaches to T-cell epitope identification exist, the quantitative data produced is seldom consistent enough to be used reliably and with confidence outside of a defined experimental protocol. Since there is no clear, lucent meaning to the concept of a partial T-cell epitope, it is often necessary to rely on working immunologists to judge what may or may not constitute a meaningful epitope.

Turning now from the experimental to the computational, much has been done to produce viable epitope prediction methods. Immunoinformatics proffers a large number of techniques and tools for the accurate and robust identification of epitopes *in silico*. Prediction methods can significantly increase the rate that epitope are discovered, with a substantial dividend for vaccine discovery.

As more and more pathogen genomes become available, computational and experimental epitope mapping are becoming pivotal, complementary issues [39–47]. Using such a strategy, deciphering and characterizing gene function in pathogen is rendered redundant. In terms of prediction, work has concentrated on predicting peptide-MHC interactions and the identification of continuous and discontinuous B-cell epitopes [48–51]. However, the most successful type of prediction has been data-driven prediction of

T-cell epitopes. In a focal retrospective study, Deavin *et al.* [52] compared a number of early T-cell epitope prediction approaches, without identifying a method of sufficient accuracy to be practically useful. Today, work focuses on the prediction class I MHC-peptide binding affinity; and for well-studied class I MHC alleles, such methods can be both accurate and reliable [53, 54].

However, predicting any other epitope data has generated only unsatisfactory and substandard results. Recently, several comparative studies have demonstrated that prediction of class II T-cell epitopes is habitually very poor and unreliable [55–57]. Moreover, data-driven and structure-driven [58, 59] prediction of B-cell epitopes is again known to be utterly unreliable. Problems abound in antibody-mediated epitope prediction, both practical and conceptual. The amino acid distribution of the antigen surface has resisted characterization: it possesses no unique features, either sequential or structural in nature, upon which one might base prediction. Likewise, due to the poorly understood recognition properties of cross-reactive antibodies, the data upon which predictions are founded remains questionable.

B-cell epitope prediction based on common patterns of binding or "motifs" or more sophisticated machine learning approaches have proved largely fruitless. More recently, structural analysis has been used as a prelude to epitope prediction. A recent review of B-cell epitope software calculated that the AROC curves for several methods were about 0.6 while protein-protein docking methods were about 65% accuracy. A key problem with antibody epitope prediction is that B-cell epitopes seem entirely context-dependent. The surface of an antigen is a continuous landscape of potential epitopes without clear boundaries. Both paratope and epitope constitute fuzzy recognition sites. They do not form a unitary invariant residue conformational ensemble but rather a succession of variable conformations. A simple binary classification between epitope and non-epitope may not capture the key characteristics of this elusive vacillating interaction. The average paratope comprises just one third of potential binding residues, consistent with the remaining amino acids potentially forming antibody-antigen complexes using an entirely different set of interactions.

9.3 Epitope Vaccines: Current Status

While no vaccine based on epitopes has yet been licensed, it is nonetheless an active research area. It is perfectly possible that during the next decade, several therapeutic epitope ensemble vaccines targeting chronic infections and other conditions will have reached the market. Indeed, due to their early success in animal studies, several peptide vaccines are being developed [60].

Epitope-based vaccines being developed include the ones targeting: HPV, HCV, CMV, HSV-2, and HIV. An anti-HCV therapeutic epitope ensemble vaccine from Intercell called IC41, which comprised four well conserved HLA-A*0201 T-cell epitopes and three helper epitopes, is currently in clinical trials [61]. IC41 is currently in Phase II and shows promise in terms of immunogenicity and decreased viral loads in chronically infected patients [62, 63]. This is the first significant efficacy seen in for an epitope ensemble anti-HCV vaccine. Another therapeutic epitope-based vaccine targets HPV, the principal cause of cervical cancer, and comprises class I and class II T-cell epitopes derived from proteins E6 and E7 together with a TLR-3 agonist-based adjuvant [64]. Several trials are being conducted involving T-cell epitopes with helper peptides [65]. Most recently, GEN-003, a multiple epitope ensemble anti-HSV-2 vaccine, has entered Phase I trials. This formulation, which includes antigens ICP4 and gD2 and a saponin-based adjuvant, has shown significant immunostimulatory properties. Extant proof-of-concept pre-clinical studies appear hopeful, showing demonstrable protection against viral disease, with clinical symptoms reduced in severity and duration.

Following few sections provide a brief but orientating narrative of recent developments in epitope-based vaccine design, complemented by Table 9.1, which seeks to give a rather broader and more synoptic overview of a representative selection of recent epitope vaccines. This then is a snapshot of the current status of peptide vaccines.

9.3.1 Alzheimer's Disease

Alzheimer's disease (AD) affects a large and growing subset of the global population, with no treatment. Pathological changes occurring in AD include the deposition of Amyloid β (Aβ) plaques leading to loss of synapses in the brain, resulting in deterioration of function. What triggers such change is controversial. One widely accepted hypothesis is the amyloid cascade hypothesis. This suggests that over production of Amyloid β peptide is the first event in the development of AD. The amyloid precursor protein (APP) is produced as an integral membrane protein, which is cleaved by β secretase to make a APPβ which is produced along with a C terminal fragment called CTFβ. This CTFβ is cleaved by γ secretase to produce a smaller C terminal fragment and Aβ peptides of various lengths; of which the Aβ_{42} fragment is considered the most dangerous. These peptides accumulate due to malfunctioning degradation enzymes, and sporadic overproduction, leading to plaque-forming peptide oligomerization.

Therapeutic epitope vaccines aim to accelerate degradation of the Aβ_{42} peptides [66]. Both B-cell and T-cell epitopes have been mapped in the Aβ_{42} peptide. The major Aβ_{42} peptide B-cell epitope comprises residues 1–15; while the major T-cell epitope is localized to amino acids 16–30 [66, 67]. Other studies have shown that the T-cell epitopes can lie in other parts of the Aβ_{42} peptide.

A trial of the vaccine AN1792, showed that antibodies raised against Aβ_{42} could slow progression of AD: that in the small number of patients who responded to the vaccine there was clearance of the Amyloid β peptide in the brain. AN1792 only reached phase II, before meningoencephalitis induced by adjuvant QS21 in about 6% of patients caused the trial to be halted [66, 68]. Long-term follow up of AN1792 has shown poor correlation between Aβ peptide clearance and the prevention of progression of Alzheimer's disease.

On-going developments include ACC – 001, a second generation epitope vaccine being developed by Elan pharmaceutical and Pfizer. This comprises an epitope representing residues 1–7 of the Aβ_{42} peptide conjugated to a protein, co-administered with adjuvant QS21. The phase II trial of ACC-001 was suspended temporarily due to vasculitis in one patient. Another vaccine, CAD106, is being developed by Novartis, and comprises amino acids 1–6 of the Aβ_{42} peptide. Again it aims to induce antibodies against Amyloid β peptide. The minimal epitope recognized by antibodies is amino acids 3–6 of the Aβ_{42} peptide. Currently in phase II trials, CAD106 was tested initially in mice [69]. IgG antibodies were induced by the vaccine, and with constant administration Amyloid plaques were reduced in mice. However once the vaccine is stopped antibody titres gradually fall.

Another important factor to consider in the development of vaccines for Alzheimer's patients is the phenomenon of immunosenescence: as patients age, the thymus shrinks, and thus production of naïve T-cells falls, reducing responses to vaccines. This may mean we will need regular immunization of those at high risk of AD [70, 71].

9.3.2 HIV

Developing a vaccine for HIV remains a challenge, and progress has been slow. Developing a safe and effective universal HIV vaccine has thus far proved confounding, despite unprecedented levels of financial and intellectual investment in the problem. Perhaps, the key challenge in devising an HIV

Table 9.1 Overview of a representative selection of recent epitope vaccines

Study	Phase	Disease	Vaccine	Antigen and Epitope sequence	adjuvant	(T) or (P)*
Fomsgaard et al, 2011	Preclinical	HIV -1	Minimal epitope	Nef 73, Pol 934, Gag 150, Gag 433, Pol 606 and others	CAF01	T
Cobb et al, 2011	preclinical	HIV-1	HIV-1 lipopep-tide antigen	Gag17–35, Gag253–284, Nef 66–97, Nef 116–145, Pol 325–355	Unknown	P
Spearman et al, 2009	I	HIV- 1	Multi epitope	MEP- 4 CTL epitopes 27–47 aa long.	RC529Se +/- Gm-CSF.	P
Gorse et al, 2007	I	HIV-1	EP1090	Ep1090-DNA vaccine(21 CTL epitopes)	None	P
Jin et al, 2009	I	HIV-1	E91043	Ep1043- 18 HTL epitopes +/- EP1090	alhydrogel	P
Salmon Ceron et al, 2010 (ANRS)	II	HIV- 1	HIV-LIPO-5	Combination of sequence see appendix	Unknown	P
Nardin et al, 2004	I (has reached phase II)	Malaria	ICC1132 malarivax	B cell epitope – $(NANP)_3$, T cell epitope NANPPN-VDPNANP	Montanide ISA 720	P
Alonso et al, 2004	II (has reached phase III)	Malaria	RTS,S	CSP 207–395 and T cell epitopes fused to HBsAg in	AS02	P
Hill et al, 2006	II	Malaria	FP9 ME-TRAP and MVA ME-TRAP	Thrombospondin related epitopes		P
Maguire et al, 1999	III	Cat allergy	CATVAX	Feld epitopes	n/a	T/P
Worm et al, 2010	III	Cat allergy	Tolero Mune cat	Feld epitopes	n/a	T/P
Counsell et al, 1996	I	Cat allergy	Feld1	Feld epitopes	n/a	T/P

Design**	Age range	Participants	Location	Duration (days)	Doses	Outcomes
n/a	n/a	Mice and rabbits	Denmark	Single dose mice 28- rabbits	1 mice 2 rabbits	The vaccine was immunogenic
A	Unknown	8	US	N/A	N/A	The vaccine was immunogenic
A	18 – 50	96	US	84	3	The vaccine was poorly immunogenic
A	18–40	42.	US and Botswana	180	4	The vaccine was poorly immunogenic
A	18–50	82	US	180	4	EP1043- immunogenic EP1090 not immunogenic
A	21–55	132	France	168	4	The vaccine was immunogenic
A	18–45	20	UK	180	3	The vaccine was immunogenic
A	1–4	2022	Mozambique	240	3	The vaccine was immunogenic
A	1–6	405	Kenya	35	3	The vaccine was moderately immunogenic
A	13–64	133	US	148	8	The vaccine was immunogenic
A	18–65	88	UK	21	5–6	The vaccine was immunogenic
B	n/a	53	US	Several weeks	3	The vaccine was immunogenic

Study	Phase	Disease	Vaccine	Antigen and Epitope sequence	adjuvant	(T) or (P)*
Weissner et al, 2011 (Novartis)	Preclinical (has reached Phase II)	Alzheimer's	CAD-106	Amylod β_{1-6}	Freund's	T
Trials number NCT00498602 (Pfizer)	II	Alzheimer's	ACC-001	Amyloid β_{1-7}	QS21	T
Odunsi et al. 2007	I	Epithelial Ovarian cancer	NY- ESO peptide	Epitope ESO 157–17	Incomplete Freunds+/- IFN	T
Miyazawaet al, 2009	I	Pancreatic cancer	Single epitope	VEGFR2–169 epitope	Montanide ISA 51	T
Staff et al, 2011	I	Colorectal cancer	DNA Vaccine	DNA vaccine with a HTL epitope	GM- CSF and cylophos-phamide	T
Kameshima et al, 2011	I	Colorectal cancer	Single epitope	Survivin 2B80–88 CTL epitope	Incomplete Freund's +/- IFNα	T
Kono et al. 2009	I	Esophageal cancer	Multi epitope	3 epitopes from cancer testis antigens	Incomplete Freund's	T
Barve et al, 2008	II	Non small cell lung cancer	Single epitope	10 amino acid CTL epitope	Incomplete freund's	T
Rezvani wt al, 2011	II	Myeloid malignan-cies	PR1 and WT1 peptides	WT1- 126–134 and PR1: 169–177 leukaemia associated antigens	Montanide + GM CSF	T
Mittendorf et al.2011	I/II	Breast cancer	Her2/neu E75	E75	GM- CSF	P
Schwartzen truber et al, 2011	III	Melanoma	GP100	GP100 epitope	IL-2	T
Slingluff et al, 2007	II	Melanoma	Multipeptide	Gp100 and tyrosinase peptides	Montanide ISA-51	T

Design**	Age range	Participants	Location	Duration (days)	Doses	Outcomes
n/a	n/a	Mice 4 rhesus Monkeys	n/a	10 - mice 110 monkeys	8-mice 3 - monkeys	The vaccine was immunogenic
A	50–85	Currently recruting participants	US	364	5	Ongoing trial
		18	US	105	5	The vaccine was immunogenic
B	20–80	18	Japan	28	3	The vaccine was immunogenic
B	≥18	10	Sweden	42	3	The vaccine was immunogenic
B	20–85	13	Japan	42	4	The vaccine was immunogenic
	20–80	10	Japan	25	5	The vaccine was immunogenic
B	≥18	63	US	728	13	The vaccine was immunogenic
B	≥18	8	US	182	2	The vaccine was poorly immunogenic
B	28–78	182	US	540	6	The vaccine was immunogenic
A	46.9–50.3 (mean age)	185	US	244 cycles	12 doses per cycle	The vaccine was immunogenic
B	>18 years old	51	US	730	6	The vaccine was immunogenic

Study	Phase	Disease	Vaccine	Antigen and Epitope sequence	adjuvant	(T) or (P)*
Scheiben bogen et al, 2000	II	Melanoma	Mutlti Peptide	Tyrosinasew peptides	GM-CSF	T
O'Rourke et al, 2002	I/II	Melanoma	Dendritic cell vaccine	Mutated epitopes	n/a	T
Kirkwood et al, 20009	II	Melanoma	Mulit epitope	Mart-1 and gp100 analogue	GM- CSF	T
Zajac et al, 2003	I/II	Melanoma	Multi epitope	Epitopes from Mart-1, gp100 and tyrosinase	GM-CSF	T
Hersey et al. 2005	I/II	Melanoma	Multi epitope	Epitopes from gp100, MAGE A3 and MART-1	Montanide ISA 720	T
Roberts et al, 2006	II	Melanoma	G209–2M	Mimics gp100 melanocyte differentiation protein	IL-2 Mon- tanide ISA 51	T
Scardino et al, 2002	Preclinical	various	Her2/neu	Cryptic epitopes	unknown	T
Waeckerle men et al, 2006	I/II	Prostate cancer	Dendritic cell	T cell epitopes	n/a	T
Matijevic et al, 2011	II	HPV	DNA vaccine	9 Epitopes from E6 and E7 proteins of HPV 16 or 18 peptides	unknown	T
Muder- spach et al, 2000	I	HPV	Single peptide	12–20 amino acids of the E7 gene	Freund's	
Tsuda et al, 2011	Preclinical	Ebola	MCMV/ZEBOV - NP_{CTL}	T cell epitope	None	P
Heathcote et al, 1999	I	Hepatitis B	CY-1899 lipopeptide	Hepatitis B core protein amino acids 18–27	Unknown	P
Elliot et al. 2007	I	EBV	CTL epitope	Synthetic peptide FLR	Montanide ISA 720	P
Ben – yedida et al. 1998	Preclinical	Infleunza	Multi epitope flagellin	NP epitopes, NP55–69 and NP 147- 158	None	P

Design**	Age range	Participants	Location	Duration (days)	Doses	Outcomes
B	unknown	18	Germany	98	5	The vaccine was immunogenic
B	21–67	19	Australia	12	8	The vaccine was immunogenic
B	23–83	115	US	356 days	6	The vaccine was immunogenic
B	Unknown	18	Switzerland	99	16	The vaccine was immunogenic
A	27–81	36	Australia	84	6	The vaccine was immunogenic
A	>18	69	US	Unknown	Unknwon	The vaccine was not immunogenic
n/a	n/a	unknown	France	Unknown	Unknown	The vaccine was immunogenic
B	58–74	6	Switzerland	Until taken off study	6 doses every other week	The vaccine was immunogenic
A		26	US	42	3	The vaccine was immunogenic
A	Median age 29	18	US	21	4	The vaccine was immunogenic
n/a	n/a	Mice	US	70	2	The vaccine was immunogenic
A	18–65	90	US	210	6	The vaccine was poorly immunogenic
A	18–50	14	Australia	56	2	The vaccine was immunogenic
n/a	Mice	n/a	Israel	42	3	The vaccine was immunogenic

vaccine is overcoming the unprecedented antigenic hyper-variability of HIV, which may require both the induction of broadly protective neutralizing antibodies, as well as inducing broad, robusT-cellular immune responses to control HIV infection. Indeed, a principal difficulty with HIV vaccine development has long been the inability to induce neutralizing antibodies to HIV. Neutralizing antibodies act early in viral infection, blocking the interaction of viruses with cell surface receptors, and thus preventing hosT-cell entry. Other complexities associated with HIV pathogenesis include the extreme multiplicity of transmission modes, a deficit of correlates of protective immunity, and HIV's ability to infect immunoregulatory cells necessary for orchestrating effective immune responses.

Epitope-based prophylactic anti-HIV vaccines have sought to identify conserved, structurally-important epitopes; conceptually, such epitopes are the least likely to mutate during viral replication as such mutations would severely compromise the ability to grow, function, and infect. A trial by Gorse and co-workers evaluated the Ep1090 vaccine injecting it intramuscularly; Ep1090 uses a DNA plasmid to encode 21 nine to eleven amino acid epitopes of HIV [72]. A long follow up to the vaccine administration indicated no clear response in the subjects. Jin and colleagues also studied Ep1090 [73], as well as Ep1043, a vaccine containing 18 helper T-cell epitopes encoded in a Baculo viral vector. Ep1043 proved more effective at inducing an immune response compared to Ep1090. ELISA results indicated that 68% of participants had memory cells to EP1043 compared to no response with EP1090; this may reflect the greater nuclear incorporation of viral vectors over plasmids.

There are, of course, many other obstacles to developing effective HIV vaccines. Even a small virus like HIV encodes thousands of potentially immunogenic peptides, yet the T-cell repertoire only responds to a small fraction of these. This is immunodominance. Many factors mediate the immunodominance of HIV epitopes. One such is the sequence of protein expression by HIV, which is known to occur in a distinct chronological order. The first proteins expressed are regulatory proteins Rev and Tat and accessory protein Nef. This is followed by expression of structural proteins Pol, Env, Gag and accessory proteins Vpr, Vif, Vpu which depend on Rev-mediated transport for translation. It has been proposed that because Rev, Tat and Nef are presented to the immune system early on they will play a prominent role in the immune response to these antigens. Another factor determining immunodominance is the variation of MHC peptide specificity across the patient population. A direct link between immunodominance and MHC type has been established. Different people have different MHC types, each with distinct peptide specificities, binding certain epitopes with greater affinity than others. Thus, immunodominant epitopes for each MHC haplotype will be different, and a vaccine only using a few epitopes is unlikely to induce an immune response acros the whole population due to the wide variety of MHC types and their different binding affinities.

A trial by Salmon-Ceron and co-workers using an immunogenic vaccine showed that the highest cytotoxic T-cell response was induced in 68% of volunteers in the low dose group [74]. The response was reached after two booster shots, and was maintained for the third and fourth booster injections. The response lasted 6 months in 45.8% of volunteers, indicating that memory T-cells had been produced. A subsequent ELISPOT assay indicated the highest response was found for the fragment Gag 253–284. This may be due to immunodominant epitopes in this section of proteins and indicates that other epitopes in the vaccine were subdominant.

The somatic MHC specificities will determine which epitopes exhibit immunodominance. Thus, having many epitopes coded by a sequence of amino acids allows for many MHC types to bind different epitopes. A trial by spearman and coworkers reported a multi-epitope vaccine aiming to target a wide spectrum of HLA specificities, using four peptides [75]: designated A,B,C and J. Each peptide contained one of three HIV Helper T-cell epitopes from Env and Gag proteins in combination with four different Cyotoxic T lymphocyte (CTL) 'hot spots' derived from Gag or Nef proteins. The CTL hot spots were regions of proteins that were identified to have a high concentration of epitopes.

In another study, epitopes from Gag p24 were combined with adjuvant GM-CSF to form vaccine Vacc-4x [76]. Phase II trials indicated high immunogenicity as well as decreased viral burden in high responders [77], and low escape mutations after immunization. Vacc-4x also exhibits a significant memory response, with patients showing high delayed-type hypersensitivity reaction four years post vaccination [78]. Although an anti-HIV vaccine based on the epitope ensemble paradigm is plausible, immunogenicity is determined by a limited number of key factors, such as the epitopes themselves, delivery system, and adjuvant.

9.3.3 Malaria

Together with TB and HIV, Malaria is one of the WHO's big three global killers. There are no effective vaccines for HIV or Malaria; and there is little hope that broadly applicable vaccines will appear in the near future. The only available vaccine for tuberculosis is of limited efficacy. Malaria is thus a major cause of global mortality particularly among children and pregnant mothers. Malaria is a mosquito-borne infectious disease of humans and other animals caused by protozoan parasites of the genus Plasmodium: primarily Plasmodium falciparum and Plasmodium vivax. The disease causes symptoms such as fever and headache, progressing in severe cases to coma or death. Malaria is prevalent in tropical and subtropical regions, including much of Sub-Saharan Africa, Asia, and the Americas. Challenges facing the development of anti-Malarial vaccine include the large size of the plasmodium genome, with over 6000 genes, and the complex multi-stage life-cycle of the plasmodium protozoa, which expresses potential antigens differently at different stages antigens. This is complicated still further by the five different species able to infect humans. Thus, producing a universal malaria vaccine is a distinct challenge.

One of the earliest approaches was to target the malaria parasite in the sporozoite stage, before it infects red blood cells. This phase of infection is normally silent and thus a successful vaccine might allow neutralization of the parasite before any signs of infection energe. This was first shown by Ruth Nussenzweig who injected mice with irradiated sporozites resulting in protection in the immunized mice. However a large number of sporozites are needed to produce immunity. This led to a search for the antigens of the sporozite that induced immunity and this was found to be the circumsporozoite protein (CSP) [79].

A further difficulty is again immunodominance. For an epitope to be immunodominant and lead to the production of high affinity neutralizing antibodies, activation of both the B-and T-helper cells is required. The malarivax vaccine comprises Hepatitis B surface antigen (HBsAG) expressing an immunodominant B-cell epitope (NANP) and one T-helper cell epitope of plasmodium falciparum adjuvanted with Alum. NANP is a 4-amino acid repeat sequence conserved in the CSP of plasmodium falciparum. Malariavax has been successful in phase I trials eliciting both a B-cell and a T-cell response as well as production of IgG antibodies specific for NANP [80].

However, the production of IgG antibodies may not necessarily be due to the presence of immunodominant epitopes but to the immunological phenomenon referred to by some as "antigenic sin". This is where an individual has previously been exposed to an antigen, thus generating memory cells. When an individual encounters a similar antigen, immunological memory is activated, producing IgG antibodies. Thus, IgG production does not necessarily imply a *de novo* reaction to an immunodominant epitope but the presence in a trial of individuals already previously exposed to similar antigens.

The RTS,S /AS02 vaccine developed by Glaxosmithkline and the Walter Reed Army institute is a notable epitope ensemble vaccine; it combines a B-cell epitope, comprising amino acids 207–395 of the circumporozite protein (containing NANP) from plasmodium falciparum, and various T-cell epitopes fused to HBsAg in AS01E, AS02a, AS02B or AS02D adjuvants. RTS,S/AS02 is safe and well tolerated, having been evaluated in a variety of clinical trials. Vaccine efficacy was 34% but protection tended to decrease from 71% at week 9 to 0% at week 6 in children aged 5 to 17 months. Some results from the phase III RTS,S/AS01E trial have been published which show a reduction in episodes of malaria by 55% after a 12 month period [81, 82].

9.3.4 Cancer

A major limitation of extant cancer therapies is their poor efficacy when treating metastasis; a limitation greatly compounded by their often extreme toxicity of cancer treatments. When compared to radio- or chemotherapy, the immune system, which can recognize and even eliminate cancer cells, may provide an alternative route to therapeutic intervention in cancer. Indeed, most patients are known to mount an immune reaction to developing neoplastic cells. Critically, at least from the current perspective, a number of tumour-associated antigens (or TAA), have been found; such TAAs can induce T-cell and B-cell mediated responses. Such so-called cancer or tumour antigens are deemed to be newly or preferentially over-expressed by newly-transformed tumour cells. However if such antigens are delivered as a vaccine, with apt ancillary adjuvants, then appropriate immune responses may be generated; inclusion of adjuvants is vital as peptide vaccines without adjuvants are typically very poorly immunogenic or may even be tolerogenic. Thus, several therapeutic strategies have arisen that seek to augment immune-responses able to circumvent tumour-induced immune suppression and thus potentially lead to tumour elimination.

As we have said, CD8 cytotoxic T-cells (CTLs) detect and respond to epitopes derived from tumour antigens as presented by MHC class I molecules on the surface of tumour cells. Thus, synthetic peptides equating to class I epitopes have been developed as therapeutic vaccines against established tumours. Early peptide vaccines comprised one or several such epitopes. Newer peptide vaccines are more often long multivalent peptides incorporating class I and class II epitopes, since peptide length and the presence of CD4 help, can modulate vaccine efficacy and safety.

There have been successful trials using whole antigens administered to animals and to cancer patients. There are also many epitope vaccination trials. Melanoma epitopes were the first cancer vaccines to be developed [83]; subsequent trials have targeted other cancers.

The anti-melanoma gp100 vaccine has reached phase III. Gp100 improved survival rates of patients when taken with interleukin 2 (IL-2) [84]. IL-2 is the only therapy licensed by the Federal Drug Administration (FDA) for metastatic melanoma. However, when gp100 was compared with monoclonal drug ipilimumab in the absence of IL-2, apparent effectiveness was much reduced [85]. Other anti-melanoma vaccines have also reached phase II. Another trial showed that a 12 peptide vaccine was immunogenic [86]. Scheinbenborgen and coworkers showed that there was an immunogenic response to a vaccine containing tyrosinase peptides [87, 88]. Another study, which used dendritic cells pulsed with epitopes, showed that three out of 12 patients developed a partial response [89].

Following the success of melanoma vaccines, other vaccines have reached phase II. For example, HER2/neu vaccines have had some success. HER2/neu is a protein over-expressed by breast cancer cells. The E75 epitope is the most commonly used in vaccines. It has been shown that an immune response develops when breast cancer patients are vaccinated with these vaccines. A trial by Mittendorf and coworkers which used dominant epitopes derived from HER2/neu indicated that the vaccine can be clinically effective if used under appropriate conditions [90–92]. However, in other trials when immunodominant self epitopes are used they induce weak immune responses to antigens, due to deletion by the thymus of self-reactive T-cells including the ones reactive to TAAs. Strategies to overcome this phenomenon include exposing individuals to non-immunodominant self epitopes. This can be achieved using the so-called cryptic epitopes; or atypical epitopes, such as partly spliced messages; mutations of a non-coding intron sequences and exon extension [93, 94].

Other studies have sought to increase the number of epitopes in an attempt to circumvent such problems. Examples include a study by Iwahashi *et al.* [95] which demonstrated that nine patients vaccinated with cancer testis antigen CpG-7909 epitope successfully exhibited immune responses. Another trial by Rezvani *et al.* showed that vaccination using PR1 and WT1 peptides for myeloid malignancies produced an immunogenic response to the vaccine but unfortunately on repeated administration there was a decrease in clinical response [96].

9.3.5 Influenza

Influenza virus is a major human pathogen. It is a severe, seasonal public health problem causing both serious illness and hospitalizations, as well as significant deaths amongst high risk populations, such as the elderly and patients with underlying disease. Influenza has major types (A, B, and C) and many subtypes and strains; it is prone to both antigenic shifts (major antigenic change) and antigenic drifts (minor antigenic change), which gives rise to the so-called seasonality, with significant new strains emerging on the scale of years rather than decades. A vaccine conferring broad, long-term protection against all or most strains remains stubbornly elusive. Influenza type A generally gives rise to moderate to severe disease across the age range, while type B is milder and primarily affects children. Currently, influenza A(H1N1) and A(H3N2) are endemic and/or seasonal viruses circulating within the human population in both hemispheres.

Vaccination is the principal preventive measure used to mitigate influenza. In the United States, historically two broad forms of Flu vaccine have been licensed: live attenuated influenza vaccine and trivalent inactivated influenza vaccine. Both are effective and safe in approved patient populations. Trivalent Inactivated vaccines (TIV) are of three types: whole virus, split-product, and subunit surface-antigen. Most influenza vaccines are currently either split-product vaccines, manufactured from highly-purified, detergent-treated influenza virus, or vaccines produced from purified surface-antigens hemagglutinin and neuraminidase.

The main limitation of current vaccines is that recall immune responses are limited to strains present in the vaccine. A putative advantage for any effective epitope vaccine over extant whole-organism or subunit vaccines would be its potential for universality: that is, its ability to act against influenza over much longer timescales, and thus avoid the economic and logistical difficulties of targeting seasonal flu. Attempts to develop universal subunit vaccines have targeted conserved antigens such as neuraminidases, hemagglutinin stalk domains, and the external domain of influenza M2. An advantage of universal epitope-based vaccine is that it can target highly conserved epitopes that are not prone to antigenic drift or antigenic shifts, such as the three highly conserved epitopes in hemagglutinin [97].

Other approaches to developing epitope-based vaccines for influenza include the use of vectors, such as naked DNA vaccines or salmonella as a carrier of epitope is a novel method that has been considered by Ben-Yedidia *et al.* [98]. Attenuated salmonella has been used previously to deliver cholera toxin epitopes. Moreover, salmonella can express multiple copies of the epitope and it can interact with Toll-like receptors as an adjuvant-like delivery system, inducing innate-like immune response. Another example of poly-epitope vaccine targeting several key immune determinants is the vaccine of Ben-Yedida and coworkers [99]: the vaccine was administered intranasally and was effective in old mice.

9.4 The *in Silico* Design of Epitope Ensemble Vaccines: A Critique

There are several compelling even persuasive reasons underlying the apparent desirability of epitope-based vaccines: certain of these are logistical in nature, and impinge on the practical problems encountered in creating effective vaccines, and others relate to general issues of design optimality. The discovery of epitope-based vaccines can potentially offer a rational and systematic approach to vaccine identification and optimization, at least as systematic and rational as reverse-vaccinology approaches seem to be for subunit vaccines.

A key pre-requisite for epitopes suitable for inclusion in vaccines is that they are immunogenic. Immunogenicity is a widely used term within immunobiology; exact definitions of immunogenicity are varied, perhaps even unique to each immunologist. Unfortunately, facile discovery of immunogenic epitopes exhibiting wide immunodominance is by no means facile. An epitope highly immunogenic in one species, or in one geographically-constrained population, or in one individual within a population, is not, of need, immunogenic in another individuals, populations, or species.

Immunogenicity at its simplest is a property characteristic of a molecular moiety – in our specific case a peptide epitope – allowing it to elicit a major response from a host immune system. However, immunogenicity is not the so-called "protective immunity", which is an enhanced immunity to re-infection, or to a first infection in the case of a successful vaccine, and comprises an enhancement of pre-formed immune reactants, including antigen-specific antibodies, and the formation of enduring memory B-cells and memory T-cells. The measurement of protective immunity is conceptually straightforward: immunized and non-immunized populations are challenged, and the incidence and severity of infection is compared between the two. Alternatively, protective immunity can be identified through the presence or absence of one or more surrogate markers correlated with protection.

Immunogenicity, however defined, manifests itself as part of both the humoral immune response (mediated by the binding of antibodies to whole antigens) and the cellular immune response (mediated through the binding of MHCs to proteolytically cleaved peptides). Immunogenicity is a clear necessity for protective immunity, but is not a sufficient condition. Other factors also help mediate protection. While it is far from easy to define immunogenicity in a comprehensive manner, understanding immunological mechanisms nonetheless underlies modern efforts to design vaccines rationally.

Within cellular immunology, peptide immunogenicity depends strongly on binding to MHC molecules; as well as to subsequent recognition by T-cell receptors. It is generally accepted that only peptides binding above some threshold will operate as T-cell epitopes and that broadly at least affinity correlates with T-cell response.

Thus, many methods for the identification of epitopes have been developed: experimental approaches, computational or predictive methods, and hybrid approaches. We have mentioned above several of the main approaches to the computational identification of epitopes, many of which are amenable to some degree of automation. However, like most high-throughput approaches, with celerity comes inaccuracy. An appealing alternative comes in the form of predictive identification of epitopes [4]. As we have also said above, prediction of B-cell epitopes is sub-optimal to say the least. As most antibodies raised against whole proteins do not cross-react with peptides derived from the said protein many believe that most epitopes are discontinuous. Around 10% are actually continuous in nature. Yet most research into B-cell epitope prediction has focussed on linear peptides since they seem easier to analyse.

The immune system is a multi-level phenomenon, exhibiting at each level from macromolecular complex to whole organism and populations' much emergent behavior, but, building on straightforward molecular interactions, can it be modeled accurately. How does immunogenicity arise? Does MHC-peptide complex lifetime govern an epitope's immunogenicity? Or is the affinity of MHC for peptide the key factor? Does immunogenicity relate rather to the population of MHC-peptide complexes displayed on the surface of APCs? There may not be a simple, single answer, but instead a complex and confounding conspiracy of causes. It should be emphasized that immunogenicity is not solely a function of peptide-MHC binding, but rather it is an emergent phenomenon that emerges as an organism recognizes bound peptide.

To address the pressing need for a fast *in silico* method able to identify vaccine epitopes, many have taken a pragmatic stance; developing approaches that rely on our emerging empirical understanding of certain aspects of immunological presentation and recognition to fast track identification identified epitopes, rather than seeking a comprehensive and cohesive molecular dissection of every contributory event. All such approaches seek to capitalize on the burgeoning wealth of genome data now available.

Sometime ago, Beverley adumbrated an over-arching paradigm for protein-focused vaccine development. He identified several factors potentially contributing to the immunogenicity of proteins, be they of pathogen-origin or another source entirely, and also other features which might make proteins particularly suitable for becoming candidate vaccines. Of these some are as yet beyond prediction, such as the attractiveness for antigen-presenting cells or the inability to down-regulate immune responses. The status of proteins as evasins is currently only possibly addressable through sequence similarity-based search methods. Likewise, attractiveness for uptake by APCs is again a difficult criterion to use, though possibly there exist motifs, structural or sequence, which could be identified. Currently, the dearth of relevant data precludes prediction of such properties; and, while it is possible to predict some of these properties with some assurance of success, others are predictable but only incidentally.

The selection of appropriate B-cell or T-cell epitopes is obvious, and dealt with by algorithms of varying accuracy: as we have said, T-cell prediction algorithms are reasonably robust and reliable for those alleles that have been well-characterized – for all other instances, accuracy is partial, poor, or non-existent. Beyond the efficacy- or affinity-based identification of epitopes, other criteria imposed in selection include both conservation (seeking epitopes and thus vaccines with broad efficacy across many pathogen strains) and epitope promiscuity (seeking epitopes and thus vaccines with broad population coverage).

There are several alternative approaches to the selection of epitopes, and we can separate most into one of two categories:

1. Genome analysis approaches

Here, the whole genome is analysed in an unbiased way, selecting the most affine epitopes from any protein. This is a conceptually clean if simplistic – even simple-minded – approach. Variants include methods which look for epitopes or continuous sequence regions culled from the so-called hotspots, where large numbers of epitopes seem to cluster.

2. Protein-focussed approaches

Here, individual proteins are prioritized prior to elucidation of epitopes, on the reasonable basis that only epitopes deriving from highly expressed proteins or proteins accessible to immune surveillance are likely candidates for vaccines.

What has not been done is to gauge whether or not either the strategy is effective, and which is the more successful, by comparing the efficacy of vaccines derived in this way with vaccines derived from epitopes selected randomly from the genome or from prioritized proteins. Such controlled studies have not been done, not least because of the confounding practical difficulties in undertaking the appropriate kind of study.

The array of approaches able to predict one or other of these quantities has led many to develop pipeline approaches to the vaccine epitope prediction problem that combine a variety of different predictors as a linear or concurrent pipeline. This in turn has led to many studies using such pipelines, and many reports of such work. The rationale for such pipelines is straightforward and draws on an accepted consensus: only secreted proteins, or those found on the cell surface, are accessible to immune surveillance, and of these, only a subset is likely to be antigenic, and are thus also likely to possess immunodominant epitopes [1]. Moutatsi *et al.* [100], amongst others, offer an experimental vindication of this approach. Overall, then, such approaches are seemingly a robust.

We ourselves have previously described the identification of an epitope ensemble comprising multiple epitopes [101]. In this work, the virtual proteome was assessed for potential epitopes using a pipeline comprising a number of servers: SignalP [102–104] (likelihood of protein secretion) followed by VaxiJen [105] (ranked antigenicity of secreted proteins) followed in parallel by MHCPred [106–108], IEDB [109], and NetMHC [110] (binding affinities of epitopes). However, this paper is one among many. An increasing – or rather an alarming – number of papers describing *in silico* analyses of genomes and proteomes, producing epitope ensembles as putative candidate vaccines, are now being published [111–115]. Technically, several are sound, and, like the present study, even rigorous; yet their value cannot easily be quantified. As a consequence, their significance is questionable. We are happy to include our own work in this category. Other studies [116, 117], combine immunoinformatic-driven vaccine design with experimental validation to give credence – rather than verisimilitude – to their computational results. For all or nearly all in computational studies addressing real-world problems, there is a pressing need for experimental validation. Continuing to publish experimentally unverified papers is almost counterproductive in this context. It is necessary to highlight the severe limitations of all such studies given the absence of proper experimental validation. Moreover, current research methodology is largely embodied in web-servers; operating such virtualized systems is facile, and the concomitant analysis of results straightforward. Indeed, the *in silico* design of epitope ensembles as candidate vaccines are well within the capacities of undergraduate researchers.

Extant methodology has many shortcomings. The next generation and the next-generation-but-one methods will need to combine more complex, less technically-naïve predictive components into more detailed and elaborate predictive schemas. Specifically, and for cellular immune response, inclusion of the following:

1. The accurate and robust prediction of peptide presentation to the T-cell
Currently, prediction of peptide presentation is very linear, following the precedent of EpiJen [118] and similar methods, and omits much of the poorly understood complexities of presentation need if we are to create quantitatively useful predictive models of cell surface populations of presented peptide [119].

2. Population coverage
Establishing more comprehensive HLA coverage will require better pan-MHC binding predictors than we currently possess. Many algorithms claim to evaluate MHC binding beyond experimental binding data, but such claims are seldom put to the sword of experimental validation.

3. Expression of epitopes
As well as accessibility to immune surveillance, pathogen side properties such as the relative expression of proteins, and thus epitopes, should prove a useful criterion in winnowing out useful epitopes from the background. Such prediction should be well advanced but strangely not, although the recent emergence of servers such as EDGE-PRO [120] promise much for the future.

4. Conservation
As we have observed above, to avoid issues of pathogen variability, highly conserved epitopes are often targeted as a key tactic within a synoptic vaccine-design strategy. HIV data from Rolland *et al.* [121] demonstrated no unequivocal link between high fitness cost, or replicative capacity, and epitope-conservation, showing that concentrating solely on conserved regions of HIV-1 is unlikely to yield an effective vaccine. However, a subset of sites proved highly function-sensitive, greatly impairing viral replication, and thus indicated they had incurred selective pressure yet proved non-mutable. So, future *in silico* targeting of epitopes that are both conserved and replication-sensitive should yield better vaccines.

5. B-cell epitope prediction

Another key element of future vaccine design will come when – finally – reliable and believable prediction methods exist for antibody mediated epitopes, since currently evidence is strongly suggestive that such existing methods are poor [58, 59, 122].

The importance of innate immune responses and the role of adjuvants should also not be forgotten. Discussion of adjuvants *per se* is beyond the present scope and intention of this chapter, as is disccuion of delivery systems adjuvaneting or otherwise, but the possibility of designing into the multilple epitope ensemble adjuvant-like peptide sequences, such as a number that has been identified over the years [123, 124], is a possibility that should certainly be factoring into emergent vaccine design strategies.

9.5 Conclusions

The vast majority of extant, licensed vaccines target viruses rather than other pathogenic microorganisms, such as bacteria, fungi, and parasites. Most of these are mediated primarily by neutralizing antibodies, rather than through cellular immunity. Successful vaccines have traditionally been developed from heat or chemically inactivated or live but attenuated microorganisms, many able to induce strong cellular as well as humoural immune responses, which in turn gives rise to long-lasting protective memory that mitigates subsequent infection. Prompted by safety concerns, and the need to address unmet medical need, successful vaccine discovery in the last decades has come to concentrate primarily on engaging the neutralizing antibody response as generated by easily synthesized, safe, easily administered subunit vaccines. However, subunit vaccines are relatively poor at inducing the cellular immune response, as they activate components that fail typically to generate broad T-cell responses. Other types of vaccines, including dendritic cell- and epitope-based, have and are also being explored. Peptide vaccines are seen to have several advantages: they are thought safe, easy to make if not create, and have long time stability [ramila81].

An epitope is the smallest part of a protein that induces an immune response; or, if you will, the immunological quantum. Vaccines based on one or more epitopes – epitope ensemble vaccines – are a relatively recent development in vaccinology. There has been a long time hope that epitope ensembles will become the *de facto* vaccines of the future. As rules governing the development and licensing of vaccines become ever more stringent, especially with regard to safety, epitope vaccines have become more appealing when compared to whole pathogen or subunit vaccines. With the increasing prevalence of infectious disease, and the lack of effective treatments of many newly emergent or newly resurgent diseases, particularly the ones which engender cellular immune responses, there is an increasing drive to research epitope vaccines. However, the vast majority of epitope ensemble vaccines have yet to progress beyond Phase III; and no epitope ensemble vaccine is currently licensed.

Like the failure to find an HIV vaccine, this is not due to lack of effort or investment. Rather it is due to a conspiracy of causes, reasons include the inability to identify conserved epitopes able to dominate the immune response and with broad population coverage, leading to distinct responses by different ethnic groups, each possessing different conserved MHC haplotypes; rapidly evolving pathogens, which can escape immune surveillance by mutating epitope sequences, a phenomenon well known from the pathology of HIV; and poor overall immunogenicity of the vaccines resulting from the small size of the ensemble even in the presence of adjuvants. Designing larger epitope ensembles with more immunogenic signals will help address immunogenicity issues, as will more extensive and complete mapping of immunodominant epitopes. Likewise, developing appropriate adjuvants and delivery systems optimal for epitope ensemble vaccines may also improve vaccine effectiveness; and pulsing dendritic cells with epitopes may, in time, prove effective in inducing immune responses. The use of site-directed mutagenesis to augment the affinity and specificity of MHC binding of individual epitope sequences may help broaden and deepen responses and allow us to engineer into the ensemble requisite immunodominance. The cell-surface abundance of the epitope also plays a role in determining the dominance shown by the epitope, and so it may also be necessary to alter any proteolytic motifs present.

Areas of greatest activity, and which thus have the greatest chance of delivering a commercially-viable epitope ensemble vaccine, include Alzheimer's disease, cancer and malaria. The CATVAX peptide vaccine has effectively stalled, mainly due to a lack of funding rather than difficulties in developing the vaccine. In Alzheimer's disease there has been much interest in epitope ensemble vaccines, and in cancer the GP100 vaccine has shown considerable promise. In malaria, promising, well-studied vaccines – such as Malarivax and a range of RTS,S vaccines – have been trialled extensively for some time. All such hindrances to developing an epitope vaccine must be overcome before epitope ensemble vaccines become a commercial reality.

The current chapter has reviewed some of the challenges faced by developers of epitope vaccines targeting different diseases. However, if these challenges are overcome, it is likely such research will lead to vaccines that target not only infectious diseases but also cancers, and possibly other chronic diseases. If this can be achieved it will have a significant impact on the health of the human population and possibly even extend average lifespan.

References

[1] Vivona, S., et al., Computer-aided biotechnology: from immuno-informatics to reverse vaccinology. Trends in Biotechnology, 2008. **26**(4): p. 190–200.

[2] Davies, M.N. and D.R. Flower, Computational Vaccinology. Bioinformatics for Immunomics, 2010. **3**: p. 1–20.

[3] Davies, M.N., et al., Using databases and data mining in vaccinology. Expert Opinion on Drug Discovery, 2007. **2**(1): p. 19–35.

[4] Flower, D.R., Databases and data mining for computational vaccinology. Current Opinion in Drug Discovery & Development, 2003. **6**(3): p. 396–400.

[5] Arnon, R. and T. Ben-Yedidia, Old and new vaccine approaches. Int Immunopharmacol, 2003. **3**(8): p. 1195–204.

[6] Davis, J.P., Experience with hepatitis A and B vaccines. Am J Med, 2005. **118 Suppl 10A**: p. 7S-15S.

[7] McShane, H., et al., Boosting BCG with MVA85A: the first candidate subunit vaccine for tuberculosis in clinical trials. Tuberculosis (Edinb), 2005. **85**(1–2): p. 47–52.

[8] Bernstein, D., Glycoprotein D adjuvant herpes simplex virus vaccine. Expert Rev Vaccines, 2005. **4**(5): p. 615–27.

[9] Bernstein, D.I., et al., Safety and immunogenicity of glycoprotein D-adjuvant genital herpes vaccine. Clin Infect Dis, 2005. **40**(9): p. 1271–81.

[10] Frey, S., et al., Comparison of the safety, tolerability, and immunogenicity of a MF59-adjuvanted influenza vaccine and a non-adjuvanted influenza vaccine in non-elderly adults. Vaccine, 2003. **21**(27–30): p. 4234–7.

[11] Huang, E.H. and H.L. Kaufman, CEA-based vaccines. Expert Rev Vaccines, 2002. **1**(1): p. 49–63.

[12] Carter, N.J., Multicomponent Meningococcal Serogroup B Vaccine (4CMenB; Bexsero((R))): A Review of its Use in Primary and Booster Vaccination. BioDrugs, 2013. **27**(3): p. 263–74.

[13] Gorringe, A.R. and R. Pajon, Bexsero: a multicomponent vaccine for prevention of meningococcal disease. Hum Vaccin Immunother, 2012. **8**(2): p. 174–83.

[14] Bittle, J.L., A New Generation of Vaccines. Veterinary Clinics of North America-Small Animal Practice, 1986. **16**(6): p. 1247–1257.

[15] Bittle, J.L., et al., Protection against Foot-and-Mouth-Disease by Immunization with a Chemically Synthesized Peptide Predicted from the Viral Nucleotide-Sequence. Nature, 1982. **298**(5869): p. 30–33.

[16] De Clercq, K., et al., Serological response to a booster foot-and-mouth disease vaccination with strains different from the primary vaccine. Vet Res Commun, 1989. **13**(3): p. 199–204.

[17] Taboga, O., et al., A large-scale evaluation of peptide vaccines against foot-and-mouth disease: lack of solid protection in cattle and isolation of escape mutants. J Virol, 1997. **71**(4): p. 2606–14.

[18] Langeveld, J.P., et al., First peptide vaccine providing protection against viral infection in the target animal: studies of canine parvovirus in dogs. J Virol, 1994. **68**(7): p. 4506–13.

[19] Ghochikyan, A., et al., Prototype Alzheimer's disease epitope vaccine induced strong Th2-type anti-Abeta antibody response with Alum to Quil A adjuvant switch. Vaccine, 2006. **24**(13): p. 2275–82.

[20] Agadjanyan, M.G., et al., Prototype Alzheimer's disease vaccine using the immunodominant B-cell epitope from beta-amyloid and promiscuous T-cell epitope pan HLA DR-binding peptide. J Immunol, 2005. **174**(3): p. 1580–6.

[21] Audran, R., et al., Phase I malaria vaccine trial with a long synthetic peptide derived from the merozoite surface protein 3 antigen. Infect Immun, 2005. **73**(12): p. 8017–26.

[22] Druilhe, P., et al., A malaria vaccine that elicits in humans antibodies able to kill Plasmodium falciparum. PLoS Med, 2005. **2**(11): p. e344.

[23] Halassy, B., et al., Immunogenicity of peptides of measles virus origin and influence of adjuvants. Vaccine, 2006. **24**(2): p. 185–94.

[24] Bouche, F.B., et al., Induction of broadly neutralizing antibodies against measles virus mutants using a polyepitope vaccine strategy. Vaccine, 2005. **23**(17–18): p. 2074–7.

[25] Jerne, N.K., Immunological speculations. Annu Rev Microbiol, 1960. **14**: p. 341–58.

[26] Lickwar, C.R., F. Mueller, and J.D. Lieb, Genome-wide measurement of protein-DNA binding dynamics using competition ChIP. Nat Protoc, 2013. **8**(7): p. 1337–53.

[27] Egger, J., et al., Nanomolar E-Selectin Antagonists with Prolonged Half-Lives by a Fragment-Based Approach. J Am Chem Soc, 2013.
[28] Mahler, M., M. Bluthner, and K.M. Pollard, Advances in B-cell epitope analysis of autoantigens in connective tissue diseases. Clin Immunol, 2003. **107**(2): p. 65–79.
[29] Geysen, H.M., S.J. Barteling, and R.H. Meloen, Small peptides induce antibodies with a sequence and structural requirement for binding antigen comparable to antibodies raised against the native protein. Proc Natl Acad Sci U S A, 1985. **82**(1): p. 178–82.
[30] Savoca, R., C. Schwab, and H.R. Bosshard, Epitope mapping employing immobilized synthetic peptides. How specific is the reactivity of these peptides with antiserum raised against the parent protein? J Immunol Methods, 1991. **141**(2): p. 245–52.
[31] Petrakou, E., et al., Evaluation of Pepscan analyses for epitope mapping of anti-MUC1 monoclonal antibodies–a comparative study and review of five antibodies. Anticancer Res, 1998. **18**(6A): p. 4419–21.
[32] Petrakou, E., A. Murray, and M.R. Price, Epitope mapping of anti-MUC1 mucin protein core monoclonal antibodies. Tumour Biol, 1998. **19 Suppl 1**: p. 21–9.
[33] Smith, G.P., Filamentous fusion phage: novel expression vectors that display cloned antigens on the virion surface. Science, 1985. **228**(4705): p. 1315–7.
[34] Williams, S.C., et al., Identification of epitopes within beta lactoglobulin recognized by polyclonal antibodies using phage display and PEPSCAN. J Immunol Methods, 1998. **213**(1): p. 1–17.
[35] Davies, M.N., et al., Using databases and data mining in vaccinology. Expert Opin Drug Discov, 2007. **2**(1): p. 19–35.
[36] Flower, D.R. and I.A. Doytchinova, Immunoinformatics and the prediction of immunogenicity. Appl Bioinformatics, 2002. **1**(4): p. 167–76.
[37] Flower, D.R., et al., Computational vaccinology: quantitative approaches. Novartis Found Symp, 2003. **254**: p. 102–20; discussion 120–5, 216–22, 250–2.
[38] Doytchinova, I.A. and D.R. Flower, Quantitative approaches to computational vaccinology. Immunology and Cell Biology, 2002. **80**(3): p. 270–9.
[39] De Groot, A.S., Immunomics: discovering new targets for vaccines and therapeutics. Drug Discov Today, 2006. **11**(5–6): p. 203–9.

[40] De Groot, A.S. and J.A. Berzofsky, From genome to vaccine–new immunoinformatics tools for vaccine design. Methods, 2004. **34**(4): p. 425–8.
[41] Serruto, D., et al., Genome-based approaches to develop vaccines against bacterial pathogens. Vaccine, 2009. **27**(25–26): p. 3245–50.
[42] Rinaudo, C.D., et al., Vaccinology in the genome era. J Clin Invest, 2009. **119**(9): p. 2515–25.
[43] Moriel, D.G., et al., Genome-based vaccine development: a short cut for the future. Adv Exp Med Biol, 2009. **655**: p. 81–9.
[44] Bambini, S. and R. Rappuoli, The use of genomics in microbial vaccine development. Drug Discov Today, 2009. **14**(5–6): p. 252–60.
[45] Barocchi, M.A., S. Censini, and R. Rappuoli, Vaccines in the era of genomics: the pneumococcal challenge. Vaccine, 2007. **25**(16): p. 2963–73.
[46] Mora, M., et al., Microbial genomes and vaccine design: refinements to the classical reverse vaccinology approach. Curr Opin Microbiol, 2006. **9**(5): p. 532–6.
[47] Serruto, D. and R. Rappuoli, Post-genomic vaccine development. FEBS Lett, 2006. **580**(12): p. 2985–92.
[48] Flower, D.R., Towards in silico prediction of immunogenic epitopes. Trends Immunol, 2003. **24**(12): p. 667–74.
[49] Korber, B., M. LaBute, and K. Yusim, Immunoinformatics comes of age. PLoS Comput Biol, 2006. **2**(6): p. e71.
[50] Flower, D., Bioinformatics for Vaccinology. 1st ed2008: Wiley.
[51] Davies, M.N. and D.R. Flower, Harnessing bioinformatics to discover new vaccines. Drug Discov Today, 2007. **12**(9–10): p. 389–95.
[52] Deavin, A.J., T.R. Auton, and P.J. Greaney, Statistical comparison of established T-cell epitope predictors against a large database of human and murine antigens. Mol Immunol, 1996. **33**(2): p. 145–55.
[53] Peters, B., et al., A community resource benchmarking predictions of peptide binding to MHC-I molecules. PLoS Comput Biol, 2006. **2**(6): p. e65.
[54] Lin, H.H., et al., Evaluation of MHC class I peptide binding prediction servers: applications for vaccine research. BMC Immunol, 2008. **9**: p. 8.
[55] El-Manzalawy, Y., D. Dobbs, and V. Honavar, On evaluating MHC-II binding peptide prediction methods. PLoS One, 2008. **3**(9): p. e3268.

[56] Lin, H.H., et al., Evaluation of MHC-II peptide binding prediction servers: applications for vaccine research. BMC Bioinformatics, 2008. **9 Suppl 12**: p. S22.

[57] Gowthaman, U. and J.N. Agrewala, In silico tools for predicting peptides binding to HLA-class II molecules: more confusion than conclusion.
J Proteome Res, 2008. **7**(1): p. 154–63.

[58] Ponomarenko, J.V. and P.E. Bourne, Antibody-protein interactions: benchmark datasets and prediction tools evaluation. BMC Struct Biol, 2007. **7**: p. 64.

[59] Blythe, M.J. and D.R. Flower, Benchmarking B-cell epitope prediction: underperformance of existing methods. Protein Sci, 2005. **14**(1): p. 246–8.

[60] Purcell, A.W., J. McCluskey, and J. Rossjohn, More than one reason to rethink the use of peptides in vaccine design. Nat Rev Drug Discov, 2007. **6**(5): p. 404–14.

[61] Firbas, C., et al., Immunogenicity and safety of a novel therapeutic hepatitis C virus (HCV) peptide vaccine: a randomized, placebo controlled trial for dose optimization in 128 healthy subjects. Vaccine, 2006. **24**(20): p. 4343–53.

[62] Klade, C.S., et al., Sustained viral load reduction in treatment-naive HCV genotype 1 infected patients after therapeutic peptide vaccination. Vaccine, 2012. **30**(19): p. 2943–50.

[63] Firbas, C., et al., Immunogenicity and safety of different injection routes and schedules of IC41, a Hepatitis C virus (HCV) peptide vaccine. Vaccine, 2010. **28**(12): p. 2397–407.

[64] Wu, C.Y., et al., Improving therapeutic HPV peptide-based vaccine potency by enhancing $CD4^{+}$ T help and dendritic cell activation. J Biomed Sci, 2010. **17**: p. 88.

[65] Lin, K., et al., Perspectives for preventive and therapeutic HPV vaccines. J Formos Med Assoc, 2010. **109**(1): p. 4–24.

[66] Ghochikyan, A., Rationale for peptide and DNA based epitope vaccines for Alzheimer's disease immunotherapy. CNS Neurol Disord Drug Targets, 2009. **8**(2): p. 128–43.

[67] Davtyan, H., et al., DNA prime-protein boost increased the titer, avidity and persistence of anti-Abeta antibodies in wild-type mice. Gene Ther, 2010. **17**(2): p. 261–71.

[68] Shah, R.S., et al., Current approaches in the treatment of Alzheimer's disease. Biomed Pharmacother, 2008. **62**(4): p. 199–207.

[69] Wiessner, C., et al., The second-generation active Abeta immunotherapy CAD106 reduces amyloid accumulation in APP transgenic mice while minimizing potential side effects. J Neurosci, 2011. **31**(25): p. 9323–31.

[70] Citron, M., Emerging Alzheimer's disease therapies: inhibition of beta-secretase. Neurobiol Aging, 2002. **23**(6): p. 1017–22.

[71] Citron, M., Beta-secretase as a target for the treatment of Alzheimer's disease. J Neurosci Res, 2002. **70**(3): p. 373–9.

[72] Gorse, G.J., et al., Safety and immunogenicity of cytotoxic T-lymphocyte poly-epitope, DNA plasmid (EP HIV-1090) vaccine in healthy, human immunodeficiency virus type 1 (HIV-1)-uninfected adults. Vaccine, 2008. **26**(2): p. 215–23.

[73] Jin, X., et al., A novel HIV T helper epitope-based vaccine elicits cytokine-secreting HIV-specific $CD4^+$ T-cells in a Phase I clinical trial in HIV-uninfected adults. Vaccine, 2009. **27**(50): p. 7080–6.

[74] Salmon-Ceron, D., et al., Immunogenicity and safety of an HIV-1 lipopeptide vaccine in healthy adults: a phase 2 placebo-controlled ANRS trial. AIDS, 2010. **24**(14): p. 2211–23.

[75] Spearman, P., et al., Safety and immunogenicity of a CTL multiepitope peptide vaccine for HIV with or without GM-CSF in a phase I trial. Vaccine, 2009. **27**(2): p. 243–9.

[76] Asjo, B., et al., Phase I trial of a therapeutic HIV type 1 vaccine, Vacc-4x, in HIV type 1-infected individuals with or without antiretroviral therapy. AIDS Res Hum Retroviruses, 2002. **18**(18): p. 1357–65.

[77] Kran, A.M., et al., Reduced viral burden amongst high responder patients following HIV-1 p24 peptide-based therapeutic immunization. Vaccine, 2005. **23**(31): p. 4011–5.

[78] Kran, A.M., et al., Delayed-type hypersensitivity responses to HIV Gag p24 relate to clinical outcome after peptide-based therapeutic immunization for chronic HIV infection. APMIS, 2012. **120**(3): p. 204–9.

[79] Crompton, P.D., S.K. Pierce, and L.H. Miller, Advances and challenges in malaria vaccine development. J Clin Invest, 2010. **120**(12): p. 4168–78.

[80] Nardin, E.H., et al., Phase I testing of a malaria vaccine composed of hepatitis B virus core particles expressing Plasmodium falciparum circumsporozoite epitopes. Infect Immun, 2004. **72**(11): p. 6519–27.

[81] Olotu, A., et al., Circumsporozoite-specific T-cell responses in children vaccinated with RTS,S/AS01E and protection against P falciparum clinical malaria. PLoS One, 2011. **6**(10): p. e25786.

[82] Olotu, A., et al., Efficacy of RTS,S/AS01E malaria vaccine and exploratory analysis on anti-circumsporozoite antibody titres and protection in children aged 5–17 months in Kenya and Tanzania: a randomised controlled trial. Lancet Infect Dis, 2011. **11**(2): p. 102–9.

[83] Bei, R. and A. Scardino, TAA polyepitope DNA-based vaccines: a potential tool for cancer therapy. Journal of Biomedicine and Biotechnology, 2010. **2010**: p. 102758.

[84] Schwartzentruber, D.J., et al., gp100 peptide vaccine and interleukin-2 in patients with advanced melanoma. N Engl J Med, 2011. **364**(22): p. 2119–27.

[85] Hodi, F.S., et al., Improved survival with ipilimumab in patients with metastatic melanoma. N Engl J Med, 2010. **363**(8): p. 711–23.

[86] Slingluff, C.L., Jr., et al., Immunologic and clinical outcomes of a randomized phase II trial of two multipeptide vaccines for melanoma in the adjuvant setting. Clin Cancer Res, 2007. **13**(21): p. 6386–95.

[87] Nagorsen, D., et al., Natural T-cell response against MHC class I epitopes of epithelial cell adhesion molecule, her-2/neu, and carcinoembryonic antigen in patients with colorectal cancer. Cancer Res, 2000. **60**(17):
p. 4850–4.

[88] Scheibenbogen, C., et al., Phase 2 trial of vaccination with tyrosinase peptides and granulocyte-macrophage colony-stimulating factor in patients with metastatic melanoma. J Immunother, 2000. **23**(2): p. 275–81.

[89] O'Rourke, M.G., et al., Durable complete clinical responses in a phase I/II trial using an autologous melanoma cell/dendritic cell vaccine. Cancer Immunol Immunother, 2003. **52**(6): p. 387–95.

[90] Mittendorf, E.A., et al., The E75 HER2/neu peptide vaccine. Cancer Immunol Immunother, 2008. **57**(10): p. 1511–21.

[91] Amin, A., et al., Assessment of immunologic response and recurrence patterns among patients with clinical recurrence after vaccination with a preventive HER2/neu peptide vaccine: from US Military Cancer Institute Clinical Trials Group Study I-01 and I-02. Cancer Immunol Immunother, 2008. **57**(12): p. 1817–25.

[92] Peoples, G.E., et al., Combined clinical trial results of a HER2/neu (E75) vaccine for the prevention of recurrence in high-risk breast cancer

patients: U.S. Military Cancer Institute Clinical Trials Group Study I-01 and I-02. Clin Cancer Res, 2008. **14**(3): p. 797–803.

[93] Graff-Dubois, S., et al., Generation of CTL recognizing an HLA-A*0201-restricted epitope shared by MAGE-A1, -A2, -A3, -A4, -A6, -A10, and -A12 tumor antigens: implication in a broad-spectrum tumor immunotherapy. J Immunol, 2002. **169**(1): p. 575–80.

[94] Scardino, A., et al., HER-2/neu and hTERT cryptic epitopes as novel targets for broad spectrum tumor immunotherapy. J Immunol, 2002. **168**(11): p. 5900–6.

[95] Iwahashi, M., et al., Vaccination with peptides derived from cancer-testis antigens in combination with CpG-7909 elicits strong specific $CD8^+$ T-cell response in patients with metastatic esophageal squamous cell carcinoma. Cancer Sci, 2010. **101**(12): p. 2510–7.

[96] Rezvani, K., et al., Repeated PR1 and WT1 peptide vaccination in Montanide-adjuvant fails to induce sustained high-avidity, epitope-specific $CD8^+$ T-cells in myeloid malignancies. Haematologica, 2011. **96**(3): p. 432–40.

[97] Ekiert, D.C. and I.A. Wilson, Broadly neutralizing antibodies against influenza virus and prospects for universal therapies. Curr Opin Virol, 2012. **2**(2): p. 134–41.

[98] Ben-Yedidia, T. and R. Arnon, Towards an epitope-based human vaccine for influenza. Hum Vaccin, 2005. **1**(3): p. 95–101.

[99] Ben-Yedidia, T., et al., Efficacy of anti-influenza peptide vaccine in aged mice. Mech Ageing Dev, 1998. **104**(1): p. 11–23.

[100] Moutaftsi, M., et al., A consensus epitope prediction approach identifies the breadth of murine T($CD8^+$)-cell responses to vaccinia virus. Nat Biotechnol, 2006. **24**(7): p. 817–9.

[101] Rai, J., et al., Immunoinformatic evaluation of multiple epitope ensembles as vaccine candidates: E coli 536. Bioinformation, 2012. **8**(6): p. 272–5.

[102] Petersen, T.N., et al., SignalP 4.0: discriminating signal peptides from transmembrane regions. Nat Methods, 2011. **8**(10): p. 785–6.

[103] Emanuelsson, O., et al., Locating proteins in the cell using TargetP, SignalP and related tools. Nat Protoc, 2007. **2**(4): p. 953–71.

[104] Bendtsen, J.D., et al., Improved prediction of signal peptides: SignalP 3.0. J Mol Biol, 2004. **340**(4): p. 783–95.

[105] Doytchinova, I.A. and D.R. Flower, VaxiJen: a server for prediction of protective antigens, tumour antigens and subunit vaccines. BMC Bioinformatics, 2007. **8**: p. 4.

[106] Guan, P., et al., MHCPred 2.0: an updated quantitative T-cell epitope prediction server. Appl Bioinformatics, 2006. **5**(1): p. 55–61.

[107] Guan, P., et al., MHCPred: bringing a quantitative dimension to the online prediction of MHC binding. Appl Bioinformatics, 2003. **2**(1): p. 63–6.

[108] Guan, P., et al., MHCPred: A server for quantitative prediction of peptide-MHC binding. Nucleic Acids Res, 2003. **31**(13): p. 3621–4.

[109] Zhang, Q., et al., Immune epitope database analysis resource (IEDB-AR). Nucleic Acids Res, 2008. **36**(Web Server issue): p. W513–8.

[110] Lundegaard, C., et al., Net MHC-3.0: accurate web accessible predictions of human, mouse and monkey MHC class I affinities for peptides of length 8–11. Nucleic Acids Res, 2008. **36**(Web Server issue): p. W509–12.

[111] Akhoon, B.A., et al., In silico identification of novel protective VSG antigens expressed by Trypanosoma brucei and an effort for designing a highly immunogenic DNA vaccine using IL-12 as adjuvant. Microb Pathog, 2011. **51**(1–2): p. 77–87.

[112] Gupta, A., D. Chaukiker, and T.R. Singh, Comparative analysis of epitope predictions: proposed library of putative vaccine candidates for HIV. Bioinformation, 2011. **5**(9): p. 386–9.

[113] Sinha, S., et al., A gp63 based vaccine candidate against Visceral Leishmaniasis. Bioinformation, 2011. **5**(8): p. 320–5.

[114] Jahangiri, A., et al., An in silico DNA vaccine against Listeria monocytogenes. Vaccine, 2011. **29**(40): p. 6948–58.

[115] John, L., G.J. John, and T. Kholia, A reverse vaccinology approach for the identification of potential vaccine candidates from Leishmania spp. Appl Biochem Biotechnol, 2012. **167**(5): p. 1340–50.

[116] Wieser, A., et al., A multiepitope subunit vaccine conveys protection against extraintestinal pathogenic Escherichia coli in mice. Infect Immun, 2010. **78**(8): p. 3432–42.

[117] Seyed, N., et al., In silico analysis of six known Leishmania major antigens and in vitro evaluation of specific epitopes eliciting HLA-A2 restricted CD8 T-cell response. PLoS Negl Trop Dis, 2011. **5**(9): p. e1295.

[118] Doytchinova, I.A., P. Guan, and D.R. Flower, EpiJen: a server for multistep T-cell epitope prediction. BMC Bioinformatics, 2006. **7**: p. 131.

[119] Dalchau, N., et al., A peptide filtering relation quantifies MHC class I peptide optimization. PLoS Comput Biol, 2011. **7**(10): p. e1002144.

[120] Magoc, T., D. Wood, and S.L. Salzberg, EDGE-pro: Estimated Degree of Gene Expression in Prokaryotic Genomes. Evol Bioinform Online, 2013. **9**: p. 127–36.

[121] Rolland, M., et al., HIV-1 conserved-element vaccines: relationship between sequence conservation and replicative capacity. J Virol, 2013. **87**(10): p. 5461–7.

[122] Greenbaum, J.A., et al., Towards a consensus on datasets and evaluation metrics for developing B-cell epitope prediction tools. J Mol Recognit, 2007. **20**(2): p. 75–82.

[123] Ferris, L.K., et al., Human beta-defensin 3 induces maturation of human langerhans cell-like dendritic cells: an antimicrobial peptide that functions as an endogenous adjuvant. J Invest Dermatol, 2013. **133**(2): p. 460–8.

[124] Saenz, R., et al., HMGB1-derived peptide acts as adjuvant inducing immune responses to peptide and protein antigen. Vaccine, 2010. **28**(47): p. 7556–62.

10

Vaccine Discovery and Translation of New Vaccine Technology

Rino Rappuoli, John Telford, Roberto Rosini and Monica Moschioni

Novartis Vaccines, via Fiorentina, 1 53100 Siena, Italy

10.1 Introduction

Vaccine development has now entered in its 4th century. Since its introduction in the 18th century, vaccination has had an impact on mortality greater than any other measure besides clean water and has profoundly improved the public health of many populations globally [1]; infectious diseases and death caused by various viruses and bacteria are now prevented by vaccination and diseases such as smallpox and poliomyelitis are globally eradicated (at least in the Western Hemisphere) [2, 3]. However, there are still no vaccines for many important diseases and new vaccines are needed to replace suboptimal existing ones or to target emerging pathogens. Although in some cases commercial priorities have had a primary inhibitory effect on vaccine development; for many of the lacking vaccines, the development has been delayed by the complexity of correlates of protection and the difficulties in constructing the right presentation of antigens to achieve an effective immune-response.

Most of the current vaccines have been developed by adaptation of living organisms to growth conditions that attenuate their virulence, by preparation of suspensions of killed microbes, or through the concentration and purification of proteins or polysaccharides from pathogens [4]. However, conventional vaccinology has proven to be inadequate for antigenically diverse or rapidly evolving pathogens, for microorganisms that cannot be cultivated in the laboratory, and for infections that lack suitable animal models and/or are controlled by mucosal or T-cell dependent immune responses [5][6]. The advent of molecular biology and of new genome–based technologies has allowed the development of multiple new tools with which novel vaccine antigens can be

Post-genomic Approaches in Drug and Vaccine Development, 245–276.

identified and also rationally designed to increase their immunogenicity and/or vaccine coverage [7]. Indeed, recombinant DNA technologies have been used to obtain rationally attenuated strains or highly purified antigenic components. In addition, the recent availability of enormous amount of information generated by whole genome sequencing projects has provided the opportunity to combine new different approaches (the so-called "-omics" approaches as they are referring to the study of the complete repertoire of genes/proteins an organisms has/is capable of expressing) to vaccine design [8]. In fact, the "*in silico*" antigen discovery based on the analysis of the genetic content of a single isolate (classical Reverse Vaccinology, RV) has been primarily flanked by a multi-genome comparative analysis of epidemiologically representative strains (comparative genomics and pan-genomics) aiming to evaluate the degree of variability of the antigens (comparative RV) but also to exclude from the analysis antigens highly conserved in the commensal flora (subtractive RV), and then by post–genomic strategies. The post-genomic strategies, referred to as *functional genomics*, complement the antigen discovery by examining the transcriptional profile of the genes (transcriptome) and the expression profile of the proteins (proteome, surfome and secretome), thus providing important information regarding the function and the regulation of the expression under different conditions. Lately, antigen identification strategies have been also flanked by: immunomic-based approaches, aiming at understanding if antigens, or specific domains thereof, selected by *in silico* and *in vitro* approaches are potentially immunogenic in humans; and structural vaccinology methodologies, which are coupling structural, variability and immunogenicity information on specific antigens to rationally engineer new immune-dominant molecules [9][10].

The objective of this chapter is to review the classical and the new genome-based vaccinology approaches by specifically highlighting how the advent of recombinant DNA technology and, more recently, the genomic information has impacted on vaccine technology development.

10.2 Classical Vaccinology Approaches

10.2.1 Live Attenuated and Inactivated Whole Organism Vaccines

Most vaccines currently licensed and available for human use are composed of killed inactivated organisms or live, attenuated organisms (Table 10.1). Historically, live attenuated vaccines were originally obtained by using animal viruses, physical attenuation of the organisms (heat, oxygenation, chemical agents), passage in animal hosts, eggs, or *in vitro* cultures, or selecting spontaneous mutants (i.e. mutants adapted to grow at low temperatures).

Table 10.1 List of whole organism and subunit/extract licensed

	Whole organism vaccines		Subunit and extract vaccines			
	Live-attenuated	Killed-inactivated	Toxoid	Conjugated capsular polysaccharide	Free capsular polysaccharide	Recombinant protein
Adenovirus	✓					
Anthrax		✓				
Cholera	✓	✓				
Diphteria			✓			
Haemophilus influenzae				typeB		
Hepatitis A		✓				
Hepatitis B						✓
Influenza virus	✓	✓				
Japanese encephalitis	✓	✓				
Lyme disease						✓
Measles	✓					
Meningococcus				serogroups ACW-135Y, A or C	serogroups ACW- 135Y, or AC	serogroup B
Mumps	✓					
Pertussis		✓	✓			✓
Plague		✓				
Pneumococcus				7, 10 or 13 serotypes	23 se rotypes	

(Continued)

Table 10.2 Continued

	Whole organism vaccines		Subunit and extract vaccines			
	Live-attenuated	Killed-inactivated	Toxoid	Conjugated capsular polysaccharide	Free capsular polysaccharide	Recombinant protein
Polio	✓	✓				
Rabies	✓	✓				
Rotavirus	✓					
Rubella	✓					
Smallpox	✓					
Tetanus			✓			
Tickborne encephalitis		✓				
Tubercolosis	✓					
Typhoid	✓	✓			✓	
Varicella	✓					
Yellow fever	✓					

They have been extensively used and have clearly demonstrated their efficacy in inducing a strong humoral immune response. However, they have been also associated with genetic instability and residual virulence and are thus not in compliance with the current regulatory requirements which call for strains that are well defined and carry precise mutations [11][12]. For dealing with these issues, and thanks to the advent of recombinant DNA technologies and to genomic sequence information, a number of strategies have been applied to obtain live attenuated organisms through genetic modifications.

The origin of the development of live attenuated and inactivated vaccines goes back to the pioneer experiments of vaccinology performed in the 18th century. The first attempts to prevent an infectious disease by means of immunoprophylaxis involved the process of "inoculation of the smallpox" or "variolation", wherein the content of smallpox vesicles was used to inoculate individuals who had not previously experienced the disease. Although the first "variolation" experiments date back to about AD1000 in China, this procedure was used in Europe and in the US only throughout the 18th century and until 1796, when Edward Jenner pioneered the first vaccination experiments by rigorously demonstrating how a prior mild cowpox infection could protect against a smallpox challenge [3]. As Jenner demonstrated with his experiments, there are instances among viruses where, because of the host specificity of virulence, an animal virus gives an attenuated infection in a human host, sometimes leading to an acceptable level of protection [13–14]. During the 19th century smallpox vaccination became popular and in the decade 1967 to 1977 smallpox became the first (and so far the only) actively eradicated communicable disease.

Following Jenner's experiments, little more was done with respect to immunization until the last quarter of the 19th century when Louis Pasteur discovered how chickens inoculated with old "attenuated" cultures of *Pasteurellaseptica* did not develop the disease and were protected when subsequently inoculated with virulent fresh cultures [15].

Subsequently, Pasteur demonstrated that farm animals were protected against lethal challenge with *Bacillus anthracis* if they had been previously immunized with a heat attenuated *B. anthracis* culture, thus leading to the development of the first preventive vaccine for animal use [16, 17]. Remarkably, Pasteur developed the first vaccine made in the laboratory, founded the terminology of vaccination and the principles established by his work, that are "isolation, inactivation and administration of disease causing microbes" have guided vaccine development throughout the 20th century.

The first bacterial vaccine used in humans was administered in 1884 by Jaime Ferran, about one year following the initial isolation of *Vibrio cholerae*. Ferran's vaccine, consisting of broth cultures containing attenuated vibrios, was given to about 30,000 individuals; the vaccine induced severe adverse reactions with apparent lack of efficacy, and afterwards resulted to be heavily contaminated with other microorganisms [18]. Years later, in 1891 Waldemar Haffkine initiated his studies to prepare *V. cholerae* strains to be used as live vaccines for parenteral immunization. Haffclines's live cholera vaccine resulted to be efficacious, but further use of the vaccine was abandoned because of difficulties in the standardization of the production process and was later on substituted by and agar-grown heat inactivated *V. cholerae,* able to confer significant short term protection in children and adults [19].

Other early successes in attenuated/inactivated bacterial vaccines have been: the BacilleCalmette-Guerin (BCG) vaccine against tuberculosis, which consisted in a tubercle Bacilli subcultured repeatedly in the presence of ox bile [20]; the vaccine agaist typhoid fever, composed of heat inactivated typhoid bacilli preserved in phenol [21]; the Sabin attenuated polio vaccine for oral administration (OPV) that has been the major approach to polio eradication [22]; and the yellow fever vaccine, developed in the late 1920s by Max Theiler by selecting an attenuated strain (named 17D) through repeated passage of a wild-type strain in minced chick embryo tissues [23].

More recently, as briefly mentioned above, to increase the safety and immunogenicity profiles of live attenuated vaccines, a number of recombinant DNA techniques such as reassortment, reverse genetics, recombination, site specific and/or deletion mutants, codon deoptimization and control of replication fidelity, have been used to obtain genetically attenuated bacteria and viruses. Genetic reassortment has been used for the creation of influenza vaccine strains that bear new hemagglutinins (HAs) and grow well *in vitro* and also for the development of a vaccine against the human rotavirus (causing infant gastroenteritis) with broad protective properties (a bovine rotavirus naturally attenuated for humans was used as donor of double strand RNA fragments coding for rotavirus proteins) [24]. Site specific and deletion mutants, instead, are being constructed by combining different mutations to obtain safer attenuated live vaccine strains that still preserve immunogenicity. As an example, attenuated *Salmonella* strains have been created by introducing specific auxotrophic mutations in the *aro* genes of the shikimate metabolic pathway, essential for the biosynthesis of aromatics, including aromatic amino acids, but also by deleting the *Salmonella* pathogenicity island type III secretion system [25]. Other innovative strategies of attenuation are codon deoptimization obtained by changing the original nucleotide triplets preferred by the virus to code for amino acids to others less frequently used, and the introduction of mutations in the RNA polymerases to enhance the fidelity of the enzyme and reduce the probability of reversion from attenuation to virulence [26].

Lately, live attenuated bacterial vaccines have been also proposed to be used as carriers (vectors) for the presentation of heterologous antigens, since they can be effectively used also for mucosal immunization. Bacterial species that are being investigated as carriers are attenuated strains of *Salmonella, V. cholerae, Listeria monocytogenes, BCG* and the Theiler's 17Dyellow fever attenuated strain [27]. As an example, the 17D yellow fever strain is in advanced clinical development as a vector for expressing dengue proteins of different serotypes but it has been also used to create a new Japanese encephalitis virus vaccine [28].

10.2.2 "Subunit" and "extract" vaccine

The second most commonly used approach for the development of the vaccines currently on the market is to use as vaccine components specific virulence factors such as toxins or polysaccharides directly isolated from the microorganisms or synthetically produced (see Table 10.1 for a list). This kind of approach is more recent in time and adequately fulfils the high regulatory standards required for commercialization of human use products.

The diphtheria and tetanus toxoids should be regarded as the pioneer subunit vaccines. The discovery at the end of the 19th century of the etiological agents of diphtheria and tetanus, *Corynebacteriumdiphteriae* and *Clostridium tetani,* was immediately followed by the observation that these two pathogens produce potent exotoxins, mostly responsible for the outcomes of the disease. Before the discovery that formalin treatment inactivates the toxins while preserving immunogenicity (1920), the active toxins were used to produce neutralizing antibodies in animals for passive protection and disease treatment and then used mixed with the animal antisera to induce active immunization in humans [29, 30].

Recombinant DNA technology was successfully applied for the first time to the production of a recombinant vaccine for hepatitis B virus (HBV), based on a purified HBV surface antigen, which was also able to self-assemble, facilitating purification and manufacture at the industrial level [31].

Another example of a toxin-based vaccine, which took advantage of the use of recombinant DNA technology, being developed during the 1980s, is the acellular pertussis vaccine. The vaccine is composed of the filamentous agglutinin and the pertactin proteins along with the pertussis toxin, which was not chemically inactivated but site-specifically detoxified by molecular engineering [32]. Production of protein antigens using recombinant DNA techniques has now become a standard for protein subunit vaccine development.

Polysaccharides extracted from liquid cultures of bacterial pathogens such as *Haemophilusinfluenzae type B* and *Streptococcus pnuemoniae* were shown to be attractive vaccine components especially because anti-polysaccharide antibodies have an important role in natural immunity. However, early observations demonstrated the limitations of plain polysaccharide as vaccine antigens, since they are poorly immunogenic during the first two years of life. The limitation was successfully overcome by using protein-conjugate (tetanus toxoid or diphtheria CRM197 carrier proteins) polysaccharides able to induce long term T-cell-based immune responses. *Haemophilusinfluenzae* (type B), *Neisseria meningitidis* (A, C, W-135, Y) and *Streptococcus pneumoniae* (7-, 10-, and 13-valent) are effective vaccines based on conjugate polysaccharides [33–35].

10.3 Identification of New Vaccine Candidates through Innovative Genome Based Technologies

10.3.1 Reverse Vaccinology

Although empiric vaccine development has served us well in the past, conventional vaccinology is currently facing a number of difficult challenges that requires us to move beyond the historical paradigm of "isolate, inactivate, and inject" [36]. The genome era, initiated with the completion of the first bacterial genome of a *Haemophilusinfluenzae* strain in 1995, catalyzed a long overdue revolution in vaccine development. Advances in sequencing technology and bioinformatics have resulted in an exponential growth of genome sequence information, and at least one genome sequence is now available for each major human pathogen. As of November 2012, 3781 bacterial genomes have been completed and 14655 are still ongoing (GOLD Genomes OnLine Database, http://www.genomesonline.org) [10]. The application of genome analysis to vaccine development, a concept termed reverse vaccinology, uses the genome sequences of viral, bacterial or parasitic pathogens of interest, rather than the cells, as starting material for the identification of novel antigens, whose activity should be subsequently confirmed by experimental biology [37]. The first step of this *in silico* approach is the application of appropriate combination of algorithms to genome analysis and the critical evaluation of the coding capacity. The predicted open reading frames (ORFs) are then used for homology searches against a database with BLASTX, BLASTN and TBLASTX programs, to identify DNA segments with potential coding regions. One of the criteria which is then applied to select the most promising ORFs is the prediction of surface localization. Indeed, secreted or extracellular proteins represent ideal vaccine candidates, being more accessible to antibodies than intracellular proteins. Since the *in silico* approach results in the identification of a large number of genes, the target genes are then cloned and expressed by means of simple high-throughput procedures. Once purified, the recombinant proteins are used to immunize mice. The post-immunization sera are analyzed to experimentally verify the computer-predicted surface localization of each polypeptide and the ability of the selected vaccine candidates to elicit an immune response [38]. However, to develop a protective immune response against an infectious agent, the selected antigens should evoke the production of functional antibodies. Hence, the protective efficacy of candidate antigens is tested using animal models of infection (Figure 10.1).

The first example of a successful application of the reverse vaccinology approach was provided by Pizza and coworkers in collaboration with The Institute for Genomic Research (TIGR) [32]. They describe the identification of vaccine candidates against *Neisseria meningitides* sero group B (MenB), the major cause of sepsis and meningitis in children and young adults. Each year, approximately 1.2 million cases of invasive meningococcal disease are recorded worldwide, 7,000 of which occur in Europe. Over 90% of cases of meningococcal meningitis and septicaemia are caused by five of the 13 meningococcal serogroups, specifically groups A, B, C, W135 and Y. In Europe, group B is the most prevalent meningococcal serogroup, with 3,406–4,819 cases reported annually between 2003 and 2007, according to a surveillance report published by the European Centre for Disease Control [39]. Conventional approaches to obtain a MenB vaccine had failed for decades, mainly for two reasons. First, the capsular polysaccharide, successfully used to make conjugate vaccines against other serogroups, resembles components of human tissues resulting in poor immunogenicity in humans and potentially able to induce an autoimmune response. Second, previously failed attempts to develop protein-based vaccines focused on variable antigens able to confer protection only against a limited number of strains.

The reverse vaccinology approach started from the determination of the complete genome sequence of a MenB pathogenic strain, MC58. Several

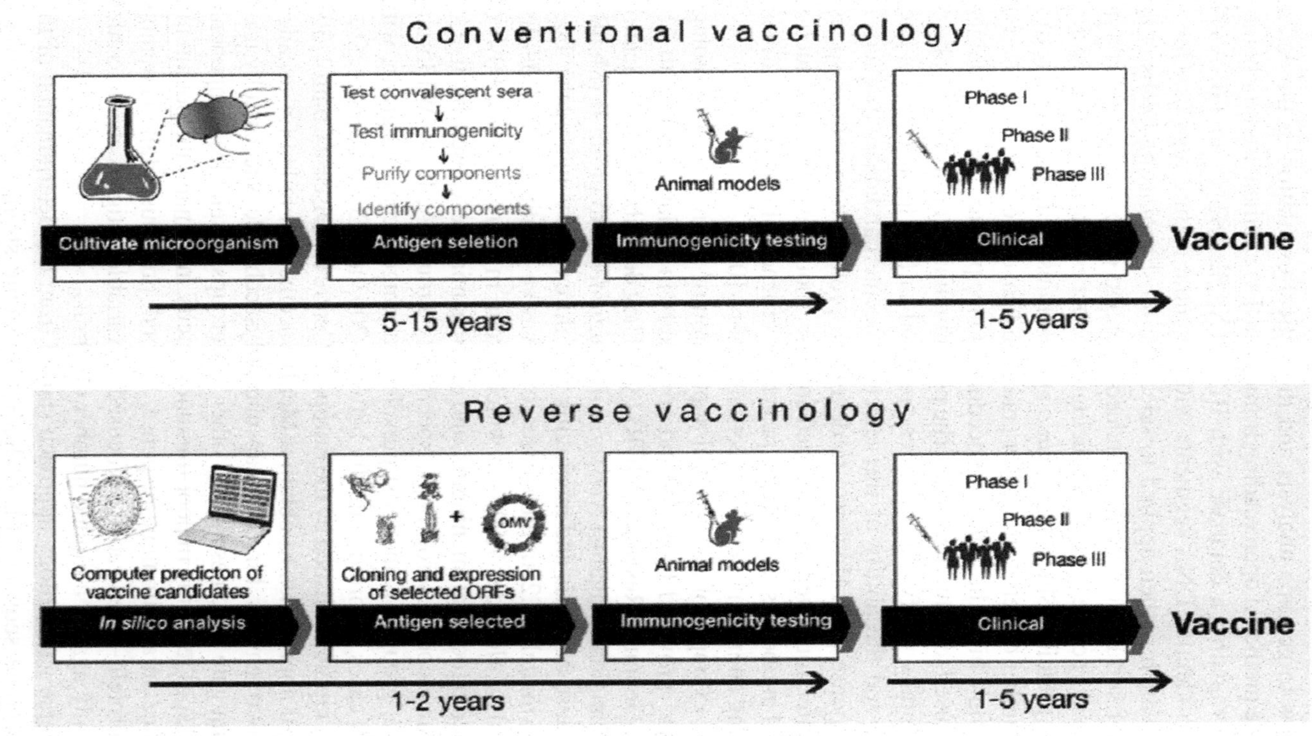

Figure 10.1 Comparison of classical (conventional) and genome-based (reverse) vaccinology approaches.

computational tools were used to find in the genome, on the basis of sequence features, the presence of amino acid motifs targeting the mature protein to the outer membrane (signal peptides), to the lipid bilayer (lipoproteins), or to the integral membrane (transmembrane domains), or potentially involved in the recognition and interaction with host structures. This analysis allowed researchers to identify 600 putative surface-exposed or secreted proteins. Three hundred and fifty proteins were expressed in a heterologous system, *Escherichia coli*, purified and used to immunize mice. Mice immune sera were tested for specificity by Western blot, used to determine antigen accessibility on the bacterial surface by flow cytometry and screened for their bactericidal activity by means of a serum bactericidal assay (SBA). In detail, SBA specifically estimates the ability of bactericidal antibodies to evoke a complement-mediated lysis of bacteria and, in the case of meningococcus, is internationally accepted as correlate of protection against the microorganism. Twenty-nine out of the about 350 surface-exposed proteins were found to elicit bactericidal immune-responses. The selected candidates were then checked for sequence conservation across a panel of strains representing the genetic diversity of meningococcus and including the clonal complexes most frequently associated to invasive disease. This analysis allowed the further selection of antigens able to elicit a cross-bactericidal response against most strains included in the panel. The results of this study were used to define the final formulation of the 4CMenB vaccine that is based on a 'cocktail' of three antigens identified by reverse vaccinology, NadA, fHbp and NHBA (fHbp and NHBA are present as fusion proteins with GNA2091 and GNA1030, respectively) in addition to an OMV preparation [40].

A number of molecular epidemiological studies revealed that the 4CMenB antigens varied both in their sequence and in protein expression levels among different strains, making the identification of a system to predict the 4CMenB vaccine coverage critical for vaccine approval [9–41]. To specifically answer to this need, a new typing system, the meningococcal antigen typing system (MATS) was developed and proposed to the regulatory bodies as a tool to predict vaccine strain coverage. Indeed, MATS is a vaccine antigen-specific sandwich ELISA that contemporaneously measures the amount of each antigen expressed by a strain and its immunological cross-reactivity with the antigen present in the 4CMenB vaccine, so that MATS scores for a specific strain are correlated with killing of the strain in the serum bactericidal activity assay. In particular, strains exceeding a threshold value in the ELISA for any of the three vaccine antigens have $\geq$80% probability of being killed by immune serum in SBA and can be estimated to be covered by the vaccine [42].

Clinical trial data from studies with >1800 infants showed that 4CMenB is safe and induces a robust immune response. Protective immune responses (human serum bactericidal activity titres ≥1:5) were observed against reference MenB strains. MATS data together with a comprehensive clinical program with 4CMenB has served as the basis for submission for regulatory approval to the European Medicines Agency [43]. Recently, the European Medicines Agency's Committee for Medicinal Products for Human Use (CHMP) has approved the marketing authorization for 4CMenB (Bexsero®), a new vaccine intended for the immunization of individuals over two months of age against invasive meningococcal disease caused by *Neisseria meningitidis* group B (http://www.ema.europa.eu) finally decreeing the first success of the reverse vaccinology approach for the development of multi-component protein-based vaccines.

10.3.2 Pan-genomic reverse vaccinology approach

While the genome sequence of a single strain reveals many aspects of the biology of a species, it fails to address how genetic variability drives pathogenesis within a bacterial species, also limiting genome-wide screens for vaccine candidates or for antimicrobial targets to a single strain. The availability of genome sequences for different isolates of a single species enables quantitative analyses of their genomic diversity through comparative genomic analyses, thus most likely reducing to a minimum the coverage analysis problems encountered for the development of the MenB vaccine, whose antigen selection was based on a single genome analysis. The advantage of multiple genome analysis in vaccine design is highlighted by the discovery of universal vaccine candidates against *S. agalactiae*, the group B Streptococcus (GBS) [44].

GBS is a Gram-positive bacterium that causes life-threatening diseases in newborns and infants. GBS is the leading cause of bacterial sepsis, pneumonia and meningitis in neonates in USA and Europe. The infection is acquired during delivery by direct mother-to baby transmission of the pathogen [45]. GBS is classified based on the polysaccharide capsular type or serotype. So far, ten different capsular serotypes have been described of which five are known to be the major disease-causing isolates in Europe and US: Ia, Ib, II, III and V. However, additional serotypes are prevalent in other areas of the world such as serotypes VI and VIII in Japan and non-typeable strains

(NT), that do not express capsule, are emerging in the GBS population [46]. Therefore, the conjugate vaccines currently under development, which are based on the most prevalent capsular polysaccharides, do not cover all possible GBS serotypes or NT strains. For these reasons, a universal protein-based GBS vaccine is highly desirable [44]. On the basis of microarray-based comparative genomic hybridizations performed by using as reference strain the first GBS genome sequence released [47], it was determined that this species' genomic diversity was fairly extensive. Eighteen percent of the genes in the reference genome were absent in at least one of the 19 isolates tested on the microarray and most of them were located in genomic islands of five or more consecutive genes. The main limitation of the microarray experiment resided in the fact that it was considering only the genes of the reference genome while the genes specific to the query strains were not identified. To circumvent this problem, additional genome sequences were generated such that a comprehensive analysis could be conducted on eight complete genomes [48]. The advantage of conducting comparative analyses on multiple genome isolates is emphasized by the discovery of universal vaccine candidates against GBS [49]. Indeed, genome analysis of the eight clinical GBS isolates, which are representative of the serogroups diversity and responsible for >90% of human infections in the United States, revealed that the global gene repertoire of the GBS bacterial species consisted of 1806 genes present in every strain (core genome), and 907 dispensable genes present in one or more but not all strains (dispensable genome). As a whole, the core genome includes all genes responsible for the survival of the bacterium, and its major phenotypic traits. By contrast, dispensable genes contribute to the species diversity and may encode supplementary biochemical pathways and functions that are not always essential for bacterial growth but which confer selective advantages, such as adaptation to different niches, antibiotic resistance, or colonization of a new host. Such genes are generally clustered on large genomic islands that are flanked by short repeated DNA sequences and are characterized by an abnormal G + C content. Investigation and functional annotation of dispensable genes reveals that hypothetical, phage- and transposon-related genes account for the vast majority of these findings. Moreover, computational predictions suggest that the more genomes are sequenced, the more new genes will be found that belong to the pan-genome of GBS. In fact, the pan-genome is predicted to grow about 33 new genes every time a new strain is sequenced. This profile is different from that observed for other microorganisms, such as *Bacillus anthracis*.

In order to identify protective antigens for GBS, computational algorithms were used to select the genes encoding putative surface-associated and secreted proteins. Among the predicted surface-exposed proteins, 396 were core genes and 193 were variable genes. Of these 589 proteins, 312 were successfully cloned and expressed in *Escherichia coli.* Each antigen was then tested in a mouse maternal immunization assay to evaluate the capacity to confer protection. No individual core proteins or a combination thereof provided high levels of immunity against a panel of GBS isolates. The best candidates, namely GBS67, GBS80, GBS104, and GBS322 identified through the reverse vaccinology combined with the multi-genome comparison approach, belonged to the dispensable genome with the exception of GBS322 which belongs to the core genome. The most significant protection results were achieved when the four best candidates were combined together. The four-protein cocktail conferred 59–100% protection against a panel of 12 GBS isolates, including the major serotypes, as well as two strains from the less common serotype VIII (81 and 94% protection). In addition, the pan-genomic reverse vaccinology approach also led to the discovery of a previously unidentified pilus structure in GBS. In fact, GBS67, GBS80, GBS104 vaccine candidates, were found to contain LPXTG motifs typical of cell wall-anchored proteins and seen to assemble into pili [50]. Further bioinformatics analysis revealed three independent loci that encode structurally distinct pilus types, each of which contains two surface-exposed antigens capable of eliciting protective immunity in mice. Because of the limited variability of GBS pili, it has been suggested that a combination of only three pilin subunits could lead to broad protective immunity [51]. The success of reverse vaccinology for meningococcus has led to the application of this approach to a variety of human and animal pathogens such as *Bacillus anthracis*, *Escherichia coli ExPEC*, *Brucella*, *Brachyspirahyodysenteriae* (Table 10.2).

10.3.3 Functional Genomics Approaches in Vaccine Development

The recent advent of high-throughput technologies has had a deep impact on the ability to analyze complex systems. These methodologies provide quantitative measurements on a global scale and provide the ability to fully interrogate more than just the genome [52]. In the vaccine discovery field, functional genomics approaches are complementary to *in silico* antigen identification. These include the large-scale analysis of gene transcription by DNA microarrays or whole transcriptome analysis (mRNA-seq), the

Table 10.2 Examples of the application of Reverse Vaccinology approaches to the development of vaccines against bacterial and protozoa pathogens

Pathogen	Disease	Approach	Refs
Neisseria meningitidisserogroup B	Bacterial meningitis and septicemia	Reverse vaccinology, Microarray, Proteomics	[32, 54, 60, 79, 80]
Bacillus anthracis	anthrax	Reverse vaccinology	[81]
Streptococcus agalactiae GBS	Bacterial sepsis, pneumonia, meningitis	Comparative reverse vaccinology	[44]
Escherichia coli ExPEC	Neonatal meningitis, sepsis, bacteremia, urinary tract infections	Subtractive reverse vaccinology	[82]
Echinococcusgranulosus	Cystic echinococciasis (zoonosis)	Reverse vaccinology	[83]
Brachyspirahyodysenteriae	Swine dysentery	Reverse vaccinology	[84]
Streptococcus suis serotype 2	Meningitis, endocarditis, septicemia, polyarthritis, polyserositis, pneumonia, and even acute death in swine	Reverse vaccinology	[85]
Schistosoma	Schistosomiasis	Reverse vaccinology	[86]
Theileriaparva	East Coast fever (theileriosis)	Reverse vaccinology	[87]
P. multocida	Hemorrhagicsepticemia in ungulates, fowl cholera in birds, atrophic rhinitis in swine, pneumonia and shipping fever in cattle, and snuffles in rabbits	Reverse vaccinology	[88]
Leishmania spp.	Spectrum of clinical manifestations ranging from cutaneous ulcers to visceral leishmaniasis	Reverse vaccinology	[89]
Rickettsial pathogens	Typhus, rickettsialpox, Boutonneuse fever, African tick bite fever, Rocky Mountain spotted fever, Flinders Island spotted fever and Queensland tick typhus	Reverse vaccinology	[90]

(Continued)

Table 10.2 Continued

Pathogen	Disease	Approach	Refs
Cryptosporidium species	Acute, persistent, and chronic diarrhoeawith life-threatening consequences in immunocompromised individuals	Reverse vaccinology	[91]
Streptococcus sanguinis	Endocarditis	Reverse vaccinology	[92]
Brucella	Zoonotic brucellosis	Reverse vaccinology	[93]
Edwardsiellatarda	Edwardsiellosis of fish	Reverse vaccinology	[94]
Ehrlichiaruminantium	Heartwater, a serious tick-borne disease of ruminants	Reverse vaccinology	[95]
Schistosomajaponicum	Schistosomiasis	Reverse vaccinology	[96]
Haemophilusparasuis	Contagious porcine Glässer's disease	Reverse vaccinology	[97]

identification of the whole set of proteins encoded by an organism (proteomics) by two-dimensional gel electrophoresis and mass spectrometry, as well as the comparative genome–proteome technologies [53]. In fact, the analysis of the role of genes and proteins is critical in order to identify genes required for survival under specific conditions. An example of the use of a microarray-based transcriptional profiling to identify potential MenB vaccine candidates was described by Grifantiniet *et al.* [54]. In this study, bacteria were incubated with human epithelial cells, cell-adhering bacteria were recovered and total RNA was purified at different times.

In parallel, RNA was prepared from non-adherent bacteria grown in the absence of epithelial cells. The two RNA preparations were comparatively analyzed on DNA microarrays carrying the entire repertoire of PCR-amplified MenB genes. Twelve proteins whose transcription was found to be particularly activated during adhesion were expressed in *E. coli*, purified and used to produce antisera in mice. Five sera showed bactericidal activity against different strains [54]. Further transcriptional analyses have then permitted a better understanding of the differential expression of genes coding for vaccine antigens against MenB in response to human blood and this can be predictive for their behavior during *in vivo* infection. The analysis of the dataset showed that several vaccine antigens, including NMB0035, tbpA, tbpB, lbpA, lbpB, NMB1030, nspA, opc, NMB1946, NMB1220, NMB2091, fHbp and porB, were differentially regulated. As reported above, the recently proposed 4CMenB vaccine contains three main antigens: NadA, fHbp and NHBA. In the vaccine, fHbp was fused with the protein GNA2091 and NHBA was fused with GNA1030. In this work, the authors found that three out of the five genes coding for vaccine included antigens were significantly up-regulated in human blood. In particular, fHbp, GNA1030 (NMB1030) and GNA2091 (NMB2091) were highly up-regulated while the other two genes encoding for NadA and NHBA, were not [55].

Other interesting applications of the microarray technology came from studies on the cholera-causing bacterium *Vibrio cholerae*. Here, genes differentially expressed during human infection were identified using bacteria directly isolated from diseased individuals. Comparison of the transcriptome of *V. cholerae* isolated from patient stool samples with that of bacteria grown *in vitro* greatly increased the understanding of the hyper infectious state that is seen after passage of bacteria through the human gastrointestinal tract [10].

Recently, pan-genome microarrays in addition to others techniques were used to study changes of genes modulated by pathogen-host interactions, which potentially encode putative vaccine targets, in Enterotoxigenic *E.coli* (ETEC) transcriptomes. Enterotoxigenic *Escherichia coli* are a leading cause of morbidity and mortality due to diarrheal illness in developing countries and there is currently no effective vaccine against these important pathogens. The reported studies suggest that pathogen-host interactions are finely orchestrated by ETEC and the elucidation of the molecular details of these interactions could highlight novel strategies for development of vaccines for these important pathogens [56]. The application of transcriptome analysis to vaccine development is expected to greatly advance with improving technologies for differentially extracting microbial RNA from tissues during *in vivo* experiments [57]. Further advances in next generation sequencing (NGS), which have enabled the rapid sequencing of cDNA and quantification of sequence reads [58, 59], should enable further transcriptome-based advances in vaccine development.

A valuable and useful tool for antigen discovery complementary to transcriptome investigations is the analysis of the proteome (the entire set of proteins expressed by a genome). The application of proteomic technologies has made significant improvement towards the understanding of the pathogenesis of many important bacterial pathogens. Advances in two-dimensional gel electrophoresis techniques and mass spectrometry analysis have enabled the separation and identification of the entire set of protein components of a cellular population. In proteomic analysis, the protein preparations are separated into their individual components using separation methods. Each protein is digested using a specific protease to generate peptide fragments and is then evaluated by molecular mass-by-mass spectrometry. The comparison between the digestion pattern thereby obtained and that predicted *in silico* enables the attribution of each peptide fragment to the cognate protein. Proteomics has been used to identify novel bacterial vaccine candidates against several human pathogens. Proteomic experiments conducted on bacterial proteomes have not only verified data obtained from genome sequencing and bioinformatic analysis but have also identified previously undescribed proteins, which could be potential new vaccine candidates.Examples include *H. pylori* [60–62], *Streptococcus pyogenes*, *Bacillus anthracis*, *Mycobacterium tuberculosis* and *Streptococcus agalactiae* [63]. Further, advanced proteomic technologies are currently employed to facilitate identification of changes in proteome accompanying tumorigenesis. Multiple tumor-related proteins currently available for clinical application are classified as tumor-specific antigens (TSAs) or tumor-associated antigens shared by tumor cells and normal cells (TAAs). Molecular characterization and identification of TSAs and TAAs rapidly evolved since van der Bruggen *et al.* [64] cloned the first human gene MAGE-1 encoding a melanoma-specific antigen. The HLA-A1 restricted nonapeptide encoded by human gene MAGE-1 was the first identified tumor-specific peptide [65]. This discovery ignited the development of peptide-based cancer vaccines. In addition, the combination of proteomics and serological analysis enabled the development of serological proteome analysis (SERPA), a technology that has been applied to screen and select new *in vivo* immunogens, potential vaccines candidates [66].

Another example of an integrated vaccine discovery approach comes from a study on *Streptococcus pyogenes* (Group A streptococcus, GAS). Three high-throughput technologies (mass spectrometry-based proteomics, protein array, and flow cytometry analysis) allowed the identification of 40 antigens and six of them were found to be protective in an *in vivo* mouse model [67].

10.4 Rational Design of Novel Vaccine Antigens Based on Integrated Genomic and Structural Information

Although by now the best and most effective vaccines have been empirically derived, the profound knowledge of the molecular details of the pathogen population structure, the pathogen-host interactions as well as the induced innate and adaptive immune-response have proven useful for the rational design of next generation vaccines. Indeed, central for the development of rationally designed vaccines and vaccines formulations are the following considerations: (i) the relevant antigen able to confer protection has to be identified; (ii) the vaccine candidate should exhibit specific immunity and broad coverage against a given pathogen; and (iii) a vaccine with a favorable safety profile able to induce strong and specific immunological response, and, if possible, immunological memory has to be formulated. Integrating insights from virology, molecular cell biology, immunology, genetics, epidemiology, biotechnology and clinical research are necessary to identify vaccine formulations adequately responding to these requirements. The analysis of whole genomes, proteomes, and transcriptomes of pathogens as well as deciphering which are the major components able to elicit a strong immune-response in humans (immunome) is now offering for the first time a powerful tool to identify and rationally design potential new antigen candidates.

Indeed, the experience accumulated in the last decade on different pathogens has demonstrated that antigens identified through RV approaches often present a certain degree of strain to strain sequence as well as expression variability, thus implying that adequate vaccine coverage can only be achieved by combining either different antigens (as in the 4CMenB) or different variants of the same antigen (in combination or as a chimeric protein). Furthermore, ongoing technological advances in protein biochemistry have allowed the development of high-throughput platforms for structural resolution. A better understanding of the molecular structure combined with the knowledge of the most immunogenic epitopes and variability information make ultimately possible the creation of broadly protective vaccine antigens by either combining in a fusion protein single protein domains or using a structure-based design of multiple immune-dominant antigenic surfaces on a single protein scaffold.

10.4.1 Design of Chimeric Vaccine Components

One of the most important features of an effective vaccine is the ability to cover a broad variety of strains with the lowest possible number of vaccine components. This last requirement is crucial to facilitate manufacturing and vaccine formulation. For this reason, vaccine antigens are chosen also based on sequence conservation across strains of a determined target species. However, many of the most immunogenic vaccine candidates are subjected to medium to high sequence variability since they are surface exposed and therefore display higher mutation rates in response to immune system pressure with respect to intracellular proteins.

One of the solutions successfully implemented to overcome sequence variability problems in designing viral vaccines, is the manufacture of chimeric virus-like particles (VLP) such as in the HPV (Human papilloma virus) vaccines. Indeed, the two HPV vaccines on the market contain chimeric VLP (self-assembling subunits of HPV viral structural antigens) of two or four most relevant HPV types, 16 and 18 or 6, 11, 16 and 18 ensuring about 70 or 90% of coverage, respectively [68, 69].

Another strategy explored for bacterial vaccines is the design of fusion proteins comprising epidemiologically relevant variants of the antigen of interest. This strategy has been successfully applied to generate a vaccine candidate, named RrgB321, comprising the three immunologically non-cross-reactive variants of RrgB, the major component of the pneumococcal pilus-1. The protein, conferring protection against *Streptococcus pneumoniae* infection in animal models of invasive disease is also able to elicit cross-reactive antibodies able to kill pneumococci in an opsonophagocytosis assay [70].

10.4.2 Structural Vaccinology Approach to Design Novel Antigens

Structural biology is increasingly being applied to vaccine development focusing on determining and understanding the structural basis of immunodominant and immunosilent antigens, on the one side to increase the understanding of previously identified vaccine candidates, and, on the other side, to enable the rational design of new peptides. Approximately 85,000 protein structures are currently available in the public databases and the current structural genomic projects are being driven with the main objectives of producing a representative set of protein folds that could be used as templates for comparative modeling purposes and providing insights into the function of still un-annotated protein sequences. To date, structural vaccinology antigen design approaches have been proposed for both viral and bacterial antigens.

Although the advent of HIV (Human Immunodeficiency Virus) medications has dramatically prolonged the lives of those infected with HIV, a vaccine targeting HIV is one of the most challenging projects as well as worldwide recognized need of the last decades. Structural and antigenic characterization of the HIV envelope has led to an increased understanding on how the viral spike (comprising the viral proteins gp120 and gp41) evades the host antibody response. Epitope mapping on the resolved structure of gp120 led to the identification of antibodies (b12, 2G12) mapping on conserved epitopes of gp120 but antibodies were also found to be induced against conformational epitopes existing only transiently during the contact with CD4. The stabilization of these latter epitopes might be obtained only through the antigen modification with the insertion of disulfide bonds or other crosslinks. However, although protective antibodies have been identified, HIV infection typically leads to an immune response against highly variable immune-dominant epitopes (loops of the envelope protein gp160), thus not providing protection against diverse strains. Taking advantage of the solved structure of gp160 Immune refocusing technology has been used to remove or dampen the immune-dominant variable gp160 epitopes in order to allow the host to respond to conserved epitopes that were previously subdominant. Engineered gp160 is able to induce neutralizing antibodies in a similar manner as wild type gp160 against homologous strains and has improved activity against heterologous strains [71–73]

Among the vaccine targets under study by means of a structural vaccinology approach is the respiratory syncytial virus (RSV). RSV is a difficult vaccine target and up to now there is no licensed RSV vaccine. The obvious target for RSV subunit vaccines is the fusion (F) glycoprotein. This trimeric protein forms a lollipop-shaped structure on the virion, with a C-terminal triple coiled-coil domain anchored in the virion membrane. During RSV entry into cells, the RSV F protein rearranges from the pre-fusion form through an intermediate extended structure to a post-fusion state. The pre-fusion form, displayed on infectious virions, flips to the post-fusion state when it is extracted from a membrane with detergent or when expressed as a non-membrane-anchored ectodomain. On the other hand, the post-fusion form can only be semipurified, as it has difficult biochemical characteristics. Structural insights from studies on the related human parainfluenza virus 3 (HPIV-3) and parainfluenza virus 5 (PIV-5) F glycoproteins have guided engineering of RSV F. Elimination of the transmembrane and cytoplasmic regions to produce an ectodomain and subsequent partial deletion of the fusion peptide from the RSV F ectodomain resulted in the formation of a hydrophilic, very stable, homogeneous post-fusion trimer. Interestingly, although structural modeling based on the known PIV-5 pre-fusion F structure predicted that the key epitope recognized by two potent neutralizing antibodies, palivizumab and motavizumab, were not accessible for antibody binding, immunization of animals with the engineered RSV post-fusion F elicits high titres of RSV-neutralizing antibodies, some of which competed for binding with motavizumab. Indeed, the crystal structure of the engineered RSV post-fusion

F antigen revealed that residues from the motavizumab (and palivizumab) epitope were out from their buried position in the intersubunit interface and onto the solvent-exposed surface and are well preserved between the pre- and post-fusion forms. The engineered post-fusion form of the F protein is now approaching clinical trials in a vaccine against RSV. [74–76].

A structural vaccinology approach has been also proposed to design a sub-type specific vaccine against the influenza A vaccine based on the hemagglutinin (HA) viral globular head (the main component of the current split influenza A virus vaccine is the viral HA). Crystallographic studies have shown that HA, which is able to induce a strong antibody-mediated immune response, forms a trimer embedded on the viral envelope surface, and each monomer consists of a globular head (HA1) and a "rod-like" stalk region (HA2), the latter being more conserved among different HA subtypes and being the primary target for universal vaccines. The rationally designed HA head based on the crystal structure of the 2009-pandemic influenza A (H1N1) virus HA as a model, was prepared by *in vitro* refolding in an *Escherichia coli* expression system, and maintained its intact structure allowing for the stimulation of a strong immune response [77].

A successful application of the design of a fully synthetic bacterial protein with multivalent protection activity is reported in Nuccitelli *et al.* The authors demonstrated that protective antibodies raised against the Group B Streptococcus (GBS) backbone subunit of type 2a pilus (BP-2a), present in six immunogenically different but structurally similar variants, were specifically directed towards one of the four domains that comprise the protein. Based on this outcome, they constructed a synthetic protein constituted by the protective domain of each one of the six variants and showed that the chimeric protein was able to protect mice against the challenge with all of the type 2a pilus-carrying strains [78].

Another example of rational design applied to a bacterial antigen specifically responding to the need to overcome sequence variability and combine in a single molecule the complete antigenic repertoire, has been the creation of a chimeric *Neisseria meningitides* fHBP protein carrying a conserved backbone and an engineered surface containing specificities for all three existing fHbp main variants. Analysis on the 3D structure of fHbp of the epitopes recognized by variant-specific monoclonal antibodies revealed that amino acids contributing to the immunogenicity of variant 1 or variants 2 and 3 were located in non-overlapping regions. To preserve folding, amino acid substitutions were introduced only for residues whose side chains were well exposed to the solvent, leaving the internal core of the protein unaltered.

Noteworthy, when injected into mice this engineered fHBP antigen was able to elicit cross-protective immunity against meningococcal strains carrying different fHBP alleles [53].

10.5 Conclusion and Future Perspectives

Vaccines are the most cost-effective method for disease prevention and represent one of the greatest successes in the history of medicine. Indeed, vaccines developed using the classical approaches described above have saved millions of human lives. However, mostly due to technical hurdles in vaccine

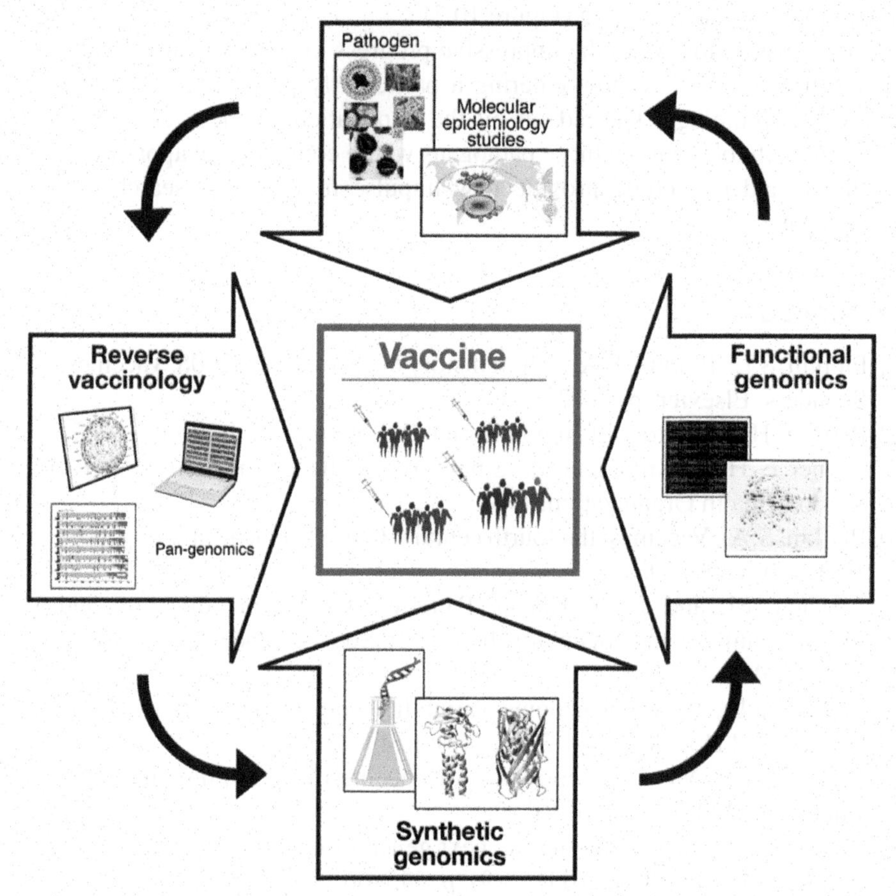

Figure 10.2 The integrated use of molecular studies, genomic, post-genomic and structural information lead to the identification and rational design of novel vaccine components.

development, many diseases can still not be controlled through vaccination and others are expected to emerge in the next future. In order to pursue the path of success, new approaches are needed to develop effective vaccines. Genomics represents the basis of the new revolution in vaccine discovery, in particular for the identification of new targets capable of defeating emerging and not yet targeted diseases. Besides, comparative genomics is also providing new insights into pathogens evolution and epidemiology, virulence mechanisms and host range specificity. In addition, through the advent and integration of new technologies such as next-generation sequencing, transcriptomics, proteomics, structural vaccinology and host genomics, it will be possible to revolutionize the approach of developing safe and effective vaccines and improve those already existing (Figure 10.2).

In conclusion, thanks to the impressive progress made in the last decade in biotechnology, we are now entering a particularly promising era in the history of vaccinology, spurred on by the application of the tools of the modern biotechnology to all those pathogens where conventional approaches have failed and that could provide new impulse in vaccine research and development.

References

[1] Plotkin, S.L., Plotkin, S.A, A short history of vaccination. 2008. Vaccines. Saunders, Elsevier: p. 1–17.
[2] Parisch, H., A History of Immunization. Edinburgh: Livingstone, 1965.
[3] Fenner F, H.D., Arita L et al., Smallpox and Its Eradication. Geneva: Worlod Health Organization, 1988.
[4] Plotkin, S.A., Vaccines: the fourth century. Clin Vaccine Immunol, 2009. 16(12): p. 1709–19.
[5] Lipsitch, M. and J.J. O'Hagan, Patterns of antigenic diversity and the mechanisms that maintain them. J R Soc Interface, 2007. 4(16): p. 787–802.
[6] Telford, J.L., Bacterial genome variability and its impact on vaccine design. Cell Host Microbe, 2008. 3(6): p. 408–16.
[7] Plotkin, S.A., Vaccines: correlates of vaccine-induced immunity. Clin Infect Dis, 2008. 47(3): p. 401–9.
[8] Bagnoli, F., et al., Designing the next generation of vaccines for global public health. OMICS, 2011. 15(9): p. 545–66.
[9] Bambini, S. and R. Rappuoli, The use of genomics in microbial vaccine development. Drug Discov Today, 2009. 14(5–6): p. 252–60.

[10] Rinaudo, C.D., et al., Vaccinology in the genome era. J Clin Invest, 2009. 119(9): p. 2515–25.
[11] Ehrenfeld, E., J. Modlin, and K. Chumakov, Future of polio vaccines. Expert Rev Vaccines, 2009. 8(7): p. 899–905.
[12] Plotkin, S.A., Six revolutions in vaccinology. Pediatr Infect Dis J, 2005. 24(1): p. 1–9.
[13] Jenner, E., An inquiry into the causes and effects of the variolae vaccinae, a disease discovered in some of the western counties of England, particularly Gloucestershire, and known by the name of the cow pox. Camac CNB, ed Classic of Medicine and Surgery. New York: Dover, 1959:213–240, 1798.
[14] Jenner, E., The Origin of the Vaccine Inoculation.London: Shury, 1801.
[15] Pasteur, L., De l'attenuation du virus du cholèra des poules. Acad Sci Paris, 1880(91): p. 673–680.
[16] Pasteur, L., Une statistique au sujet de la vaccination préventive contre le harbon, portant sur quatre vignt-cinq-mille animaux. CR Acad Si Paris, 1882(95): p. 1250–1252.
[17] Pasteur L, C.C.-E., Roux E., Sur la vaccination charbonneuse. CR Acad Si Paris, 1881(92): p. 1378–1383.
[18] Ferran, J., Sur la prophylaxie du choléera au moygen d'injections hypodermiques de cultures pures du bacille-virgule. CR Acad Si Paris, 1885(101): p. 147–149.
[19] Haffkine, W., Inoculation de vacins anticholériques à l'homme. CR Seanc. Soc Biol, 1892(44): p. 740–746.
[20] Calmette A. Guérin C, W.H.B.E.d.i.c.l.i.t., Essais d'immunisation contre l'infection tuberculeuse.Bull Acad Natl Med, 1924(91).
[21] Groschel, D., Hornick, RB., Who introduced typhoid vaccination: Almoth Wright or Richard Pfeiffer? Rev. Infect. Dis., 1891(3): p. 1251–1254.
[22] Sabin, A.B., Oral poliovirus vaccine. Recent results and recommendations for optimum use. R Soc Health J, 1962. 82: p. 51–9.
[23] Theiler, M. and H.H. Smith, The Effect of Prolonged Cultivation in Vitro Upon the Pathogenicity of Yellow Fever Virus. J Exp Med, 1937. 65(6): p. 767–86.
[24] Clark, H.F., et al., The new pentavalent rotavirus vaccine composed of bovine (strain WC3) -human rotavirus reassortants. Pediatr Infect Dis J, 2006. 25(7): p. 577–83.
[25] Hone, D.M., et al., Construction of genetically defined double aro mutants of Salmonella typhi. Vaccine, 1991. 9(11): p. 810–6.

[26] Coffin, J.M., Attenuation by a thousand cuts. N Engl J Med, 2008. 359(21): p. 2283–5.
[27] Daudel, D., G. Weidinger, and S. Spreng, Use of attenuated bacteria as delivery vectors for DNA vaccines. Expert Rev Vaccines, 2007. 6(1): p. 97–110.
[28] Chambers, T.J., et al., Yellow fever/Japanese encephalitis chimeric viruses: construction and biological properties. J Virol, 1999. 73(4): p. 3095–101.
[29] Fitzgerald, J.G., Diphtheria Toxoid as an Immunizing Agent. Can Med Assoc J, 1927. 17(5): p. 524–9.
[30] Park, W.H., et al., Observations on Diphtheria Toxoid as an Immunizing Agent. Am J Public Health (N Y), 1924. 14(12): p. 1047–9.
[31] Valenzuela, P., et al., Synthesis and assembly of hepatitis B virus surface antigen particles in yeast. Nature, 1982. 298(5872): p. 347–50.
[32] Pizza, M., et al., Identification of vaccine candidates against serogroup B meningococcus by whole-genome sequencing. Science, 2000. 287(5459): p. 1816–20.
[33] Kelly, D.F., E.R. Moxon, and A.J. Pollard, Haemophilus influenzae type b conjugate vaccines. Immunology, 2004. 113(2): p. 163–74.
[34] Hausdorff, W.P. and R. Dagan, Serotypes and pathogens in paediatric pneumonia. Vaccine, 2008. 26 Suppl 2: p. B19–23.
[35] Pichichero, M., et al., Comparative trial of the safety and immunogenicity of quadrivalent (A, C, Y, W-135) meningococcal polysaccharide-diphtheria conjugate vaccine versus quadrivalent polysaccharide vaccine in two- to ten-year-old children. Pediatr Infect Dis J, 2005. 24(1): p. 57–62.
[36] Kennedy, R.B. and G.A. Poland, The top five "game changers" in vaccinology: toward rational and directed vaccine development. OMICS, 2011. 15(9): p. 533–7.
[37] Rappuoli, R., Reverse vaccinology, a genome-based approach to vaccine development. Vaccine, 2001. 19(17–19): p. 2688–91.
[38] Mora, M., et al., Reverse vaccinology. Drug Discov Today, 2003. 8(10): p. 459–64.
[39] Thigpen, M.C., et al., Bacterial meningitis in the United States, 1998–2007. N Engl J Med, 2011. 364(21): p. 2016–25.
[40] Serruto, D., et al., The new multicomponent vaccine against meningococcal serogroup B, 4CMenB: immunological, functional and structural characterization of the antigens. Vaccine, 2012. 30 Suppl 2: p. B87–97.

[41] Serruto, D., et al., Neisseria meningitidis GNA2132, a heparin-binding protein that induces protective immunity in humans. Proc Natl Acad Sci U S A, 2010. 107(8): p. 3770–5.

[42] Donnelly, J., et al., Qualitative and quantitative assessment of meningococcal antigens to evaluate the potential strain coverage of protein-based vaccines. Proc Natl Acad Sci U S A, 2010. 107(45): p. 19490–5.

[43] Seib, K.L., X. Zhao, and R. Rappuoli, Developing vaccines in the era of genomics: a decade of reverse vaccinology. Clinical Microbiology and Infection, 2012. 18: p. 109–116.

[44] Maione, D., et al., Identification of a Universal Group B Streptococcus Vaccine by Multiple Genome Screen. Science, 2005. 309(5731): p. 148–150.

[45] Schuchat, A. and J.D. Wenger, Epidemiology of group B streptococcal disease. Risk factors, prevention strategies, and vaccine development. Epidemiol Rev, 1994. 16(2): p. 374–402.

[46] Ippolito, D.L., et al., Group B streptococcus serotype prevalence in reproductive-age women at a tertiary care military medical center relative to global serotype distribution. BMC Infect Dis, 2010. 10: p. 336.

[47] Tettelin, H., et al., Complete genome sequence and comparative genomic analysis of an emerging human pathogen, serotype V Streptococcus agalactiae. Proc Natl Acad Sci U S A, 2002. 99(19): p. 12391–6.

[48] Tettelin, H., et al., Genome analysis of multiple pathogenic isolates of Streptococcus agalactiae: implications for the microbial "pan-genome". Proc Natl Acad Sci U S A, 2005. 102(39): p. 13950–5.

[49] Maione, D., et al., Identification of a universal Group B streptococcus vaccine by multiple genome screen. Science, 2005. 309(5731): p. 148–50.

[50] Lauer, P., et al., Genome Analysis Reveals Pili in Group B Streptococcus. Science, 2005. 309(5731): p. 105-.

[51] Margarit, I., et al., Preventing bacterial infections with pilus-based vaccines: the group B streptococcus paradigm. J Infect Dis, 2009. 199(1): p. 108–15.

[52] Poland, G.A., et al., Vaccinomics and a new paradigm for the development of preventive vaccines against viral infections. OMICS, 2011. 15(9): p. 625–36.

[53] Scarselli, M., et al., The impact of genomics on vaccine design. Trends Biotechnol, 2005. 23(2): p. 84–91.

[54] Grifantini, R., et al., Previously unrecognized vaccine candidates against group B meningococcus identified by DNA microarrays. Nat Biotechnol, 2002. 20(9): p. 914–21.

[55] Echenique-Rivera, H., et al., Transcriptome analysis of Neisseria meningitidis in human whole blood and mutagenesis studies identify virulence factors involved in blood survival. PLoS Pathog, 2011. 7(5): p. e1002027.

[56] Kansal, R., et al., Transcriptional Modulation of Enterotoxigenic Escherichia coli Virulence Genes in Response to Epithelial Cell Interactions. Infect Immun, 2012.

[57] Hinton, J.C., et al., Benefits and pitfalls of using microarrays to monitor bacterial gene expression during infection. Curr Opin Microbiol, 2004. 7(3): p. 277–82.

[58] Nielsen, K.L., DeepSAGE: higher sensitivity and multiplexing of samples using a simpler experimental protocol. Methods Mol Biol, 2008. 387: p. 81–94.

[59] Morozova, O. and M.A. Marra, Applications of next-generation sequencing technologies in functional genomics. Genomics, 2008. 92(5): p. 255–64.

[60] Bernardini, G., et al., Helicobacter pylori: immunoproteomics related to different pathologies. Expert Rev Proteomics, 2007. 4(5): p. 679–89.

[61] Del Giudice, G., P. Malfertheiner, and R. Rappuoli, Development of vaccines against Helicobacter pylori. Expert Rev Vaccines, 2009. 8(8): p. 1037–49.

[62] Jagusztyn-Krynicka, E.K., et al., Proteomic technology in the design of new effective antibacterial vaccines. Expert Rev Proteomics, 2009. 6(3): p. 315–30.

[63] Doro, F., et al., Surfome analysis as a fast track to vaccine discovery: identification of a novel protective antigen for Group B Streptococcus hypervirulent strain COH1. Mol Cell Proteomics, 2009. 8(7): p. 1728–37.

[64] van der Bruggen, P., et al., A gene encoding an antigen recognized by cytolytic T lymphocytes on a human melanoma. Science, 1991. 254(5038): p. 1643–7.

[65] Traversari, C., et al., A nonapeptide encoded by human gene MAGE-1 is recognized on HLA-A1 by cytolytic T lymphocytes directed against tumor antigen MZ2-E. J Exp Med, 1992. 176(5): p. 1453–7.

[66] Suzuki, A., et al., Identification of melanoma antigens using a Serological Proteome Approach (SERPA). Cancer Genomics Proteomics, 2010. 7(1): p. 17–23.

[67] Bensi, G., et al., Multi high-throughput approach for highly selective identification of vaccine candidates: the Group A Streptococcus case. Mol Cell Proteomics, 2012. 11(6): p. M111 015693.

[68] Golden, O., et al., Protection of cattle against a natural infection of Fasciola hepatica by vaccination with recombinant cathepsin L1 (rFhCL1). Vaccine, 2010. 28(34): p. 5551–7.

[69] Munoz, N., et al., Against which human papillomavirus types shall we vaccinate and screen? The international perspective. Int J Cancer, 2004. 111(2): p. 278–85.

[70] Moschioni, M., et al., Immunization with the RrgB321 fusion protein protects mice against both high and low pilus-expressing Streptococcus pneumoniae populations. Vaccine, 2012. 30(7): p. 1349–56.

[71] Zhou, T., et al., Structural definition of a conserved neutralization epitope on HIV-1 gp120. Nature, 2007. 445(7129): p. 732–7.

[72] West, A.P., Jr., et al., Design and expression of a dimeric form of human immunodeficiency virus type 1 antibody 2G12 with increased neutralization potency. J Virol, 2009. 83(1): p. 98–104.

[73] Garrity, R.R., et al., Refocusing neutralizing antibody response by targeted dampening of an immunodominant epitope. J Immunol, 1997. 159(1): p. 279–89.

[74] Kim, H.W., et al., Respiratory syncytial virus disease in infants despite prior administration of antigenic inactivated vaccine. Am J Epidemiol, 1969. 89(4): p. 422–34.

[75] Swanson, K.A., et al., Structural basis for immunization with postfusion respiratory syncytial virus fusion F glycoprotein (RSV F) to elicit high neutralizing antibody titers. Proc Natl Acad Sci U S A, 2011. 108(23): p. 9619–24.

[76] Magro, M., et al., Neutralizing antibodies against the preactive form of respiratory syncytial virus fusion protein offer unique possibilities for clinical intervention. Proc Natl Acad Sci U S A, 2012. 109(8): p. 3089–94.

[77] Xuan, C., et al., Structural vaccinology: structure-based design of influenza A virus hemagglutinin subtype-specific subunit vaccines. Protein Cell, 2011. 2(12): p. 997–1005.

[78] Nuccitelli, A., et al., Structure-based approach to rationally design a chimeric protein for an effective vaccine against Group B Streptococcus infections. Proc Natl Acad Sci U S A, 2011. 108(25): p. 10278–83.

[79] Tettelin, H., et al., Complete genome sequence of Neisseria meningitidis serogroup B strain MC58. Science, 2000. 287(5459): p. 1809–15.
[80] Giuliani, M.M., et al., A universal vaccine for serogroup B meningococcus. Proc Natl Acad Sci U S A, 2006. 103(29): p. 10834–9.
[81] Ariel, N., et al., Search for potential vaccine candidate open reading frames in the Bacillus anthracis virulence plasmid pXO1: in silico and in vitro screening. Infect Immun, 2002. 70(12): p. 6817–27.
[82] Moriel, D.G., et al., Identification of protective and broadly conserved vaccine antigens from the genome of extraintestinal pathogenic Escherichia coli. Proc Natl Acad Sci U S A, 2010. 107(20): p. 9072–7.
[83] Gan, W., et al., Reverse vaccinology approach identify an Echinococcus granulosus tegumental membrane protein enolase as vaccine candidate. Parasitol Res, 2010. 106(4): p. 873–82.
[84] Song, Y., et al., A reverse vaccinology approach to swine dysentery vaccine development. Vet Microbiol, 2009. 137(1–2): p. 111–9.
[85] Liu, L., et al., Identification and experimental verification of protective antigens against Streptococcus suis serotype 2 based on genome sequence analysis. Curr Microbiol, 2009. 58(1): p. 11–7.
[86] Feng, X.G., et al., [Application of reverse vaccinology in Schistosoma vaccine development: advances and prospects]. Zhongguo Ji Sheng Chong Xue Yu Ji Sheng Chong Bing Za Zhi, 2007. 25(3): p. 237–47.
[87] Graham, S.P., et al., A novel strategy for the identification of antigens that are recognised by bovine MHC class I restricted cytotoxic T cells in a protozoan infection using reverse vaccinology. Immunome Res, 2007. 3: p. 2.
[88] Hatfaludi, T., et al., Screening of 71 P. multocida proteins for protective efficacy in a fowl cholera infection model and characterization of the protective antigen PlpE. PLoS One, 2012. 7(7): p. e39973.
[89] John, L., G.J. John, and T. Kholia, A reverse vaccinology approach for the identification of potential vaccine candidates from Leishmania spp. Appl Biochem Biotechnol, 2012. 167(5): p. 1340–50.
[90] Palmer, G.H., et al., Genome-wide screening and identification of antigens for rickettsial vaccine development. FEMS Immunol Med Microbiol, 2012. 64(1): p. 115–9.
[91] Manque, P.A., et al., Identification and immunological characterization of three potential vaccinogens against Cryptosporidium species. Clin Vaccine Immunol, 2011. 18(11): p. 1796–802.

[92] Ge, X., et al., Pooled protein immunization for identification of cell surface antigens in Streptococcus sanguinis. PLoS One, 2010. 5(7): p. e11666.

[93] He, Y., Analyses of Brucella Pathogenesis, Host Immunity, and Vaccine Targets using Systems Biology and Bioinformatics. Front Cell Infect Microbiol, 2012. 2: p. 2.

[94] Zhang, M., et al., Edwardsiella tarda flagellar protein FlgD: a protective immunogen against edwardsiellosis. Vaccine, 2012. 30(26): p. 3849–56.

[95] Liebenberg, J., et al., Identification of Ehrlichia ruminantium proteins that activate cellular immune responses using a reverse vaccinology strategy. Vet Immunol Immunopathol, 2012. 145(1–2): p. 340–9.

[96] Zhao, B.P., et al., In silico prediction of binding of promiscuous peptides to multiple MHC class-II molecules identifies the Th1 cell epitopes from secreted and transmembrane proteins of Schistosoma japonicum in BALB/c mice. Microbes Infect, 2011. 13(7): p. 709–19.

[97] Hong, M., et al., Identification of novel immunogenic proteins in pathogenic Haemophilus parasuis based on genome sequence analysis. Vet Microbiol, 2011. 148(1): p. 89–92.

11

Outer Membrane Proteins as Potential Candidate Vaccine Targets

Yuka Hara[1] and Sheila Nathan[2]

[1]Post-Doctoral Research Fellow
Universiti Kebangsaan Malaysia
[2]Professor
Faculty of Science and Technology
Universiti Kebangsaan Malaysia

11.1 Introduction

Infectious diseases caused by bacterial pathogens are responsible for a vast number of deaths globally. Although most pathogenic bacteria have previously been successfully eradicated with the use of narrow and broad range antibiotics, clinical management of these infections is increasingly hampered by the emergence of multi-drug-resistant strains. Therefore, prevention through vaccination currently represents the best course of action to combat these pathogens. However, immune escape and evasion by pathogens often render vaccine development difficult. Selecting the optimal antigen represents the cornerstone in vaccine design. With the advent of genomics, the traditional process of selecting candidate antigens one-by-one has been replaced by reverse vaccinology approaches. The use of strategies and approaches such as functional genomics, structural biology/vaccinology and proteomics has enabled the exploitation of a pathogen's genome to define promising antigens in relation to *in vivo* expressed genes and clonal variation [1–3]. Systems vaccinology adopts a global approach to correlate successful vaccination with the specific antigen-induced immune response, early vaccine efficacy and provide suggestions on mechanisms of successful immunogenicity. Functional genomics allows for the identification of specific molecular signatures of

Post-genomic Approaches in Drug and Vaccine Development, 277–322.

individual vaccines to be used as predictors of vaccination efficiency [4]. The selected choice of antigen should contain appropriate B-cell receptor epitopes and peptides that can be recognized by the T-cell receptor in a complex with MHC molecules.

Identification of specific antigens responsible for the ability of complex immunogens to induce protection is a major goal in the development of bacterial vaccines. Much of the investigation has focused on highly abundant and highly immunodominant outer membrane proteins. The structure and functional variety of outer membrane proteins on the cell surface are presented as potent immunogens based on the structure of the cell envelope. These proteins participate in the stabilization of the membrane structure and adhesion to other cells, are receptors for bacteriophages and play a key role in signal transduction, intracellular transport and energy transformation processes to ensure proper cell functioning. More importantly, these proteins have a protective function against immune reactions of the infected organism. This chapter aims to provide a non-exhaustive review on bacterial outer membrane proteins as candidate vaccines towards bacterial diseases.

11.1.1 Outer Membrane Proteins

The outer membrane (OM) of Gram-negative bacteria is a complex structure with a major role in bacterial adaptation to various external environments while passively and selectively controlling influx and efflux of important solutes, peptides or proteins, nucleic acids and other organic compounds such as lipids and polysaccharides. OM primarily consists of phospholipids, lipopolysaccharide (LPS), peptidoglycan and a group of outer membrane proteins (OMPs) that account for approximately 50% of the OM mass [5] (Figure 11.1). As a highly selective permeability barrier, the OM insulates the bacterial cell against a variety of potentially cytotoxic agents within the extracellular environment. The targeting and assembly of membrane proteins in general is a complex process that requires multiple folding factors. OM biogenesis is further complicated by the fact that these proteins must traverse one lipid bilayer and then integrate specifically into another; the cell must be capable of discriminating between inner and outer membrane proteins in addition to coordinating the assembly of each [6].

OMPs include integral membrane proteins as well as lipoproteins anchored to the outer membrane via attachment to N-terminal lipids. Gram-negative

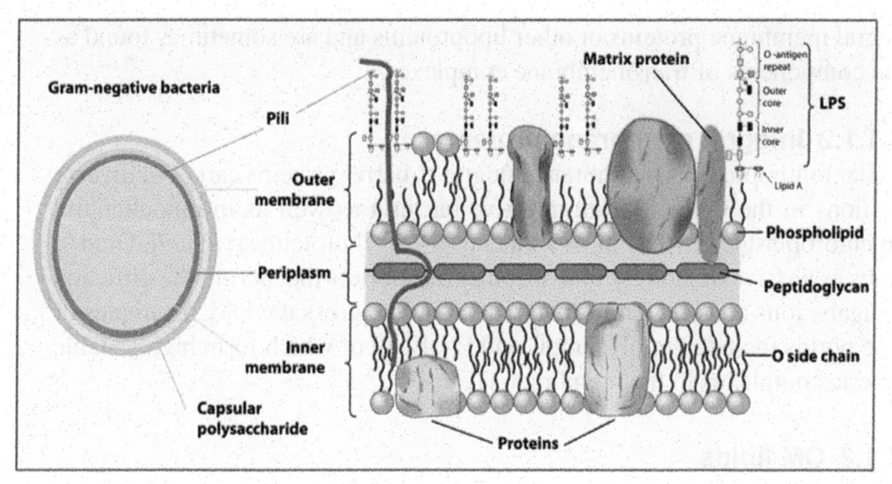

Figure 11.1 Schematic diagram of the outer membrane (OM), cytoplasmic or inner membrane (IM) and the intermediate periplasmic layer containing peptidoglycan. The IM consists of the two phospholipid leaflets and different lipoproteins. The OM consists of two leaflets, the inner leaflet being composed of one phospholipid layer and the outer leaflet of lipid-A, core polysaccharide and the O-antigen polysaccharide chains (LPS) projecting outward.

bacteria contain several classes of OMPs that form monomeric or trimeric barrels to facilitate the transport of nutrients into the cell [7]. Integral OMPs are characterized by β-barrel structures and are essential for maintaining integrity as well as serving as passive, relatively non-specific transporters that permit the diffusion of sugars, ions and small hydrophilic molecules [8]. In addition, they also participate in a wide array of cellular processes including energy dependent efflux, active transport, secretion and adhesion [9]. Most OMPs are surface-exposed and are important during the interaction of the bacteria with its mammalian host and defense mechanisms, bacteriophages and other microbes [7].

11.1.1.1 Lipoproteins

Bacterial lipoproteins are membrane attached via an N-terminal N-acryl-diacylglyceryl cysteine. They serve to spatially restrict their activity, presumably enabling them to function more efficiently. Over 90 lipoproteins have been reported in *Escherichia coli* [10] and they participate in a number of processes including envelop biogenesis [11], cell division [12], secretion [13] and signaling [14]. Lipoproteins may interact with periplasmic proteins, integral membrane proteins or other lipoproteins and are sometimes found as vital components of transmembrane complexes.

11.1.1.2 Integral membrane proteins

Similar to lipoproteins, membrane integral β-barrel proteins carry out diverse functions in the OM of Gram-negative bacteria as well as in mitochondria and chloroplast OMs. The most abundant β-barrel proteins in *E. coli*, OmpA, serves as passive, relatively non-specific transporters that permit the diffusion of sugars, ions and small hydrophilic molecules across the OM. Examples of these porins include OmpF, OmpC and PhoE, all of which form highly stable trimeric complexes in the membrane [15].

11.1.2 OM lipids

Most Gram-negative bacteria in OM are composed of two major lipidic species, phospholipids and LPS, which are asymmetrically distributed in the membrane. They have unique properties that substantially influence the fluidity and permeability of the OM.

11.1.2.1 Phospholipids

The major OM phospholipid of *E. coli* is phophatidylethanolamine (PEth), which accounts for roughly 70–80% of the total phospholipid content in the cell while phosphatidylglycerol (PG) comprises the remaining 20–30% [8]. Both PEth and PG are synthesized on the cytoplasmic side of the inner membrane (IM) [16]. In order to reach the OM, both phospholipids first need to rotate the membrane followed by transportation into the OM which is mediated by the LPS transporter MsbA [17]. Various α-helical membrane-spanning peptides also induce phospholipid translocation *in vitro* [18], suggesting that the membrane rotation and localization of the phospholipids is not totally transporter-dependent as typical α-helical membrane-spanning segments of some IM proteins can play a similar function. The next steps in phospholipid biogenesis, i.e. transport through the periplasm and incorporation into the inner leaflet of the OM, remain obscure. It has been reported that phospholipid shuttling does not require ATP or protein synthesis, but is heavily dependent on the proton motive force [19]. However, the mechanism that requires the proton motive force in the transport of phospholipids is not obvious. Numerous mechanisms of lipid transport [20] have been observed or proposed for eukaryotic systems, however, lipid trafficking employed by bacteria clearly requires further clarification.

11.1.2.2 Lipopolysaccharides (LPS)

The barrier property of the OM can largely be attributed to the presence and asymmetrical distribution of the well-known LPS. It consists of three distinctive fractions: a hydrophobic membrane anchor, lipid A, which secures LPS in the OM, oligosaccharide core region, and the distal O-antigen, which is not produced in derivatives of *E. coli* K12 [8]. The lipid A moiety of LPS is rather conserved among Gram-negative bacteria whereas the core oligosaccharide is much more variable between bacterial species [6]. The O-antigen, if present, is the most variable part of LPS and shows high degree of variation between different species of bacteria and is thus subjected to intense selection by the host immune system [21]. Synthesis of lipid A and addition of the core oligosaccharide occur at the cytoplasmic face of the IM, whereas O-antigen ligation only takes place after the nascent glycolipid is flipped to the periplasmic face of the IM [22].

It is clear that the lipid A-core moiety and O-antigen subunits, if present, are transported separately over the IM (Figure 11.2). The transport pathway consists of two major membrane proteins, the flippaseMsbA in the IM and

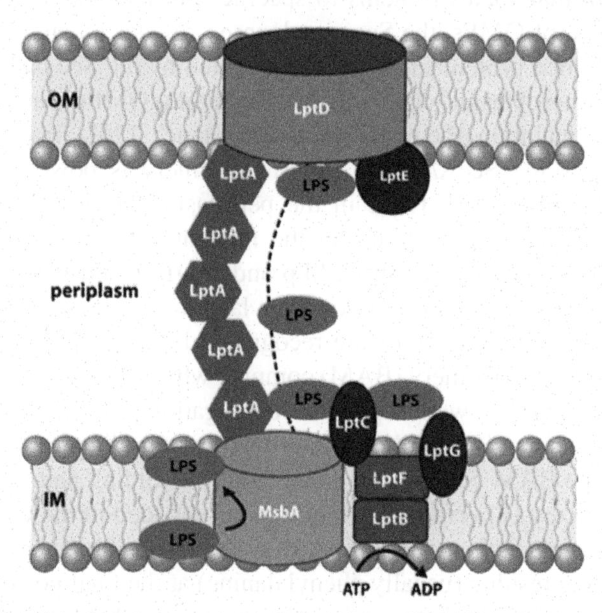

Figure 11.2 Model of LPS transport through the bacterial cell envelop. After synthesis at the inner leaflet IM, LPS is flipped to the outer leaflet of the IM. MsbA is required for the translocation of LPS but helical transmembrane domains have also been shown to translocate phospholipids. ABC transporter consisting of LptBFG recruits LPS to traverse the periplasmic space via LptA to the OM LptDE complex.

the β-barrel IMP/LptD in the OM [23]. Rough LPS (lipid A+ core) is first flipped towards the periplasmic face of the IM by the ATP-dependent translocaseMsbA, which can promote the translocation of phospholipids whereas O-antigen is translocated by another ATP binding cassette (ABC) transporter, Wzm/Wzt [24]. Once in the periplasmic side of the IM, ABC transporters consisting of two integral membrane proteins (LptF and LptG) as well as additional IM proteins located solely on the cytoplasmic face of the membrane (LptB) are involved in the recruitment of LPS to the OM [25]. Once in the periplasm, O-antigen is ligated onto rough LPS to form mature (smooth) LPS, then LPS moves from the IM LptBFG complex to the OM LptDE complex, traversing the periplasm via LptA [24, 25].

11.1.3 OMP Biogenesis and Assembly

Like all cellular proteins, OM proteins are synthesized within the cytoplasm. Since they are destined for the cell envelope, they are synthesized as nascent proteins with an N-terminal signal sequence [6]. The signal sequence is a targeting element that routes proteins to specific translocases embedded in the IM. In the case of OMPs, the Sec translocase provides a path across the IM and only unfolded protein can be accommodated by the Sec pathway [8] (Figure 11.3). The cytoplasmic chaperon SecB maintains OMPs in an unfolded conformation to maintain their transport-competence [26]. Translocation through the IM by the SecA/Y/E/G export machinery is powered by the essential ATPase, SecA [27]. Once in the periplasm, the signal sequence is processed by a signal peptidase and the nascent OMP associates with periplasmic chaperones such as SurA, Skp and DegP which guide nascent OMPs across the periplasm and peptidoglycan layer to the inner leaflet of the OM [27]. Here, the nascent OMPs are recognized by a complex known as the β-barrel assembly machinery (BAM) complex which folds and inserts the new OMPs into the OM as well as prevents aggregation [28]. Interestingly, a subset of OMPs also requires lipid synthesis for correct folding and assembly, suggesting that the BAM complex mediates multiple folding pathways [29]. The BAM recognizes a specific recognition motif encoded in the C-terminal β-strands of OMPs and this motif consists of several hydrophobic residues as well as an aromatic residue (usually phenylalanine) at the final position [30]. These targeting motifs differ between bacteria suggesting that OMP sorting is species-specific [31].

The *E. coli* BAM complex consists of five subunits named BamA (an OMP itself), BamB, BamC, BamD and BamE [8]. BamA and BamD are essential

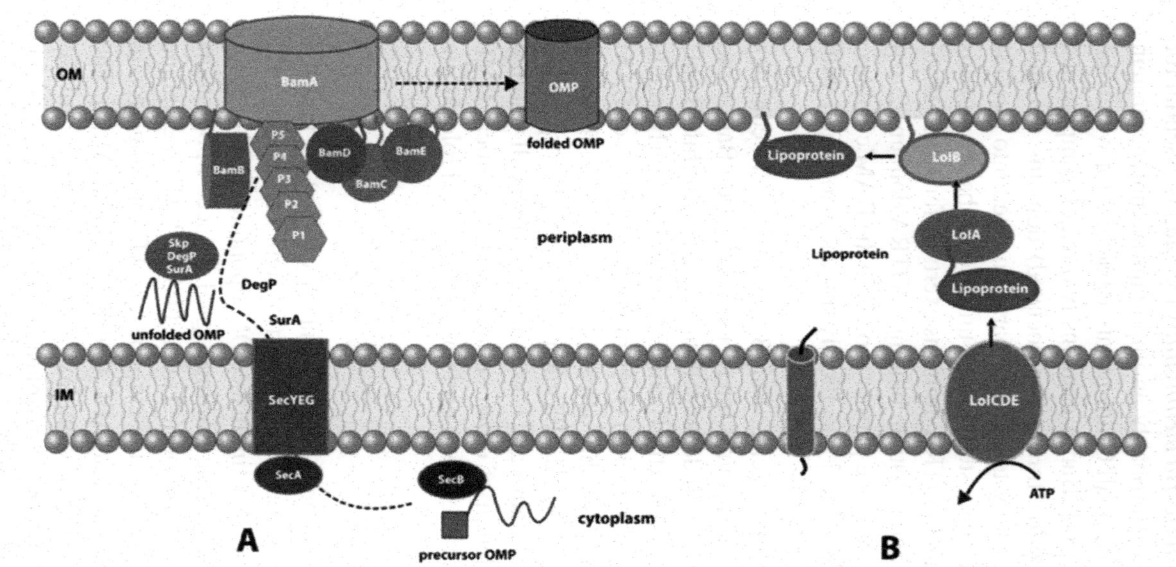

Figure 11.3 Biogenesis of bacterial (A) outer membrane proteins and (B) lipoproteins. (A) OMPs are synthesised in the cytoplasm as precursors with an N-terminal signal sequence. Next, they are transported across the inner membrane (IM) to the periplasm via the Sec translocase. The holding chaperones SecB and Skp prevent premature folding and aggregation of the OMPs in the cytoplasm and periplasm, respectively. The OMPs are then targeted to the Bam complex in the outer membrane (OM), which consists of the integral membrane protein BamA (Omp85) and four membrane-associated lipoproteins BamB–E. BamA consists of a C-terminal b-barrel embedded in the membrane and an N-terminal part consisting of five polypeptide-transport-associated (POTRA) domains (P1–P5) extending in the periplasm. The periplasmic chaperone SurA assists in the folding of the OMPs at the Bam complex that assembles them into the outer membrane. (B) OM lipoproteins interact with the ATP-binding cassette (ABC) transporter LolCDE, which hands them over to the periplasmic chaperone LolA. LolA escorts OM lipoproteins across the periplasm and delivers them to the OM-assembly site, the OM lipoprotein LolB. LolB helps lipoprotein to dissociate from LolA complex and transfer to the inner leaflet of OM.

for cell viability and OMP biogenesis, while BamB, BamC, and BamE serve regulatory roles [32]. BamA, also known as Omp85, is the first component required for OMP insertion in *Neiserria meningitidis* [33] and is essential for bacteria viability. Omp85 is most probably an important membrane protein as its gene homologues are found in all Gram-negative bacteria for which the genome sequence is available [34]. BamA contains two major components; a set of five POTRA (polypeptide transport-associated) domains oriented towards the periplasm and a carboxy-terminal β-barrel inserted into the OM [35]. These repeat domains are suggested to bind unfolded β-barrel proteins [35] and the deletion of individual POTRA domains exerts different effects to stability of the BAM complex and substrate binding. POTRA domains 2 to 4 are important in the interaction with the partner protein in the BAM complex, BamB, whereas deletion of POTRA 5 specifically affects association with other lipoprotein partners [6].

The four BAM accessory lipoproteins, BamB–E, form a tight complex with *E. coli* BamA [36]. BamA and BamB interact directly, as do BamA and BamD. Both of these interactions are independent of one another, while BamC and BamE form contact with BamD and the BamBE sub-complex can be stably isolated *in vitro* [37]. They all have roles in OMP biogenesis, as loss of BAM leads to varying degrees of OMP assembly defects, including accumulation of unfolded protein substrates and reduced LPS and phospholipid incorporation into the OM [38]. However, only BamA and BamD are crucial for cell viability and OMP biogenesis and depletion of either subunit causes identical OMP assembly defects, indicating that these two proteins play a central role during OMP assembly [38].

11.1.4 Lipoprotein Transport

Some lipoproteins are retained in the IM but the majority (90%) are targeted to the OM by a dedicated periplasmic transport system called Lol [39] (Figure 11.3). Lol transport system distinguishes OM lipoproteins from those that function at the IM based on the sorting signal, hence, lipoproteins lacking an IM insertion signal are transported to the OM [40]. The membrane associated heterodimer LolCE recruits the cytoplasmic ATP binding protein (LolD) to from a complex that acts as an ABC transporter in the IM [6]. The binding of lipoproteins to this LolCDE complex followed by ATP hydrolysis leads to the transfer of lipoproteins to the periplasmic chaperone LolA, which subsequently forms a complex with lipoproteins and traverses the periplasmic space [41]. The LolA-lipoprotein complex then interacts with an OM receptor, LolB, itself an OM lipoprotein. Once the LolA-lipoprotein comes in contact with LolB, it dissociates and the lipoprotein is transferred to the inner leaflet of the OM via LolB [39]. Thus, the localization of LolB at the OM and LolCDE at the IM ensures that lipoprotein transport is unidirectional and irreversible.

11.2 Evaluation of Outer Membrane Proteins as Vaccine Candidates

11.2.1 OMP in Bacterial Pathogenesis and Virulence

OmpA is a multifunctional major OMP of *E. coli* and other *Enterobacteriaceae* and is highly conserved among Gram-negative bacteria. In pathogenic bacteria, OmpA-like proteins are surface-exposed and exist at high copy numbers. These proteins actively participate in a variety of pathogenic roles, including adhesion to mucosal surfaces, serum resistance, invasion, antimicrobial peptide resistance and host cell activation [42]. In addition, the surface-exposed Omp-A like proteins can activate both innate and adaptive immune mechanisms in the infected host. These inherent characteristics of surface exposure and high copy number make Omp-A like proteins ideal candidates for vaccine development.

The importance of OmpA in pathogenesis is well established using a number of model systems. The first report describing a role for OmpA in virulence was of a neonatal rat model of *E. coli* K1 meningitis. *E. coli* K1 lacking OmpA expression was inefficient in inducing bacteraemia in neonatal rats and embryonated chick embryos and was more sensitive to serum complement-mediated bactericidal effects [43]. OmpA is also capable of interacting with host immune cells to manipulate the host immune system in order to avoid killing by polymorphonuclear leukocytes (PMNs) [44]. In addition, OmpA plays a crucial role in invading the central nervous system capillary endothelium where OmpA interacts with its receptor, Ecgp96, and upon binding, a cascade of cellular signalling occurs resulting in the successful invasion of the bacterium into the blood brain barrier [45].

Pseudomonas aeruginosa(*P. aeruginosa*) OprF is widely considered an ortholog of OmpA with significant amino acid similarity in both OprF and OmpA C-terminal domains [46]. A recent report demonstrated the involvement of OprF in *P. aeruginosa* virulence, at least partly through modulation of the quorum-sensing network [47]. The absence of OprF results in impaired adhesion to animal cells, secretion of ExoT and ExoS toxins through the type III secretion system (T3SS) and production of the quorum-sensing-dependent virulence factors pyocyanin, elastase, lectin PA-1L and exotoxin A.

OmpA of pathogenic strains of *E. coli* [48] and *Klebsiella pneumoniae* [49] mediates adhesion to leukocytes and macrophages whilst *Pasteurella multocida* OmpA facilitates binding to extracellular matrix proteins such as heparin and fibronectin [50]. In addition, *K. pneumoniae* OmpA (KpOmpA) triggers cytokine production, dendritic cell maturation, modulates the inflammatory response and is responsible for immune evasion [49]. These strategies utilized by KpOmpA may facilitate pathogen survival in the hostile environment of the lung.

11.2.2 Vaccines Against Intracellular and Extracellular Bacteria – Induction of Different Types of Protective Immunity

In general, the success in developing vaccines against extracellular bacteria can be attributed to several factors. In many cases of extracellular pathogens, protective immunity is often elicited towards either a single antigen or a limited number of antigens which can be isolated from bacteria *in vitro*. Thus, the protective immunity is predominantly dependent on the induction of antibody responses to these antigens or antibodies having opsonizing or/and bactericidal activities [51, 52].

In the case of intracellular pathogens, the nature of protective immunity is much less clearly defined due to the complex nature of their life cycle to avoid the host immune defense. Intracellular pathogens usually cause disease after invading and growing in the host cells and utilize an array of virulence factors acting *in concert*. Thus, to identify a single gene product with a predominant role in pathogenesis as a potential vaccine candidate is not feasible [53]. Although antibodies against defined toxins and virulence factors play a role in protection against intracellular pathogens, the level of such protection can be limited. For example, antibodies can protect low but not high virulence strains of *Francisella tularensis* [54] or can provide protection against low challenge doses of *Burkholderia pseudomallei* given by the intraperitoneal route, but not by the aerosol route of infection [55, 56]. Thus, although antibodies might protect against intracellular pathogens, the consensus is that cellular immune responses, particularly those involving $CD4^+$ and $CD8^+$ cells, are crucial for protective immunity against intracellular pathogens [57]. These typical T-cell types play distinct and complementary roles in protective immunity. For example, $CD4^+$ T-cells (T helper cells) produce a range of inflammatory cytokines such as interferon (IFN)-γ and tumour necrosis factor (TNF)-α that orchestrate the immune response and may activate host cells such as macrophages to kill the pathogen while $CD8^+$ T-cells (cytotoxic T-cells) are able to directly kill infected cells [57]. IFN-γ and TNF-α are important mediators of protection against a number of Gram negative bacterial infections such as ehrlichioses, *Brucella* and *P. aeruginosa* infections [58–60]. Similarly, similar classes of bacterial OMPs may invoke different immune responses during infection by different pathogens. For example, administration of LPS protected mice against challenge by *K. pneumoniae*[61], *P. aeruginosa* and *B. melitensis.* However, LPS vaccines are unable to induce a CMI response and fail to protect against intracellular pathogens such as *F. tularensis* [54].

A new subset of CD4$^+$ cells, Th17, which secretes interleukin 17 (IL-17) has been reported to play a role in protective immunity against extracellular pathogens as well as enhancing host defense against intracellular pathogens [62]. IL-17 reduced bacterial burden in the host as well as increased host resistance to *K. penumoniae* and *Porphyromonas gingivalis* [63–66]. IL-17 most likely protects against extracellular pathogens by coordinating early recruitment of neutrophils to the local site of infection and may increase neutrophil bactericidal activity [67, 68]. In contrast, IL-17-mediated protective effect against intracellular pathogens is modest. Both IL-17 deficient mice and control mice survived infection with sub-lethal doses of *Salmonella enteric* although consistently higher bacterial burden was observed in the organs of IL-17 deficient mice compared to control mice [69].

11.2.3 Leptospirosis

Leptospirosis is a widespread disease caused by the spirochete bacterium *Leptospira* and affects almost all mammals [70–72]. Infection is facilitated through the mucosa or open wounds and after gaining entry into the host, the bacterium causes serious disease [73]. The symptoms of leptospirosis are extremely diverse, from meningitis, pneumonitis or hepatitis to death [74–76]. Pathogenic *leptospires* are able to translocate through cell monolayers and rapidly gain access to the bloodstream allowing them to disseminate to multiple organs [77].

The current available leptospiral vaccines, either live attenuated whole cell vaccines or vaccines using LPS components, only provide short-term immunity and are not effective at preventing disease [78]. Leptospiral OMPs are an important alternative and in contrast to LPS, leptospiral OMPs are relatively well conserved and those that are surface-exposed represent potential targets for immune-mediated defence mechanisms [79]. Such OMPs include lipoproteins (LipL32, LipL21 and LipL41 [80]), the leptospiral immunoglobulin-like proteins LigA and B [81], and porins such as OmpL1 [82]. OmpL1 and

LipL41 are antigenically conserved among pathogenic *Leptospira* species; their promise as vaccine candidates is enhanced by the finding that OmpL1 and LipL41 are expressed during infection of the mammalian host [83]. Recombinant outer membrane proteins OmpL1 and LipL41 used as subunit vaccines demonstrated synergistic protective effects in a hamster model of leptospirosis [84]. Recently, Omp1 and LipL41 conserved immunodominant T- and B-cell epitopes were identified and shown to collaborate in the production of antibodies against *Leptospira* and induction of lymphocyte differentiation [85].

11.2.4 *Porphyromonas gingivalis*

Chronic periodontitis is a common oral inflammatory disease that causes breakdown of periodontal tissue, including the resorption of alveolar bone, leading to tooth loss [86]. Chronic periodontitis has also been reported to complicate systemic conditions such as cardiovascular diseases and presents a higher risk of preterm low birth-weight babies [87]. Hence, the prevention of periodontitis is important for both oral and systemic health. *P. gingivalis*, a Gram-negative anaerobic bacterium, is one of the major causative pathogens of chronic periodontitis.

A 40 kDa *P. gingivalis* OMP (40-kDa OMP) is a key virulence factor involved in the coaggregation activity of *P. gingivalis* [88]. Furthermore, this OMP resides both on the cell surface and in extracellular vesicles and is found in many strains of *P. gingivalis* [88]. Previous studies demonstrated that monoclonal antibodies against the 40-kDa OMP provide an inhibitory effect on the coaggregation activity and possess complement-mediated bactericidal activity against *P. gingivalis* [88, 89]. Moreover, human monoclonal antibodies against the 40-kDa OMP provide protection against bone loss caused by *P. gingivalis* in rats [90]. Nasal administration of this protein in a non-toxic chimeric mutant Cholera toxin (CT) adjuvant elicited long-term OMP specific secretory IgA in saliva and serum IgG in immunized mice [91]. Furthermore, immunized mice were significantly protected against alveolar bone loss caused by oral infection with *P. gingivalis* [91]. However, whilst genetically detoxified CT mutants appear to be non-toxic in animal models, safety evaluation in humans is warranted. An alternative synthetic adjuvant, oligodeoxynucleotides containing unmethylated CpGd dinucleotides (CpG ODN) conjugated with the 40-kDa OMP, is effective in inducing mucosal 40-kDa OMP-specific IgA compared to CT-conjugated vaccines when administered orally [92]. This immunization protocol elicits T helper 1 and 2 cytokines for enhanced protective immunity in mice as well as protective immunity against alveolar bone loss caused by a *P. gingivalis* infection [92].

11.2.5 *Pseudomonas aeruginosa*

P. aeruginosa, an environmental Gram-negative microorganism, is one of the most important opportunistic pathogens in hospital-acquired infections [93]. Pulmonary infection with *P. aeruginosa* is a frequent problem for patients with cystic fibrosis (CF), immunodeficiency or bronchiectasis [94]. Nosocomial *P. aeruginosa* strains are characterized by an intrinsic resistance to various antimicrobial agents and common antibiotic therapies. The low permeability of the major outer membrane porins and the presence of multiple drug efflux pumps are factors contributing to the mechanisms of drug resistance [95]. This high antibiotic resistance often leads to therapeutic complications and is associated with treatment failure and death. Since there is no current vaccine available against *P. aeruginosa,* the development of a vaccine is necessary, particularly in CF patients prior to their lungs being colonized by this pathogen.

The *P. aeruginosa* OprF and OprI, OMPs that are surface-exposed and antigenically conserved in various strains of *P. aeruginosa*, are promising antigens as vaccines [96]. As the organism is prone to phenotypic variation due to changing environmental conditions such as in the airways of CF patients [97], highly conserved antigens such as OMP represent ideal vaccine candidates. OprF and OprI cross-reactivity against all serotypes of *P. aeruginosa* makes them promising candidate vaccines for further evaluation. OprF appears to be a key player in the adaptation of *P. aeruginosa* to the host immune defense. *P. aeruginosa* senses activation of the immune response by binding to IFN-γ through OprF, which leads to the development of a more virulent phenotype [98]. Ding *et al.* demonstrated that OprF-OprI vaccinated serum from human volunteers inhibited *P. aeruginosa* binding to IFN-γ, suggesting an additional effector mechanism for OprF containing vaccines [99]. OprI adheres to the mucosal surface of the respiratory and intestinal tracts and may act as a mucosal carrier to facilitate antigen delivery to antigen-presenting cells [100].

The *P. aeruginosa* recombinant fusion OprF and OprI proteins were shown to afford protection in various animal models and to be safe and immunogenic following several clinical trials [101–103]. Nasal vaccination of this recombinant vaccine induces a long lasting antibody response in serum as well as in the lower airways in immunized healthy individuals [103]. The observation that antibodies were detectable at high levels in both compartments after 1 year suggests a lasting immunogenicity and a high responder rate for this vaccination strategy [103]. Intranasal immunization followed by systemic boost induced higher levels of systemic anti-*P. aeruginosa* IgG compared with intranasal priming and intranasal boosts [104]. Subsequent phase I/II clinical trials in patients with chronic pulmonary disease demonstrated that nasal OprF-OprI vaccination followed by systemic boost was well tolerated and induced airway mucosal *P. aeruginosa*-specific IgG and IgA up to 6 months in more than 90% of those vaccinated [105].

A multivalent vaccine containing type A and B flagellin, OprI and OprF was developed and its efficacy was evaluated in mice. Flagellin serves as a potent adjuvant and generates high titer antigen-specific IgG that exhibits a high degree of functional activity and a robust memory response in immunized mice compared to control [106]. Furthermore, the vaccine promotes enhanced clearance of non-mucoid *P. aeruginosa* without secondary tissue damage [106], an important implication for CF patients as the initial *P. aeruginosa* infection in CF patients is mediated by non-mucoid bacteria [107]. In addition, immunized mice had significantly less inflammation and lung damage throughout the infection, demonstrating the substantial potential of this multivalent vaccine against *P. aeruginosa* [106]. An Opr vaccine (CFC-101), composed of Opr extracts from four *P. aeruginosa* strains, induced Opr-specific antibody titres with opsonophagocytic activity and increased *P. aeruginosa* blood clearance rates in healthy volunteers and burn patients [108, 109]. Replication-deficient adenoviral (Ad) vectors are attractive platforms as vaccines due to their immunogenicity and their function as adjuvants [110]. AdC7-based vector expressing OprF induces long-term anti-*P. aeruginosa* systemic, mucosal and protective immunity following systemic and airway mucosal immunization in mice [59]. Furthermore, mucosal immunization with AdC7-OprF led to sustained high levels of mucosal anti-Opr FIgG, IgA and lung T-cell immunity [59].

While many experimental vaccines have been tested in preclinical trials as well as in the clinical phase, it is difficult to study CF patients as improved antibiotic therapy impaired a proper evaluation of the vaccine's efficacy [95]. New promising perspectives for the development of vaccination strategies against various types of pathogens include the use of antigen-pulsed dendritic cells (DCs) as biological immunizing agents [111]. DCs are specialized antigen-presenting cells that play a role in inducing adaptive immune responses to foreign antigens and in maintaining T-cell tolerance to self [111]. Recombinant OprF-pulsed DCs were evaluated as a potential vaccine against *P. aeruginosa* in mice. The cells conferred protection against both the conventional *P. aeruginosa* O1 strain as well as a more virulent mucoid strain by inducing Th1 resistance to the infection [112]. OMPs of *P. aeruginosa* continue to be one of the most promising vaccine antigens to provide protection against a broad range of *P. aeruginosa* strains and controlled clinical trials are required for further validation of the efficacy of OMP vaccines.

11.2.6 *Burkholderia pseudomallei*

Melioidosis is a disease prevalent in subtropical and tropical climates [113]. In northeast Thailand, melioidosis is the most common cause of severe community-acquired septicaemia with a documented prevalence of 4.4 cases per 100,000 [113]. The reported mortality rate ranges from 19% in Australia to 50% in Northeast Thailand [114]. The worldwide incidence of melioidosis appears to be increasing, partly as a result of increased travel to endemic areas and partly because of increased reporting as a result of improved diagnosis [115].

Melioidosis is caused by the Gram-negative bacterium *Burkholderia pseudomallei* which is inherently resistant to many antibiotics, including β-lactams, amino-glycosides, macrolides and polymyxins [116]. Therefore, treatment may involve intensive and prolonged regimens that are often unsuccessful and may lead to recurrence; in fact, 75% of these cases are a consequence of relapse rather than re-infection [113]. There is currently no approved vaccine against melioidosis. A hurdle in the production of vaccines for melioidosis is the need to protect against both acute and chronic infection.

Live attenuated mutants of *B. pseudomallei* have been found to be the most effective immunogens in mice [117, 118], although it is unlikely that such mutants would be appropriate for a vaccine against melioidosis in humans. LPS and capsular polysaccharide are the main surface-associated antigens of *B. pseudomallei.*When tested as potential vaccines, both proteins resulted in a strong IgM and IgG_3 or IgG_{2b} response [56]. The combination of both antigens provided better protection [119] although the antigen combination failed to provide sterilizing immunity. Nevertheless, IgG antibodies to LPS were significantly higher in patients who survived melioidosis compared to those who succumbed to the infection [120]. The ability of anti-LPS and anti-capsular polysaccharide (CPS) antibodies to provide passive protection from challenge with *Burkholderia* has also been evaluated [55]; however, whilst mice treated with the mAbs against these proteins were protected against an intraperitoneal challenge, the mAbs were less protective at higher challenge doses or when the bacteria were introduced through an aerosol challenge.

In two subsequent studies, three OMPs were evaluated as potential candidates for melioidosis: BPSL2522, BPSL2765 and BPSL2151 [121, 122]. BPSL 2522 and BPSL2765 were identified as OMPs by evaluating the *B. pseudomallei* whole genome sequence for potential OMP CDSs whilst BPSL2151 was identified through screening of a small insert expression library with melioidosis patient sera. Immunization with BPSL2522 or BPSL2765 protected 50% of mice challenged over a period of 21 days while 70% of BPSL2151 immunized mice survived 14 days post-challenge. However, all surviving mice had evidence of bacterial colonization in multiple organs, indicating the failure of immunization with these OMPs to provide sterile immunity [121, 122]. It is clear that a single component vaccine, at least delivered as purified protein, will not provide complete protection against this disease. Development of an OMP-based vaccine which delays the onset of bacteremia could extend the window of opportunity for antibiotic treatment. In contrast, almost all the candidate vaccines evaluated to date deal with challenge models of acute infection whilst challenge models that mimic natural routes of infection leading to a chronic infection have yet to be fully developed to assist in the discovery of OMP-based vaccines to address chronic infections of *B. pseudomallei.*

11.2.7 *Neiserria meningitidis*

N. meningitidis is a Gram-negative, oxidase-positive, aerobic diplococcus and is a common harmless commensal inhabitant of the mucosal membranes of the human nasopharynx [123]. Invasive infections caused by *N. meningitidis* are a serious public health problem worldwide and have a heavy economic impact, not only in epidemic areas, but also in areas where sporadic forms occur and it is estimated that approximately 500,000 cases and 50,000 to 135,000 deaths occur every year worldwide [124].

Among the 13 distinct *N. meningitidis* serogroups that have been defined on the basis of their CPS immunochemistry, groups A, B, C, W135 and Y are responsible for over 90% of severe meningitis and septicaemia [123]. Group B *N. meningitidis* (MenB) which causes about 50% of meningococcal disease cases worldwide is the only serogroup whose infection cannot be prevented by CPS-based vaccines because group B CPS is poorly immunogenic in humans [125]. Furthermore, the CPS structure of MenB mimics the sialylated structure in human neutral tissue posing potential safety concerns for a vaccine formulation containing these polymers [126]. Therefore, most of the vaccine research for MenB has focused on outer membrane vesicles (OMV)–based vaccines. OMV–based vaccine candidates containing a predominantly single sero-subtype OMP antigen, PorA, which is expressed by almost all meningococci or multiple sero-subtype antigens, have been evaluated extensively [127]. PorA-based vaccines have shown promise due to their ability to induce sero-subtype specific and low levels of cross-reactive bactericidal antibodies in humans [128]. The efficacy of ProA vaccines was extensively evaluated in several countries in response to national outbreaks and the vaccine successfully controlled clonal MenB epidemics in Cuba [129] and Brazil [130]. Since these vaccines are based on a single meningococcal isolate, they can provide only partial protection against virulent heterologous meningococci. Furthermore, it is clear that OMV vaccines can be useful in controlling localized epidemics through the administration of a specific "tailor-made" vaccine MeNZB [131] and the prospective study to assess the effectiveness of MeNZB showed an estimated disease rate of 3.7 times higher in the unvaccinated group compared to the vaccinated group with a vaccine effectiveness of 73% [132]. Combined evidence from Phase I and Phase II clinical trials together with an observational cohort study on the effectiveness of MeNZB in infants in New Zealand supported the protective efficacy of MeNZB [133].

One limitation of conventional OMV vaccines is that serum bactericidal antibody responses in children are largely directed against surface-accessible loops on PorA [134], which is antigenically variable. To broaden protection, several multi-valent OMV vaccines were developed. A hexavalent PorA-based OMV, Hexamen™ developed in the Netherlands was prepared from two *N. meningitidis* strains that were each engineered to express three different PorA proteins [135, 136]. The vaccine has been proven to be effective after a four-dose regimen in children [137]. In order to provide even broader protection, a nanovalent PorA OMV vaccine, Nanomen™ was developed to cover the nine most frequently occurring subtypes. This nanovalent vaccine was immunogenic in mice and rabbits and has the potential to offer protection against 80% of disease across Europe [138, 139]. Another OMV-based vaccine was composed of six PorA variants, three PorB variants, two factor H binding protein (fhbp) variants, OpcA (Outer membrane adhesion protein) and NadA (an antigen which promotes adhesion to and invasion of epithelial cells) resulting in a 13-valent protein vaccine. This vaccine has been used in phase I trials and elicits a bactericidal response in humans [140].

11.2.8 *Haemophilus influenzae*

H. influenzae is a Gram-negative bacterium that commonly colonizes the human respiratory tract. This bacterium can be differentiated into typeable and non-typeable strains based on the presence or absence of a polysaccharide capsule [141]. Among encapsulated strains, serotype b (Hib) is the most commonly associated strain with human diseases but since the introduction of the Hib vaccine in the late 1980s, invasive Hib diseases have dramatically reduced in many developed countries [142]. In contrast, non-encapsulated and therefore non-typeable *H. influenzae* (NTHi) is a common cause of respiratory tract infections and otitis media in humans [143]. At present, there is no vaccine available for prevention of infection by NTHi. As NTHi strains lack a conserved capsule, vaccine development has focused on highly conserved OMPs like Omp26, P1, P2, P6 and Protein D which have been shown to elicit immune responses in animal models [143].

P6 is one of the vaccine candidates for NTHi and its most attractive feature is the conservation among all *H. influenzae* strains studied [144]. Nasal immunization is an effective therapeutic regimen for prevention of upper respiratory infection. When a plasmid DNA vaccine encoding P6 protein was used to immunize mice, it induced NTHi specific long-term immune responses and enhanced bacteria clearance [145]. In addition to antigen-specific secretory-IgA and IgG, specific Th17 cells induced by nasal vaccination contributed to protection against NTHi [146]. Another OMP, Omp26, is present on all NTHi strains tested and is highly conserved between strains [147]. Rats immunized with Omp26 enhanced the clearance of both homologous and heterologous strains of NTHi post-pulmonary challenge and more recently, Omp26 has been shown to be effective against NTHi otitis media and nasopharyngeal carriage in the chinchilla model [147, 148]. The bacterial ghost platform technology has been proposed as an advanced delivery system for potential vaccine candidates [149] and recently, the *E. coli* ghost harboring Omp26 was evaluated as a candidate vaccine using different mucosal immunization regimens in rats [150]. Animals received the gut/lung regime followed by recombinant Omp26 boost demonstrated enhanced pulmonary clearance of NTHi, high levels of Omp26 specific antibodies and significant cellular immune response. This confirmed the capacity of Omp26 to induce specific and protective immune response in a rodent model of acute lung infection [150].

Chang *et al.* have shown that P6 is not conserved in all NTHi strains and may not be surface–exposed [151]. This variation in P6 protein structure was observed in only 10% of NTHi studied and though it does not eliminate this antigen as a potential vaccine candidate, focusing on one antigen may not be the best approach to develop an effective NTHi vaccine. Instead, a combination of multiple heterogeneous antigens may increase the efficacy of a vaccine against heterologous bacterial strains. In this aspect, OMVs could be considered as a new promising vaccine candidate. OMVs of *H. influenzae* mainly consist of OMPs including P1, P2, P5 and P6 among others as well as phospholipids and LPS [152]. Intranasal immunization in mice induced strong humoral and mucosal immune responses [152]. The induced immune response conferred not only protection against colonization with a homologous NTHi strain which served as a OMV donor for the immunization mixture, but also against a heterologous NTHi strain, whose OMVs were not part of the immunization mixture, indicating the potential of OMVs as potential vaccine against heterologous NTHi infection.

11.2.9 Ehrlichiosis

Members of the genus *Ehrlichia* are tick-borne obligate intracellular Gram-negative bacteria that cause persistent infections in natural animal hosts, but are also associated with emerging human zoonoses of public health importance. *E. ruminantium* causes heartwater in cattle, sheep and goats and is one of the most economically important ehrlichioses, as heartwater is a devastating endemic disease of livestock in sub-Saharan regions of Africa and a few eastern Caribbean islands [153]. *E. chaffeensis* resides in mononuclear phagocytes and is the etiologic agent for human monocytropic ehrlichiosis (HME), the most frequent cause of severe and potentially fatal ehrlichiosis in humans whereas *E. ewingii* is associated with disease primarily in immuno-compromised individuals and *E. cans* has been identified as the etiologic agent of human infections in Venezuela [154–156].

There is no commercially available vaccine for these tick-transmitted pathogens and this is partly due to the unusual nature of these pathogens that lack the inflammatory pathogen-associated molecular pattern molecules, LPS and peptidoglycans [157, 158]. One of the most extensively studied vaccine candidates are a family of OMPs (MAP, OMP, P28 and P30) ranging from 16 *E. ruminantium* to 25 *E. canis* members that are present in all species. Immunodominant B-cell epitopes on OMPs located in hypervariable regions of these antigens have been reported [159]. Recently, three principal T-cell epitopes of OMP19 were mapped and are conserved across *E. muris*, *E. chaffeensis* and highly virulent *Ehrlichia* from *Ixodesovatus* (IOE) and exhibited a high degree of homology with a family of P28 OMPs in *E. chaffeensis* [160]. These epitopes are associated with protective immunity against IOE in mice and may induce cross-protective T-cell immunity against related Ehrlichial infections. Moreover, P28–19 of *E. chaffeensis* has been identified as an effective target mediating clearance of the bacteria and additionally a *p28* gene-based naked DNA vaccine (MAP1) was found to protect mice against challenge with a lethal dose of *E. ruminantium* [161, 162]. Furthermore, boosting DNA vaccine-primed mice with recombinant MAP1 protein significantly augmented protection against a homologous challenge [163]. MAP1 and MAP2 also elicited strong T-cell and antibody-mediated responses in cattle infected with *E. ruminantium* [164, 165].

11.2.10 *Brucella*

Brucellosis is a globally widespread zoonotic disease that is transmitted from domestic animals to humans. It is mostly caused by *Brucella abortus* and *B. melitensis* and is frequently acquired by ingestion, inhalation or direct contact of conjunctiva or skin lesions with infected animal products [166]. Bacteria spread from the site of entry to different organs causing acute disease symptoms and develop into localized foci of infection. Here, the pathogen survives intracellularly in the mononuclear phagocytic system leading to chronic disease [167]. The human disease represents an important cause of morbidity worldwide whereas animal brucellosis is associated with serious economic losses caused mainly by abortion and infertility in ruminants [166]. While a human vaccine would be valuable for individuals who may be occupationally exposed to *Brucella*, human brucellosis incidence can be reduced by control of the infected domestic animals [60]. Currently, live attenuated *B. abortus* S19 is used to immunize cattle whereas *B. melitensis* Rev1 is used to immunize goats and sheep [60]. Despite their effectiveness, these vaccines have inherent disadvantages such as being too infectious for human application, interfering with diagnosis and resulting in abortion when administered to pregnant animals [60]. Thus, improved vaccines which combine safety and efficacy in all species at risk need to be designed.

It is well established that the production of IFN-γ by Th1 cells as well as $CD8^+$ T-cell-mediated responses are key mediators of protective immunity against *Brucella* infections, whereas Th2 responses are minor contributors in host resistance [168, 169]. *Brucella* OMP antigens are categorized according to their molecular weight into three groups: groups 1, 2, and 3 with approximate molecular masses of 94, 41 to 43, and 25 to 30 kDa, respectively [170]. OMPs of *Brucella spp.* have been extensively characterized as potential immunogenic and protective antigens [171]. Passive protection experiments in mice have shown that mixtures of monoclonal Abs (mAbs) previously shown to bind individually to several OMPs, were protective against rough *B. ovis* [172] but conferred no or poor protection against smooth *Brucella* strains in mice [173]. However, protection was improved when the anti-Omp31 mAb was present in the mixture of mAbs [174]. This could be attributed to the presence of the O-polysaccharide-bearing LPS on smooth strains, which has been shown to hinder buried OMP epitopes [174]. Indeed, vaccination with Omp31-enriched preparations, which induced a strong antibody response but a poor cellular response, provided protection against *B. ovis* [175] but not against *B. melitensis* [176] challenge. These findings have added support to the contention that *Brucella* OMPs seem to have little relevance as protective antigens in smooth *B. melitensis* infections, whereas these proteins, in particular, Omp31 appear as important protective antigens against *B. ovis* infection. Nonetheless, Omp31 could be potentially protective against smooth *Brucella* strains if an appropriate cell-mediated immunity could be achieved. DNA vaccination is a relatively novel and powerful method of immunization that induces both humoral and cellular (Th1 and CTL) immune responses and protection against a wide range of pathogens [177]. A DNA vaccine coding for Omp31 demonstrated significant protection against *B. ovis* and *B. melitensis* infection in mice [178]. Protection was observed in the spleen and also in the liver, an important site for the control of *Brucella* infection. Levels of protection afforded in the spleen and the liver after *B. ovis* challenge were comparable to the ones achieved by the H38 control vaccine. Thus, the Omp31 DNA vaccine could be included in the development of a multi-subunit vaccine in the immunoprophylaxis of brucellosis.

Omp28, also a group 3 antigen, elicits a protective response in mice against *B. abortus* and *B. melitensis* infections whereby immunization with Omp28 induced both Th1 and Th2 responses in the host, which are critical for promoting bacterial clearance and mounting immune responses [179, 180]. *B. abortus* Omp16 and Omp19 are lipoproteins expressed broadly within the *Brucella genus* [181]. Both proteins are able to induce protective immunity against virulent *B. abortus* in mice when delivered by the intraperitoneal or oral route [182, 183] and the level of protection induced was similar to that elicited by live *B. abortus* S19 immunization, which is commonly used to vaccinate cattle. Immunization with Omp16 or Omp19 induced a protective Th1 response, without the addition of adjuvants. Furthermore, the elicited systemic protective response was protective against *B. abortus, B. melitensis* and *B. suis* infections [183, 184]. The self-adjuvant property of Omp19 and Omp16 is not unique to these proteins, because several bacterial proteins such as KpOmpA have intrinsic adjuvant activity that allows them to induce specific and effective immune responses without the help of external adjuvants [49].

11.3 Genomics and Functional Genomics in OMP-Based Vaccine Development

The capacity to identify immunogens, particularly OMPs for vaccine development by genome-wide screening has been clearly enhanced by the availability of microbial genome sequences coupled to proteomic and genomic analysis. Sequencing of the first bacterial genome, *H. influenzae*, in 1995, ushered vaccine development into a new era [185]. The continuous advances in genome sequencing technologies and bioinformatics have enabled scientists to identify vaccine candidates without relying on conventional vaccinology principles. This new approach, termed reverse vaccinology, provides full access to all the proteins that microorganisms can encode and by computer analysis, it is possible to identify potential surface-exposed proteins in a reverse manner beginning from the genome rather than the microorganism. Thus, problems related to non-culturable microorganisms and also to antigens that are not expressed under *in vitro* conditions, the most important obstacles for vaccine development, could be avoided by adopting the reverse vaccinology approach.

The first classic example of the use of reverse vaccinology to identify vaccine candidates is the development of a *N. meningitidis* serogroup B (MenB) vaccine [186]. Researchers were able to select nearly 600 potential vaccine candidates by *in silico* analysis. Following this, 350 recombinant proteins were successfully expressed in *E. coli* and used to immunize mice, from which 91 novel proteins were confirmed to be surface exposed and 28 proteins were found to induce bactericidal antibodies [187]. After further analysis, a final four-component MenB (4CMenB) vaccine was formulated, which consists of three recombinant proteins and OMV containing PorA, an OMP antigen [186]. Clinical trial data from studies with >1800 infants showed that 4CMenB (commercial name Bexsero^{TM}) is safe and induces a robust immune response [188].

The success obtained with the *N. meningitidis* experience prompted the application of reverse vaccinology to other pathogens including *P. gingivalis*, *Anaplasma marginale* and many others. As mentioned above, *P. ginvivalis* is a key periodontal pathogen which has been implicated in the etiology of chronic adult periodontitis. The genome contains ~ 15,000 CDSs and various bioinformatics tools have predicted that 120 proteins are surface exposed. Of these, two OMPs, PG32 and PG33 are highly conserved in both clinical and laboratory strains of *P. gingivalis* [189, 190]. When murine models of *P. gingivalis* were immunized with both proteins, significant protection from lesion development was observed [189]. For rickettsial pathogens, bottlenecks in vaccine development have led to the unavailability of potential vaccines to prevent human infection and/or disease. Nevertheless, the availability of more than 30 complete rickettsial genomes has provided new resources to accelerate identification of potential vaccine candidates. Bioinformatics identification of OMPs in the natural rickettsial pathogen *A. marginale* genome using PSORT, PSORTB and homology to experimentally identified surface proteins predicted 62 OMPs within 949 proteins predicted to represent the complete proteome [191]. Subsequent two-independent genome-wide screens using IgG_2 from vaccinates protected from challenge following vaccination with individual OMPs or bacterial surface complexes, resulted in 10 OMP vaccine candidates [192]. To further narrow down the choice of vaccine candidates, OMPs with broad conserved epitopes were identified by immunization with a live heterologous vaccine, *A. marginale ss. central* vaccine strain, and three OMPs were selected as final vaccine candidates [193]. More recently, the reverse vaccinology approach has also identified nine *B. melitensis* OMPs that reacted to pooled sera from exposed goats, mice and humans and protective efficacy studies are currently underway [194].

Complementary to *in silico* antigen discovery approaches, a functional genomics approach combining genomics, transcriptional profiling and proteomics has emerged to accelerate the vaccine discovery research process. The application of functional genomics revealed potential vaccine candidate genes for leptospirosis. Currently available vaccines for leptospirosis are of low efficacy, have unacceptable side-effects and do not induce long-term protection [79, 195, 196]. Functional genomics studies revealed 226 genes as vaccine candidates from 4727 ORFs in the genome, of which 26 were OMPs. These genes are not only potentially surface-exposed in the bacterium, but also conserved in two sequenced *L. interrogans*. Moreover, these genes are conserved among ten epidemic serovars in China and have high transcriptional levels *in vitro*. For the case of *Actinobacillus pleuropneumoniae*, the causative agent of porcine pleuropneumonia, microarray-based comparative genomic hybridizations revealed 58 genes coding for OMPs which were conserved in 15 reference strains as well as 21 fresh field isolates [197]. Many of these OMPs were identified as vaccine candidates since homologs of them are already under investigation as vaccine components in other bacteria. In a complementary study, Chen *et al.* [198] performed a genome-wide strategy combined with bioinformatics and experimental work described an OmpW that could induce high titres of antibodies but vaccination elicited low protective immunity against *A. pleuropneumoniae*. Montigian *et al.* (2002) reported the first systematic attempt to define surface protein organization in *Chlamydia pneumoniae*, a human pathogen causing respiratory infections [199]. A combined genomics–proteomics approach identified 53 surface-exposed proteins from 141 ORFs selected *in silico* from the genome of *C. pneumoniae*. Among 53 surface-exposed proteins, 22 are OMPs and they are proposed as potential vaccine candidates against C. *pneumoniae* infection, paving the way to a rational selection of new vaccine candidates.

11.4 Conclusions

Infectious diseases caused by Gram–negative bacteria that are transmissible from one person to another or between inter- and intra-species, are responsible for the majority of deaths of children and young adults and cause significant loss to the agriculture industry. The first massive immunization campaign against some of the more contagious bacterial and viral infections, for instance, diphtheria, pertussis, polio and measles was initiated in the mid-twentieth century and since then, vaccines have greatly diminished the morbidity and mortality stemming from multiple infectious diseases and remain the safest, most effective, cost-beneficial method to control infectious disease spread. However, despite the innovative technologies in the vaccine industry, safe and effective vaccines are still unavailable for many infectious diseases which continue to be a global problem.

OMPs are unique to Gram-negative bacteria and their strategic location on the cell surface allows for direct interaction with the various environments encountered by pathogenic organisms. Thus, OMPs represent important virulence factors and play essential roles in bacterial adaptation to host niches which are by and large hostile to invading pathogens. Accumulating evidence suggests that OMPs are potentially promising vaccine candidates and they are able to stimulate both humoral and cellular immune responses to combat both extracellular and intracellular pathogens, respectively. The availability of complete genome sequences together with progress in high-throughput technologies such as functional and structural genomics, has led to a new paradigm in vaccine development. Although many are still at the preclinical stage, OMP vaccination often demonstrates protection levels comparable to currently used vaccines which usually comprise of attenuated pathogens. Some OMP-based vaccines for *P. aeruginosa* and *N. meningitidis* are already at phase I/II clinical trials and have shown promising results. Although there is more to learn in order to improve their effectiveness, OMP-based vaccines will contribute significantly for the better management of infectious diseases.

References

[1] Sette, A., and R. Rappuoli, "Reverse vaccinology: developing vaccines in the era of genomics," Immunity, Vol. 33, No. 4, Oct 29 2010, pp. 530–41.

[2] Bagnoli, F., B. Baudner, R. P. Mishra, E. Bartolini, L. Fiaschi, P. Mariotti, V. Nardi-Dei, P. Boucher, and R. Rappuoli, "Designing the next generation of vaccines for global public health," OMICS, Vol. 15, No. 9, Sep 2011, pp. 545–66.

[3] Poland, G. A., R. B. Kennedy, and I. G. Ovsyannikova, "Vaccinomics and personalized vaccinology: is science leading us toward a new path of directed vaccine development and discovery?," PLoS Pathog, Vol. 7, No. 12, Dec 2011, pp. e1002344.

[4] Six, A., B. Bellier, V. Thomas-Vaslin, and D. Klatzmann, "Systems biology in vaccine design," Microb Biotechnol, Vol. 5, No. 2, Mar 2012, pp. 295–304.

[5] Koebnik, R., K. P. Locher, and P. Van Gelder, "Structure and function of bacterial outer membrane proteins: barrels in a nutshell," Mol Microbiol, Vol. 37, No. 2, Jul 2000, pp. 239–53.

[6] Bos, M. P., V. Robert, and J. Tommassen, "Biogenesis of the gram-negative bacterial outer membrane," Annu Rev Microbiol, Vol. 61, 2007, pp. 191–214.

[7] Krishnan, S., and N. V. Prasadarao, "Outer membrane protein A and OprF: versatile roles in Gram-negative bacterial infections," FEBS J, Jan 12 2012, pp.

[8] Ricci, D. P., and T. J. Silhavy, "The Bam machine: A molecular cooper," Biochim Biophys Acta, Vol. 1818, No. 4, Apr 2012, pp. 1067–84.

[9] Buchanan, S. K., "β-barrel proteins from bacterial outer membranes: structure, function and refolding," Curr Opin Struct Biol, Vol. 9, No. 4, Aug 1999, pp. 455–61.

[10] Miyadai, H., K. Tanaka-Masuda, S. Matsuyama, and H. Tokuda, "Effects of lipoprotein overproduction on the induction of DegP (HtrA) involved in quality control in the Escherichia coli periplasm," J Biol Chem, Vol. 279, No. 38, Sep 17 2004, pp. 39807–13.

[11] Malinverni, J. C., J. Werner, S. Kim, J. G. Sklar, D. Kahne, R. Misra, and T. J. Silhavy, "YfiO stabilizes the YaeT complex and is essential for outer membrane protein assembly in Escherichia coli," Mol Microbiol, Vol. 61, No. 1, Jul 2006, pp. 151–64.

[12] Uehara, T., T. Dinh, and T. G. Bernhardt, "LytM-domain factors are required for daughter cell separation and rapid ampicillin-induced lysis in Escherichia coli," J Bacteriol, Vol. 191, No. 16, Aug 2009, pp. 5094–107.

[13] Collin, S., I. Guilvout, N. N. Nickerson, and A. P. Pugsley, "Sorting of an integral outer membrane protein via the lipoprotein-specific Lol pathway and a dedicated lipoprotein pilotin," Mol Microbiol, Vol. 80, No. 3, May 2011, pp. 655–65.

[14] Laubacher, M. E., and S. E. Ades, "The Rcs phosphorelay is a cell envelope stress response activated by peptidoglycan stress and contributes to intrinsic antibiotic resistance," J Bacteriol, Vol. 190, No. 6, Mar 2008, pp. 2065–74.

[15] Wexler, H. M., "Pore-forming molecules in gram-negative anaerobic bacteria," Clin Infect Dis, Vol. 25 Suppl 2, Sep 1997, pp. S284–6.

[16] Cronan, J. E., "Bacterial membrane lipids: where do we stand?," Annu Rev Microbiol, Vol. 57, 2003, pp. 203–24.

[17] Zhou, Z., K. A. White, A. Polissi, C. Georgopoulos, and C. R. Raetz, "Function of Escherichia coli MsbA, an essential ABC family transporter, in lipid A and phospholipid biosynthesis," J Biol Chem, Vol. 273, No. 20, May 15 1998, pp. 12466–75.

[18] Kol, M. A., A. van Dalen, A. I. de Kroon, and B. de Kruijff, "Translocation of phospholipids is facilitated by a subset of membrane-spanning proteins of the bacterial cytoplasmic membrane," J Biol Chem, Vol. 278, No. 27, Jul 4 2003, pp. 24586–93.

[19] Donohue-Rolfe, A. M., and M. Schaechter, "Translocation of phospholipids from the inner to the outer membrane of Escherichia coli," Proc Natl Acad Sci U S A, Vol. 77, No. 4, Apr 1980, pp. 1867–71.

[20] Prinz, W. A., "Lipid trafficking sans vesicles: where, why, how?," Cell, Vol. 143, No. 6, Dec 10 2010, pp. 870–4.

[21] Lerouge, I., and J. Vanderleyden, "O-antigen structural variation: mechanisms and possible roles in animal/plant-microbe interactions," FEMS Microbiol Rev, Vol. 26, No. 1, Mar 2002, pp. 17–47.

[22] Raetz, C. R., and C. Whitfield, "Lipopolysaccharide endotoxins," Annu Rev Biochem, Vol. 71, 2002, pp. 635–700.

[23] Liechti, G., and J. B. Goldberg, "Outer membrane biogenesis in Escherichia coli, Neisseria meningitidis, and Helicobacter pylori: paradigm deviations in H. pylori," Front Cell Infect Microbiol, Vol. 2, 2012, pp. 29.

[24] Narita, S., "ABC transporters involved in the biogenesis of the outer membrane in gram-negative bacteria," Biosci Biotechnol Biochem, Vol. 75, No. 6, 2011, pp. 1044–54.

[25] Ma, B., C. M. Reynolds, and C. R. Raetz, "Periplasmic orientation of nascent lipid A in the inner membrane of an Escherichia coli LptA mutant," Proc Natl Acad Sci U S A, Vol. 105, No. 37, Sep 16 2008, pp. 13823–8.

[26] Rigel, N. W., and T. J. Silhavy, "Making a beta-barrel: assembly of outer membrane proteins in Gram-negative bacteria," Curr Opin Microbiol, Vol. 15, No. 2, Apr 2012, pp. 189–93.

[27] Rapoport, T. A., "Protein translocation across the eukaryotic endoplasmic reticulum and bacterial plasma membranes," Nature, Vol. 450, No. 7170, Nov 29 2007, pp. 663–9.

[28] Knowles, T. J., A. Scott-Tucker, M. Overduin, and I. R. Henderson, "Membrane protein architects: the role of the BAM complex in outer membrane protein assembly," Nat Rev Microbiol, Vol. 7, No. 3, Mar 2009, pp. 206–14.

[29] Bolla, J. M., C. Lazdunski, and J. M. Pages, "The assembly of the major outer membrane protein OmpF of Escherichia coli depends on lipid synthesis," EMBO J, Vol. 7, No. 11, Nov 1988, pp. 3595–9.

[30] de Cock, H., M. Struyve, M. Kleerebezem, T. van der Krift, and J. Tommassen, "Role of the carboxy-terminal phenylalanine in the biogenesis of outer membrane protein PhoE of Escherichia coli K-12," J Mol Biol, Vol. 269, No. 4, Jun 20 1997, pp. 473–8.

[31] Robert, V., E. B. Volokhina, F. Senf, M. P. Bos, P. Van Gelder, and J. Tommassen, "Assembly factor Omp85 recognizes its outer membrane protein substrates by a species-specific C-terminal motif," PLoS Biol, Vol. 4, No. 11, Nov 2006, pp. e377.

[32] Jiang, J. H., J. Tong, K. S. Tan, and K. Gabriel, "From Evolution to Pathogenesis: The Link Between β-Barrel Assembly Machineries in the Outer Membrane of Mitochondria and Gram-Negative Bacteria," Int J Mol Sci, Vol. 13, No. 7, 2012, pp. 8038–50.

[33] Voulhoux, R., M. P. Bos, J. Geurtsen, M. Mols, and J. Tommassen, "Role of a highly conserved bacterial protein in outer membrane protein assembly," Science, Vol. 299, No. 5604, Jan 10 2003, pp. 262–5.

[34] Voulhoux, R., and J. Tommassen, "Omp85, an evolutionarily conserved bacterial protein involved in outer-membrane-protein assembly," Res Microbiol, Vol. 155, No. 3, Apr 2004, pp. 129–35.

[35] Sanchez-Pulido, L., D. Devos, S. Genevrois, M. Vicente, and A. Valencia, "POTRA: a conserved domain in the FtsQ family and a class of β-barrel outer membrane proteins," Trends Biochem Sci, Vol. 28, No. 10, Oct 2003, pp. 523–6.

[36] Wu, T., J. Malinverni, N. Ruiz, S. Kim, T. J. Silhavy, and D. Kahne, "Identification of a multicomponent complex required for outer membrane biogenesis in Escherichia coli," Cell, Vol. 121, No. 2, Apr 22 2005, pp. 235–45.

[37] Hagan, C. L., T. J. Silhavy, and D. Kahne, "β-Barrel membrane protein assembly by the Bam complex," Annu Rev Biochem, Vol. 80, Jun 7 2011, pp. 189–210.

[38] Genevrois, S., L. Steeghs, P. Roholl, J. J. Letesson, and P. van der Ley, "The Omp85 protein of Neisseria meningitidis is required for lipid export to the outer membrane," EMBO J, Vol. 22, No. 8, Apr 15 2003, pp. 1780–9.

[39] Okuda, S., and H. Tokuda, "Lipoprotein sorting in bacteria," Annu Rev Microbiol, Vol. 65, 2011, pp. 239–59.

[40] Terada, M., T. Kuroda, S. I. Matsuyama, and H. Tokuda, "Lipoprotein sorting signals evaluated as the LolA-dependent release of lipoproteins from the cytoplasmic membrane of Escherichia coli," J Biol Chem, Vol. 276, No. 50, Dec 14 2001, pp. 47690–4.

[41] Okuda, S., and H. Tokuda, "Model of mouth-to-mouth transfer of bacterial lipoproteins through inner membrane LolC, periplasmic LolA, and outer membrane LolB," Proc Natl Acad Sci U S A, Vol. 106, No. 14, Apr 7 2009, pp. 5877–82.

[42] Smith, S. G., V. Mahon, M. A. Lambert, and R. P. Fagan, "A molecular Swiss army knife: OmpA structure, function and expression," FEMS Microbiol Lett, Vol. 273, No. 1, Aug 2007, pp. 1–11.

[43] Weiser, J. N., and E. C. Gotschlich, "Outer membrane protein A (OmpA) contributes to serum resistance and pathogenicity of Escherichia coli K-1," Infect Immun, Vol. 59, No. 7, Jul 1991, pp. 2252–8.

[44] Mayer-Scholl, A., P. Averhoff, and A. Zychlinsky, "How do neutrophils and pathogens interact?," Curr Opin Microbiol, Vol. 7, No. 1, Feb 2004, pp. 62–6.

[45] Sukumaran, S. K., and N. V. Prasadarao, "Escherichia coli K1 invasion increases human brain microvascular endothelial cell monolayer permeability by disassembling vascular-endothelial cadherins at tight junctions," J Infect Dis, Vol. 188, No. 9, Nov 1 2003, pp. 1295–309.

[46] Brinkman, F. S., M. Bains, and R. E. Hancock, "The amino terminus of Pseudomonas aeruginosa outer membrane protein OprF forms channels in lipid bilayer membranes: correlation with a three-dimensional model," J Bacteriol, Vol. 182, No. 18, Sep 2000, pp. 5251–5.

[47] Fito-Boncompte, L., A. Chapalain, E. Bouffartigues, H. Chaker, O. Lesouhaitier, G. Gicquel, A. Bazire, A. Madi, N. Connil, W. Veron, L. Taupin, B. Toussaint, P. Cornelis, Q. Wei, K. Shioya, E. Deziel, M. G. Feuilloley, N. Orange, A. Dufour, and S. Chevalier, "Full virulence of Pseudomonas aeruginosa requires OprF," Infect Immun, Vol. 79, No. 3, Mar 2011, pp. 1176–86.

[48] Mittal, R., S. K. Sukumaran, S. K. Selvaraj, D. G. Wooster, M. M. Babu, A. D. Schreiber, J. S. Verbeek, and N. V. Prasadarao, "Fcγ receptor I alpha chain (CD64) expression in macrophages is critical for the onset of meningitis by Escherichia coli K1," PLoS Pathog, Vol. 6, No. 11, 2010, pp. e1001203.

[49] Jeannin, P., G. Magistrelli, L. Goetsch, J. F. Haeuw, N. Thieblemont, J. Y. Bonnefoy, and Y. Delneste, "Outer membrane protein A (OmpA): a new pathogen-associated molecular pattern that interacts with antigen presenting cells-impact on vaccine strategies," Vaccine, Vol. 20 Suppl 4, Dec 19 2002, pp. A23–7.

[50] Dabo, S. M., A. W. Confer, and R. A. Quijano-Blas, "Molecular and immunological characterization of Pasteurella multocida serotype A:3 OmpA: evidence of its role in P. multocida interaction with extracellular matrix molecules," Microb Pathog, Vol. 35, No. 4, Oct 2003, pp. 147–57.

[51] Grabenstein, J. D., "Vaccines: countering anthrax: vaccines and immunoglobulins," Clin Infect Dis, Vol. 46, No. 1, Jan 1 2008, pp. 129–36.

[52] Gordon, S. B., G. R. Irving, R. A. Lawson, M. E. Lee, and R. C. Read, "Intracellular trafficking and killing of Streptococcus pneumoniae by human alveolar macrophages are influenced by opsonins," Infect Immun, Vol. 68, No. 4, Apr 2000, pp. 2286–93.

[53] Titball, R. W., "Vaccines against intracellular bacterial pathogens," Drug Discov Today, Vol. 13, No. 13–14, Jul 2008, pp. 596–600.

[54] Fulop, M., P. Mastroeni, M. Green, and R. W. Titball, "Role of antibody to lipopolysaccharide in protection against low- and high-virulence strains of Francisella tularensis," Vaccine, Vol. 19, No. 31, Aug 14 2001, pp. 4465–72.

[55] Jones, S. M., J. F. Ellis, P. Russell, K. F. Griffin, and P. C. Oyston, "Passive protection against Burkholderia pseudomallei infection in mice by monoclonal antibodies against capsular polysaccharide, lipopolysaccharide or proteins," J Med Microbiol, Vol. 51, No. 12, Dec 2002, pp. 1055–62.
[56] Nelson, M., J. L. Prior, M. S. Lever, H. E. Jones, T. P. Atkins, and R. W. Titball, "Evaluation of lipopolysaccharide and capsular polysaccharide as subunit vaccines against experimental melioidosis," J Med Microbiol, Vol. 53, No. Pt 12, Dec 2004, pp. 1177–82.
[57] Seder, R. A., and A. V. Hill, "Vaccines against intracellular infections requiring cellular immunity," Nature, Vol. 406, No. 6797, Aug 17 2000, pp. 793–8.
[58] McBride, J. W., and D. H. Walker, "Progress and obstacles in vaccine development for the ehrlichioses," Expert Rev Vaccines, Vol. 9, No. 9, Sep 2010, pp. 1071–82.
[59] Krause, A., W. Z. Whu, Y. Xu, J. Joh, R. G. Crystal, and S. Worgall, "Protective anti-Pseudomonas aeruginosa humoral and cellular mucosal immunity by AdC7-mediated expression of the P. aeruginosa protein OprF," Vaccine, Vol. 29, No. 11, Mar 3 2011, pp. 2131–9.
[60] Schurig, G. G., N. Sriranganathan, and M. J. Corbel, "Brucellosis vaccines: past, present and future," Vet Microbiol, Vol. 90, No. 1–4, Dec 20 2002, pp. 479–96.
[61] Clements, A., A. W. Jenney, J. L. Farn, L. E. Brown, G. Deliyannis, E. L. Hartland, M. J. Pearse, M. B. Maloney, S. L. Wesselingh, O. L. Wijburg, and R. A. Strugnell, "Targeting subcapsular antigens for prevention of Klebsiella pneumoniae infections," Vaccine, Vol. 26, No. 44, Oct 16 2008, pp. 5649–53.
[62] Khader, S. A., and R. Gopal, "IL-17 in protective immunity to intracellular pathogens," Virulence, Vol. 1, No. 5, Sep-Oct 2010, pp. 423–7.
[63] Ye, P., P. B. Garvey, P. Zhang, S. Nelson, G. Bagby, W. R. Summer, P. Schwarzenberger, J. E. Shellito, and J. K. Kolls, "Interleukin-17 and lung host defense against Klebsiella pneumoniae infection," Am J Respir Cell Mol Biol, Vol. 25, No. 3, Sep 2001, pp. 335–40.
[64] Happel, K. I., P. J. Dubin, M. Zheng, N. Ghilardi, C. Lockhart, L. J. Quinton, A. R. Odden, J. E. Shellito, G. J. Bagby, S. Nelson, and J. K. Kolls, "Divergent roles of IL-23 and IL-12 in host defense against Klebsiella pneumoniae," J Exp Med, Vol. 202, No. 6, Sep 19 2005, pp. 761–9.

[65] Yu, J. J., M. J. Ruddy, G. C. Wong, C. Sfintescu, P. J. Baker, J. B. Smith, R. T. Evans, and S. L. Gaffen, "An essential role for IL-17 in preventing pathogen-initiated bone destruction: recruitment of neutrophils to inflamed bone requires IL-17 receptor-dependent signals," Blood, Vol. 109, No. 9, May 1 2007, pp. 3794–802.

[66] Shibata, K., H. Yamada, H. Hara, K. Kishihara, and Y. Yoshikai, "Resident V·1+··T-cells control early infiltration of neutrophils after Escherichia coli infection via IL-17 production," J Immunol, Vol. 178, No. 7, Apr 1 2007, pp. 4466–72.

[67] Lu, Y. J., J. Gross, D. Bogaert, A. Finn, L. Bagrade, Q. Zhang, J. K. Kolls, A. Srivastava, A. Lundgren, S. Forte, C. M. Thompson, K. F. Harney, P. W. Anderson, M. Lipsitch, and R. Malley, "Interleukin-17A mediates acquired immunity to pneumococcal colonization," PLoS Pathog, Vol. 4, No. 9, 2008, pp. e1000159.

[68] Curtis, M. M., and S. S. Way, "Interleukin-17 in host defence against bacterial, mycobacterial and fungal pathogens," Immunology, Vol. 126, No. 2, Feb 2009, pp. 177–85.

[69] Schulz, S. M., G. Kohler, C. Holscher, Y. Iwakura, and G. Alber, "IL-17A is produced by Th17, ·· T-cells and other CD4- lymphocytes during infection with Salmonella enterica serovar Enteritidis and has a mild effect in bacterial clearance," Int Immunol, Vol. 20, No. 9, Sep 2008, pp. 1129–38.

[70] Vinetz, J. M., "Leptospirosis," Curr Opin Infect Dis, Vol. 14, No. 5, Oct 2001, pp. 527–38.

[71] Holk, K., S. V. Nielsen, and T. Ronne, "Human leptospirosis in Denmark 1970–1996: an epidemiological and clinical study," Scand J Infect Dis, Vol. 32, No. 5, 2000, pp. 533–8.

[72] Hernandez, M. S., J. M. Ferrer, C. D. Oval, and J. M. Sanchez, "Outbreaks of animal and human leptospirosis in the province of Ciego de Avila," Rev Cubana Med Trop, Vol. 57, No. 1, Jan-Apr 2005, pp. 79–80.

[73] Schmid, G. P., A. C. Steere, A. N. Kornblatt, A. F. Kaufmann, C. W. Moss, R. C. Johnson, K. Hovind-Hougen, and D. J. Brenner, "Newly recognized Leptospira species ("Leptospira inadai" serovar lyme) isolated from human skin," J Clin Microbiol, Vol. 24, No. 3, Sep 1986, pp. 484–6.

[74] de Souza, A. L., J. Sztajnbok, S. R. Marques, and A. C. Seguro, "Leptospirosis-induced meningitis and acute renal failure in a

19-month-old male child," J Med Microbiol, Vol. 55, No. Pt 6, Jun 2006, pp. 795–7.

[75] Adamus, C., M. Buggin-Daubie, A. Izembart, C. Sonrier-Pierre, L. Guigand, M. T. Masson, G. Andre-Fontaine, and M. Wyers, "Chronic hepatitis associated with leptospiral infection in vaccinated beagles," J Comp Pathol, Vol. 117, No. 4, Nov 1997, pp. 311–28.

[76] Katz, A. R., V. E. Ansdell, P. V. Effler, C. R. Middleton, and D. M. Sasaki, "Assessment of the clinical presentation and treatment of 353 cases of laboratory-confirmed leptospirosis in Hawaii, 1974–1998," Clin Infect Dis, Vol. 33, No. 11, Dec 1 2001, pp. 1834–41.

[77] Barocchi, M. A., A. I. Ko, M. G. Reis, K. L. McDonald, and L. W. Riley, "Rapid translocation of polarized MDCK cell monolayers by Leptospira interrogans, an invasive but nonintracellular pathogen," Infect Immun, Vol. 70, No. 12, Dec 2002, pp. 6926–32.

[78] Gebriel, A. M., G. Subramaniam, and S. D. Sekaran, "The detection and characterization of pathogenic Leptospira and the use of OMPs as potential antigens and immunogens," Trop Biomed, Vol. 23, No. 2, Dec 2006, pp. 194–207.

[79] Sonrier, C., C. Branger, V. Michel, N. Ruvoen-Clouet, J. P. Ganiere, and G. Andre-Fontaine, "Evidence of cross-protection within Leptospira interrogans in an experimental model," Vaccine, Vol. 19, No. 1, Aug 15 2000, pp. 86–94.

[80] Cullen, P. A., X. Xu, J. Matsunaga, Y. Sanchez, A. I. Ko, D. A. Haake, and B. Adler, "Surfaceome of Leptospira spp," Infect Immun, Vol. 73, No. 8, Aug 2005, pp. 4853–63.

[81] Matsunaga, J., M. A. Barocchi, J. Croda, T. A. Young, Y. Sanchez, I. Siqueira, C. A. Bolin, M. G. Reis, L. W. Riley, D. A. Haake, and A. I. Ko, "Pathogenic Leptospira species express surface-exposed proteins belonging to the bacterial immunoglobulin superfamily," Mol Microbiol, Vol. 49, No. 4, Aug 2003, pp. 929–45.

[82] Shang, E. S., M. M. Exner, T. A. Summers, C. Martinich, C. I. Champion, R. E. Hancock, and D. A. Haake, "The rare outer membrane protein, OmpL1, of pathogenic Leptospira species is a heat-modifiable porin," Infect Immun, Vol. 63, No. 8, Aug 1995, pp. 3174–81.

[83] Dong, H., Y. Hu, F. Xue, D. Sun, D. M. Ojcius, Y. Mao, and J. Yan, "Characterization of the ompL1 gene of pathogenic Leptospira species in China and cross-immunogenicity of the OmpL1 protein," BMC Microbiol, Vol. 8, 2008, pp. 223.

[84] Haake, D. A., M. K. Mazel, A. M. McCoy, F. Milward, G. Chao, J. Matsunaga, and E. A. Wagar, "Leptospiral outer membrane proteins OmpL1 and LipL41 exhibit synergistic immunoprotection," Infect Immun, Vol. 67, No. 12, Dec 1999, pp. 6572–82.

[85] Lin, X., A. Sun, P. Ruan, Z. Zhang, and J. Yan, "Characterization of conserved combined T- and B-cell epitopes in Leptospira interrogans major outer membrane proteins OmpL1 and LipL41," BMC Microbiol, Vol. 11, No. 1, 2011, pp. 21.

[86] Cutler, C. W., J. R. Kalmar, and C. A. Genco, "Pathogenic strategies of the oral anaerobe, Porphyromonas gingivalis," Trends Microbiol, Vol. 3, No. 2, Feb 1995, pp. 45–51.

[87] Seymour, G. J., P. J. Ford, M. P. Cullinan, S. Leishman, and K. Yamazaki, "Relationship between periodontal infections and systemic disease," Clin Microbiol Infect, Vol. 13 Suppl 4, Oct 2007, pp. 3–10.

[88] Hiratsuka, K., Y. Abiko, M. Hayakawa, T. Ito, H. Sasahara, and H. Takiguchi, "Role of Porphyromonas gingivalis 40-kDa outer membrane protein in the aggregation of P. gingivalis vesicles and Actinomyces viscosus," Arch Oral Biol, Vol. 37, No. 9, Sep 1992, pp. 717–24.

[89] Saito, S., M. Hayakawa, K. Hiratsuka, H. Takiguchi, and Y. Abiko, "Complement-mediated killing of Porphyromonas gingivalis 381 by the immunoglobulin G induced by recombinant 40-kDa outer membrane protein," Biochem Mol Med, Vol. 58, No. 2, Aug 1996, pp. 184–91.

[90] Hamada, N., K. Watanabe, T. Tahara, K. Nakazawa, I. Ishida, Y. Shibata, T. Kobayashi, H. Yoshie, Y. Abiko, and T. Umemoto, "The r40-kDa outer membrane protein human monoclonal antibody protects against Porphyromonas gingivalis-induced bone loss in rats," J Periodontol, Vol. 78, No. 5, May 2007, pp. 933–9.

[91] Momoi, F., T. Hashizume, T. Kurita-Ochiai, Y. Yuki, H. Kiyono, and M. Yamamoto, "Nasal vaccination with the 40-kilodalton outer membrane protein of Porphyromonas gingivalis and a nontoxic chimeric enterotoxin adjuvant induces long-term protective immunity with reduced levels of immunoglobulin E antibodies," Infect Immun, Vol. 76, No. 6, Jun 2008, pp. 2777–84.

[92] Liu, C., T. Hashizume, T. Kurita-Ochiai, K. Fujihashi, and M. Yamamoto, "Oral immunization with Porphyromonas gingivalis outer membrane protein and CpGoligodeoxynucleotides elicits T helper 1 and 2 cytokines for enhanced protective immunity," Mol Oral Microbiol, Vol. 25, No. 3, Jun 2010, pp. 178–89.

[93] Gaynes, R., and J. R. Edwards, "Overview of nosocomial infections caused by gram-negative bacilli," Clin Infect Dis, Vol. 41, No. 6, Sep 15 2005, pp. 848–54.
[94] Lyczak, J. B., C. L. Cannon, and G. B. Pier, "Lung infections associated with cystic fibrosis," Clin Microbiol Rev, Vol. 15, No. 2, Apr 2002, pp. 194–222.
[95] Mesaros, N., P. Nordmann, P. Plesiat, M. Roussel-Delvallez, J. Van Eldere, Y. Glupczynski, Y. Van Laethem, F. Jacobs, P. Lebecque, A. Malfroot, P. M. Tulkens, and F. Van Bambeke, "Pseudomonas aeruginosa: resistance and therapeutic options at the turn of the new millennium," Clin Microbiol Infect, Vol. 13, No. 6, Jun 2007, pp. 560–78.
[96] Stanislavsky, E. S., and J. S. Lam, "Pseudomonas aeruginosa antigens as potential vaccines," FEMS Microbiol Rev, Vol. 21, No. 3, Nov 1997, pp. 243–77.
[97] Oliver, A., R. Canton, P. Campo, F. Baquero, and J. Blazquez, "High frequency of hypermutable Pseudomonas aeruginosa in cystic fibrosis lung infection," Science, Vol. 288, No. 5469, May 19 2000, pp. 1251–4.
[98] Wu, L., O. Estrada, O. Zaborina, M. Bains, L. Shen, J. E. Kohler, N. Patel, M. W. Musch, E. B. Chang, Y. X. Fu, M. A. Jacobs, M. I. Nishimura, R. E. Hancock, J. R. Turner, and J. C. Alverdy, "Recognition of host immune activation by Pseudomonas aeruginosa," Science, Vol. 309, No. 5735, Jul 29 2005, pp. 774–7.
[99] Ding, B., B. U. von Specht, and Y. Li, "OprF/I-vaccinated sera inhibit binding of human interferon-gamma to Pseudomonas aeruginosa," Vaccine, Vol. 28, No. 25, Jun 7 2010, pp. 4119–22.
[100] Loots, K., H. Revets, and B. M. Goddeeris, "Attachment of the outer membrane lipoprotein (OprI) of Pseudomonas aeruginosa to the mucosal surfaces of the respiratory and digestive tract of chickens," Vaccine, Vol. 26, No. 4, Jan 24 2008, pp. 546–51.
[101] Knapp, B., E. Hundt, U. Lenz, K. D. Hungerer, J. Gabelsberger, H. Domdey, E. Mansouri, Y. Li, and B. U. von Specht, "A recombinant hybrid outer membrane protein for vaccination against Pseudomonas aeruginosa," Vaccine, Vol. 17, No. 13–14, Mar 26 1999, pp. 1663–6.
[102] Larbig, M., E. Mansouri, J. Freihorst, B. Tummler, G. Kohler, H. Domdey, B. Knapp, K. D. Hungerer, E. Hundt, J. Gabelsberger, and B. U. von Specht, "Safety and immunogenicity of an intranasal Pseudomonas aeruginosa hybrid outer membrane protein F-I vaccine

in human volunteers," Vaccine, Vol. 19, No. 17–19, Mar 21 2001, pp. 2291–7.

[103] Baumann, U., K. Gocke, B. Gewecke, J. Freihorst, and B. U. von Specht, "Assessment of pulmonary antibodies with induced sputum and bronchoalveolar lavage induced by nasal vaccination against Pseudomonas aeruginosa: a clinical phase I/II study," Respir Res, Vol. 8, 2007, pp. 57.

[104] Berry, L. J., D. K. Hickey, K. A. Skelding, S. Bao, A. M. Rendina, P. M. Hansbro, C. M. Gockel, and K. W. Beagley, "Transcutaneous immunization with combined cholera toxin and CpG adjuvant protects against Chlamydia muridarum genital tract infection," Infect Immun, Vol. 72, No. 2, Feb 2004, pp. 1019–28.

[105] Sorichter, S., U. Baumann, A. Baumgart, S. Walterspacher, and B. U. von Specht, "Immune responses in the airways by nasal vaccination with systemic boosting against Pseudomonas aeruginosa in chronic lung disease," Vaccine, Vol. 27, No. 21, May 11 2009, pp. 2755–9.

[106] Weimer, E. T., H. Lu, N. D. Kock, D. J. Wozniak, and S. B. Mizel, "A fusion protein vaccine containing OprF epitope 8, OprI, and type A and B flagellins promotes enhanced clearance of nonmucoid Pseudomonas aeruginosa," Infect Immun, Vol. 77, No. 6, Jun 2009, pp. 2356–66.

[107] Burns, J. L., R. L. Gibson, S. McNamara, D. Yim, J. Emerson, M. Rosenfeld, P. Hiatt, K. McCoy, R. Castile, A. L. Smith, and B. W. Ramsey, "Longitudinal assessment of Pseudomonas aeruginosa in young children with cystic fibrosis," J Infect Dis, Vol. 183, No. 3, Feb 1 2001, pp. 444–52.

[108] Jang, I. J., I. S. Kim, W. J. Park, K. S. Yoo, D. S. Yim, H. K. Kim, S. G. Shin, W. H. Chang, N. G. Lee, S. B. Jung, D. H. Ahn, Y. J. Cho, B. Y. Ahn, Y. Lee, Y. G. Kim, S. W. Nam, and H. S. Kim, "Human immune response to a Pseudomonas aeruginosa outer membrane protein vaccine," Vaccine, Vol. 17, No. 2, Jan 1999, pp. 158–68.

[109] Lee, N. G., S. B. Jung, B. Y. Ahn, Y. H. Kim, J. J. Kim, D. K. Kim, I. S. Kim, S. M. Yoon, S. W. Nam, H. S. Kim, and W. J. Park, "Immunization of burn-patients with a Pseudomonas aeruginosa outer membrane protein vaccine elicits antibodies with protective efficacy," Vaccine, Vol. 18, No. 18, Mar 17 2000, pp. 1952–61.

[110] Hackett, N. R., S. M. Kaminsky, D. Sondhi, and R. G. Crystal, "Antivector and antitransgene host responses in gene therapy," Curr Opin Mol Ther, Vol. 2, No. 4, Aug 2000, pp. 376–82.

[111] Worgall, S., T. Kikuchi, R. Singh, K. Martushova, L. Lande, and R. G. Crystal, "Protection against pulmonary infection with Pseudomonas aeruginosa following immunization with P. aeruginosa-pulsed dendritic cells," Infect Immun, Vol. 69, No. 7, Jul 2001, pp. 4521–7.

[112] Peluso, L., C. de Luca, S. Bozza, A. Leonardi, G. Giovannini, A. Lavorgna, G. De Rosa, M. Mascolo, L. Ortega De Luna, M. R. Catania, L. Romani, and F. Rossano, "Protection against Pseudomonas aeruginosa lung infection in mice by recombinant OprF-pulsed dendritic cell immunization," BMC Microbiol, Vol. 10, 2010, pp. 9.

[113] Cheng, A. C., and B. J. Currie, "Melioidosis: epidemiology, pathophysiology, and management," Clin Microbiol Rev, Vol. 18, No. 2, Apr 2005, pp. 383–416.

[114] Wuthiekanun, V., and S. J. Peacock, "Management of melioidosis," Expert Rev Anti Infect Ther, Vol. 4, No. 3, Jun 2006, pp. 445–55.

[115] White, N. J., "Melioidosis," Lancet, Vol. 361, No. 9370, May 17 2003, pp. 1715–22.

[116] Kenny, D. J., P. Russell, D. Rogers, S. M. Eley, and R. W. Titball, "In vitro susceptibilities of Burkholderia mallei in comparison to those of other pathogenic Burkholderia spp," Antimicrob Agents Chemother, Vol. 43, No. 11, Nov 1999, pp. 2773–5.

[117] Stevens, M. P., A. Haque, T. Atkins, J. Hill, M. W. Wood, A. Easton, M. Nelson, C. Underwood-Fowler, R. W. Titball, G. J. Bancroft, and E. E. Galyov, "Attenuated virulence and protective efficacy of a Burkholderia pseudomallei bsa type III secretion mutant in murine models of melioidosis," Microbiology, Vol. 150, No. Pt 8, Aug 2004, pp. 2669–76.

[118] Srilunchang, T., T. Proungvitaya, S. Wongratanacheewin, R. Strugnell, and P. Homchampa, "Construction and characterization of an unmarked aroC deletion mutant of Burkholderia pseudomallei strain A2," Southeast Asian J Trop Med Public Health, Vol. 40, No. 1, Jan 2009, pp. 123–30.

[119] Zhang, S., S. H. Feng, B. Li, H. Y. Kim, J. Rodriguez, S. Tsai, and S. C. Lo, "In Vitro and In Vivo studies of monoclonal antibodies with prominent bactericidal activity against Burkholderia pseudomallei and Burkholderia mallei," Clin Vaccine Immunol, Vol. 18, No. 5, May 2011, pp. 825–34.

[120] Charuchaimontri, C., Y. Suputtamongkol, C. Nilakul, W. Chaowagul, P. Chetchotisakd, N. Lertpatanasuwun, S. Intaranongpai, P. J. Brett, and D. E. Woods, "Antilipopolysaccharide II: an antibody protective against fatal melioidosis," Clin Infect Dis, Vol. 29, No. 4, Oct 1999, pp. 813–8.

[121] Hara, Y., R. Mohamed, and S. Nathan, "Immunogenic Burkholderia pseudomallei outer membrane proteins as potential candidate vaccine targets," PLoS One, Vol. 4, No. 8, 2009, pp. e6496.

[122] Su, Y. C., K. L. Wan, R. Mohamed, and S. Nathan, "Immunization with the recombinant Burkholderia pseudomallei outer membrane protein Omp85 induces protective immunity in mice," Vaccine, Vol. 28, No. 31, Jul 12 2010, pp. 5005–11.

[123] Girard, M. P., M. P. Preziosi, M. T. Aguado, and M. P. Kieny, "A review of vaccine research and development: meningococcal disease," Vaccine, Vol. 24, No. 22, May 29 2006, pp. 4692–700.

[124] Reisinger, K. S., S. Black, and J. J. Stoddard, "Optimizing protection against meningococcal disease," Clin Pediatr (Phila), Vol. 49, No. 6, Jun 2010, pp. 586–97.

[125] Ala'Aldeen, D. A., H. A. Davies, and S. P. Borriello, "Vaccine potential of meningococcal FrpB: studies on surface exposure and functional attributes of common epitopes," Vaccine, Vol. 12, No. 6, May 1994, pp. 535–41.

[126] Pillai, S., A. Howell, K. Alexander, B. E. Bentley, H. Q. Jiang, K. Ambrose, D. Zhu, and G. Zlotnick, "Outer membrane protein (OMP) based vaccine for Neisseria meningitidis serogroup B," Vaccine, Vol. 23, No. 17–18, Mar 18 2005, pp. 2206–9.

[127] Pollard, A. J., and C. Frasch, "Development of natural immunity to Neisseria meningitidis," Vaccine, Vol. 19, No. 11–12, Jan 8 2001, pp. 1327–46.

[128] Vermont, C. L., H. H. van Dijken, A. J. Kuipers, C. J. van Limpt, W. C. Keijzers, A. van der Ende, R. de Groot, L. van Alphen, and G. P. van den Dobbelsteen, "Cross-reactivity of antibodies against PorA after vaccination with a meningococcal B outer membrane vesicle vaccine," Infect Immun, Vol. 71, No. 4, Apr 2003, pp. 1650–5.

[129] Sierra, G. V., H. C. Campa, N. M. Varcacel, I. L. Garcia, P. L. Izquierdo, P. F. Sotolongo, G. V. Casanueva, C. O. Rico, C. R. Rodriguez, and M. H. Terry, "Vaccine against group B Neisseria meningitidis: protection trial and mass vaccination results in Cuba," NIPH Ann, Vol. 14, No. 2, Dec 1991, pp. 195–207; discussion 208–10.

[130] de Moraes, J. C., B. A. Perkins, M. C. Camargo, N. T. Hidalgo, H. A. Barbosa, C. T. Sacchi, I. M. Landgraf, V. L. Gattas, G. Vasconcelos Hde, and *et al.*, "Protective efficacy of a serogroup B meningococcal vaccine in Sao Paulo, Brazil," Lancet, Vol. 340, No. 8827, Oct 31 1992, pp. 1074–8.

[131] Oster, P., D. Lennon, J. O'Hallahan, K. Mulholland, S. Reid, and D. Martin, "MeNZB: a safe and highly immunogenic tailor-made vaccine against the New Zealand Neisseria meningitidis serogroup B disease epidemic strain," Vaccine, Vol. 23, No. 17–18, Mar 18 2005, pp. 2191–6.

[132] Kelly, C., R. Arnold, Y. Galloway, and J. O'Hallahan, "A prospective study of the effectiveness of the New Zealand meningococcal B vaccine," Am J Epidemiol, Vol. 166, No. 7, Oct 1 2007, pp. 817–23.

[133] Galloway, Y., P. Stehr-Green, A. McNicholas, and J. O'Hallahan, "Use of an observational cohort study to estimate the effectiveness of the New Zealand group B meningococcal vaccine in children aged under 5 years," Int J Epidemiol, Vol. 38, No. 2, Apr 2009, pp. 413–8.

[134] Tappero, J. W., R. Lagos, A. M. Ballesteros, B. Plikaytis, D. Williams, J. Dykes, L. L. Gheesling, G. M. Carlone, E. A. Hoiby, J. Holst, H. Nokleby, E. Rosenqvist, G. Sierra, C. Campa, F. Sotolongo, J. Vega, J. Garcia, P. Herrera, J. T. Poolman, and B. A. Perkins, "Immunogenicity of 2 serogroup B outer-membrane protein meningococcal vaccines: a randomized controlled trial in Chile," JAMA, Vol. 281, No. 16, Apr 28 1999, pp. 1520–7.

[135] van der Ley, P., J. van der Biezen, and J. T. Poolman, "Construction of Neisseria meningitidis strains carrying multiple chromosomal copies of the porA gene for use in the production of a multivalent outer membrane vesicle vaccine," Vaccine, Vol. 13, No. 4, Mar 1995, pp. 401–7.

[136] Claassen, I., J. Meylis, P. van der Ley, C. Peeters, H. Brons, J. Robert, D. Borsboom, A. van der Ark, I. van Straaten, P. Roholl, B. Kuipers, and J. Poolman, "Production, characterization and control of a Neisseria meningitidis hexavalent class 1 outer membrane protein containing vesicle vaccine," Vaccine, Vol. 14, No. 10, Jul 1996, pp. 1001–8.

[137] Cartwright, K., R. Morris, H. Rumke, A. Fox, R. Borrow, N. Begg, P. Richmond, and J. Poolman, "Immunogenicity and reactogenicity in UK infants of a novel meningococcal vesicle vaccine containing multiple class 1 (PorA) outer membrane proteins," Vaccine, Vol. 17, No. 20–21, Jun 4 1999, pp. 2612–9.

[138] van den Dobbelsteen, G. P., H. H. van Dijken, S. Pillai, and L. van Alphen, "Immunogenicity of a combination vaccine containing pneumococcal conjugates and meningococcal PorA OMVs," Vaccine, Vol. 25, No. 13, Mar 22 2007, pp. 2491–6.
[139] Trotter, C. L., and M. E. Ramsay, "Vaccination against meningococcal disease in Europe: review and recommendations for the use of conjugate vaccines," FEMS Microbiol Rev, Vol. 31, No. 1, Jan 2007, pp. 101–7.
[140] Keiser, P. B., S. Biggs-Cicatelli, E. E. Moran, D. H. Schmiel, V. B. Pinto, R. E. Burden, L. B. Miller, J. E. Moon, R. A. Bowden, J. F. Cummings, and W. D. Zollinger, "A phase 1 study of a meningococcal native outer membrane vesicle vaccine made from a group B strain with deleted lpxL1 and synX, over-expressed factor H binding protein, two PorAs and stabilized OpcA expression," Vaccine, Vol. 29, No. 7, Feb 4 2011, pp. 1413–20.
[141] Erwin, A. L., and A. L. Smith, "Nontypeable Haemophilus influenzae: understanding virulence and commensal behavior," Trends Microbiol, Vol. 15, No. 8, Aug 2007, pp. 355–62.
[142] Morris, S. K., W. J. Moss, and N. Halsey, "Haemophilus influenzae type b conjugate vaccine use and effectiveness," Lancet Infect Dis, Vol. 8, No. 7, Jul 2008, pp. 435–43.
[143] Foxwell, A. R., J. M. Kyd, and A. W. Cripps, "Nontypeable Haemophilus influenzae: pathogenesis and prevention," Microbiol Mol Biol Rev, Vol. 62, No. 2, Jun 1998, pp. 294–308.
[144] Nelson, M. B., M. A. Apicella, T. F. Murphy, H. Vankeulen, L. D. Spotila, and D. Rekosh, "Cloning and sequencing of Haemophilus influenzae outer membrane protein P6," Infect Immun, Vol. 56, No. 1, Jan 1988, pp. 128–34.
[145] Kodama, S., T. Hirano, K. Noda, S. Umemoto, and M. Suzuki, "Nasal immunization with plasmid DNA encoding P6 protein and immunostimulatory complexes elicits nontypeable Haemophilus influenzae-specific long-term mucosal immune responses in the nasopharynx," Vaccine, Vol. 29, No. 10, Feb 24 2011, pp. 1881–90.
[146] Noda, K., S. Kodama, S. Umemoto, N. Nomi, T. Hirano, and M. Suzuki, "Th17 cells contribute to nontypeable Haemophilus influenzae-specific protective immunity induced by nasal vaccination with P6 outer membrane protein and alpha-galactosylceramide," Microbiol Immunol, Vol. 55, No. 8, Aug 2011, pp. 574–81.

[147] El-Adhami, W., J. M. Kyd, D. A. Bastin, and A. W. Cripps, "Characterization of the gene encoding a 26-kilodalton protein (OMP26) from nontypeable Haemophilus influenzae and immune responses to the recombinant protein," Infect Immun, Vol. 67, No. 4, Apr 1999, pp. 1935–42.
[148] Kyd, J. M., A. W. Cripps, L. A. Novotny, and L. O. Bakaletz, "Efficacy of the 26-kilodalton outer membrane protein and two P5 fimbrin-derived immunogens to induce clearance of nontypeable Haemophilus influenzae from the rat middle ear and lungs as well as from the chinchilla middle ear and nasopharynx," Infect Immun, Vol. 71, No. 8, Aug 2003, pp. 4691–9.
[149] Lubitz, W., "Bacterial ghosts as carrier and targeting systems," Expert Opin Biol Ther, Vol. 1, No. 5, Sep 2001, pp. 765–71.
[150] Riedmann, E. M., W. Lubitz, J. McGrath, J. M. Kyd, and A. W. Cripps, "Effectiveness of engineering the nontypeable Haemophilus influenzae antigen Omp26 as an S-layer fusion in bacterial ghosts as a mucosal vaccine delivery," Hum Vaccin, Vol. 7 Suppl, Jan-Feb 2011, pp. 99–107.
[151] Chang, A., R. Kaur, L. V. Michel, J. R. Casey, and M. Pichichero, "Haemophilus influenzae vaccine candidate outer membrane protein P6 is not conserved in all strains," Hum Vaccin, Vol. 7, No. 1, Jan 1 2011, pp. 102–5.
[152] Roier, S., D. R. Leitner, J. Iwashkiw, K. Schild-Prufert, M. F. Feldman, G. Krohne, J. Reidl, and S. Schild, "Intranasal Immunization with Non-typeable Haemophilus influenzae Outer Membrane Vesicles Induces Cross-Protective Immunity in Mice," PLoS One, Vol. 7, No. 8, 2012, pp. e42664.
[153] Provost, A., and J. D. Bezuidenhout, "The historical background and global importance of heartwater," Onderstepoort J Vet Res, Vol. 54, No. 3, Sep 1987, pp. 165–9.
[154] Perez, M., Y. Rikihisa, and B. Wen, "Ehrlichia canis-like agent isolated from a man in Venezuela: antigenic and genetic characterization," J Clin Microbiol, Vol. 34, No. 9, Sep 1996, pp. 2133–9.
[155] McQuiston, J. H., C. D. Paddock, R. C. Holman, and J. E. Childs, "The human ehrlichioses in the United States," Emerg Infect Dis, Vol. 5, No. 5, Sep-Oct 1999, pp. 635–42.
[156] Paddock, C. D., S. M. Folk, G. M. Shore, L. J. Machado, M. M. Huycke, L. N. Slater, A. M. Liddell, R. S. Buller, G. A. Storch, T. P. Monson, D. Rimland, J. W. Sumner, J. Singleton, K. C. Bloch, Y. W. Tang, S. M.

Standaert, and J. E. Childs, "Infections with Ehrlichia chaffeensis and Ehrlichia ewingii in persons coinfected with human immunodeficiency virus," Clin Infect Dis, Vol. 33, No. 9, Nov 1 2001, pp. 1586–94.

[157] Lin, M., and Y. Rikihisa, "Ehrlichia chaffeensis and Anaplasma phagocytophilum lack genes for lipid A biosynthesis and incorporate cholesterol for their survival," Infect Immun, Vol. 71, No. 9, Sep 2003, pp. 5324–31.

[158] Thomas, S., N. Thirumalapura, E. C. Crossley, N. Ismail, and D. H. Walker, "Antigenic protein modifications in Ehrlichia," Parasite Immunol, Vol. 31, No. 6, Jun 2009, pp. 296–303.

[159] Li, J. S., E. Yager, M. Reilly, C. Freeman, G. R. Reddy, A. A. Reilly, F. K. Chu, and G. M. Winslow, "Outer membrane protein-specific monoclonal antibodies protect SCID mice from fatal infection by the obligate intracellular bacterial pathogen Ehrlichia chaffeensis," J Immunol, Vol. 166, No. 3, Feb 1 2001, pp. 1855–62.

[160] Nandi, B., K. Hogle, N. Vitko, and G. M. Winslow, "CD4 T-cell epitopes associated with protective immunity induced following vaccination of mice with an ehrlichial variable outer membrane protein," Infect Immun, Vol. 75, No. 11, Nov 2007, pp. 5453–9.

[161] Ohashi, N., N. Zhi, Y. Zhang, and Y. Rikihisa, "Immunodominant major outer membrane proteins of Ehrlichia chaffeensis are encoded by a polymorphic multigene family," Infect Immun, Vol. 66, No. 1, Jan 1998, pp. 132–9.

[162] Nyika, A., S. M. Mahan, M. J. Burridge, T. C. McGuire, F. Rurangirwa, and A. F. Barbet, "A DNA vaccine protects mice against the rickettsial agent Cowdria ruminantium," Parasite Immunol, Vol. 20, No. 3, Mar 1998, pp. 111–9.

[163] Nyika, A., A. F. Barbet, M. J. Burridge, and S. M. Mahan, "DNA vaccination with map1 gene followed by protein boost augments protection against challenge with Cowdria ruminantium, the agent of heartwater," Vaccine, Vol. 20, No. 7–8, Jan 15 2002, pp. 1215–25.

[164] Jongejan, F., and M. J. Thielemans, "Identification of an immunodominant antigenically conserved 32-kilodalton protein from Cowdria ruminantium," Infect Immun, Vol. 57, No. 10, Oct 1989, pp. 3243–6.

[165] Mwangi, D. M., D. J. McKeever, J. K. Nyanjui, A. F. Barbet, and S. M. Mahan, "Major antigenic proteins 1 and 2 of Cowdria ruminantium are targets for T-lymphocyte responses of immune cattle," Ann N Y Acad Sci, Vol. 849, Jun 29 1998, pp. 372–4.

[166] Gorvel, J. P., "Brucella: a Mr "Hide" converted into Dr Jekyll," Microbes Infect, Vol. 10, No. 9, Jul 2008, pp. 1010–3.
[167] Kohler, S., S. Michaux-Charachon, F. Porte, M. Ramuz, and J. P. Liautard, "What is the nature of the replicative niche of a stealthy bug named Brucella?," Trends Microbiol, Vol. 11, No. 5, May 2003, pp. 215–9.
[168] Baldwin, C. L., and M. Parent, "Fundamentals of host immune response against Brucella abortus: what the mouse model has revealed about control of infection," Vet Microbiol, Vol. 90, No. 1–4, Dec 20 2002, pp. 367–82.
[169] Perkins, S. D., S. J. Smither, and H. S. Atkins, "Towards a Brucella vaccine for humans," FEMS Microbiol Rev, Vol. 34, No. 3, May 2010, pp. 379–94.
[170] Verstreate, D. R., M. T. Creasy, N. T. Caveney, C. L. Baldwin, M. W. Blab, and A. J. Winter, "Outer membrane proteins of Brucella abortus: isolation and characterization," Infect Immun, Vol. 35, No. 3, Mar 1982, pp. 979–89.
[171] Cloeckaert, A., N. Vizcaino, J. Y. Paquet, R. A. Bowden, and P. H. Elzer, "Major outer membrane proteins of Brucella spp.: past, present and future," Vet Microbiol, Vol. 90, No. 1–4, Dec 20 2002, pp. 229–47.
[172] Bowden, R. A., S. M. Estein, M. S. Zygmunt, G. Dubray, and A. Cloeckaert, "Identification of protective outer membrane antigens of Brucella ovis by passive immunization of mice with monoclonal antibodies," Microbes Infect, Vol. 2, No. 5, Apr 2000, pp. 481–8.
[173] Jacques, I., A. Cloeckaert, J. N. Limet, and G. Dubray, "Protection conferred on mice by combinations of monoclonal antibodies directed against outer-membrane proteins or smooth lipopolysaccharide of Brucella," J Med Microbiol, Vol. 37, No. 2, Aug 1992, pp. 100–3.
[174] Bowden, R. A., A. Cloeckaert, M. S. Zygmunt, and G. Dubray, "Outer-membrane protein- and rough lipopolysaccharide-specific monoclonal antibodies protect mice against Brucella ovis," J Med Microbiol, Vol. 43, No. 5, Nov 1995, pp. 344–7.
[175] Estein, S. M., J. Cassataro, N. Vizcaino, M. S. Zygmunt, A. Cloeckaert, and R. A. Bowden, "The recombinant Omp31 from Brucella melitensis alone or associated with rough lipopolysaccharide induces protection against Brucella ovis infection in BALB/c mice," Microbes Infect, Vol. 5, No. 2, Feb 2003, pp. 85–93.
[176] Guilloteau, L. A., K. Laroucau, N. Vizcaino, I. Jacques, and G. Dubray, "Immunogenicity of recombinant Escherichia coli expressing

the omp31 gene of Brucella melitensis in BALB/c mice," Vaccine, Vol. 17, No. 4, Jan 28 1999, pp. 353–61.

[177] Gurunathan, S., C. Y. Wu, B. L. Freidag, and R. A. Seder, "DNA vaccines: a key for inducing long-term cellular immunity," Curr Opin Immunol, Vol. 12, No. 4, Aug 2000, pp. 442–7.

[178] Cassataro, J., C. A. Velikovsky, S. de la Barrera, S. M. Estein, L. Bruno, R. Bowden, K. A. Pasquevich, C. A. Fossati, and G. H. Giambartolomei, "A DNA vaccine coding for the Brucella outer membrane protein 31 confers protection against B. melitensis and B. ovis infection by eliciting a specific cytotoxic response," Infect Immun, Vol. 73, No. 10, Oct 2005, pp. 6537–46.

[179] Kaushik, P., D. K. Singh, S. V. Kumar, A. K. Tiwari, G. Shukla, S. Dayal, and P. Chaudhuri, "Protection of mice against Brucella abortus 544 challenge by vaccination with recombinant OMP28 adjuvanted with CpG oligonucleotides," Vet Res Commun, Vol. 34, No. 2, Feb 2010, pp. 119–32.

[180] Lim, J. J., D. H. Kim, J. J. Lee, D. G. Kim, W. Min, H. J. Lee, M. H. Rhee, and S. Kim, "Protective effects of recombinant Brucella abortus Omp28 against infection with a virulent strain of Brucella abortus 544 in mice," J Vet Sci, Vol. 13, No. 3, Sep 2012, pp. 287–92.

[181] Tibor, A., B. Decelle, and J. J. Letesson, "Outer membrane proteins Omp10, Omp16, and Omp19 of Brucella spp. are lipoproteins," Infect Immun, Vol. 67, No. 9, Sep 1999, pp. 4960–2.

[182] Pasquevich, K. A., S. M. Estein, C. Garcia Samartino, A. Zwerdling, L. M. Coria, P. Barrionuevo, C. A. Fossati, G. H. Giambartolomei, and J. Cassataro, "Immunization with recombinant Brucella species outer membrane protein Omp16 or Omp19 in adjuvant induces specific $CD4^+$ and $CD8^+$ T-cells as well as systemic and oral protection against Brucella abortus infection," Infect Immun, Vol. 77, No. 1, Jan 2009, pp. 436–45.

[183] Pasquevich, K. A., A. E. Ibanez, L. M. Coria, C. Garcia Samartino, S. M. Estein, A. Zwerdling, P. Barrionuevo, F. S. Oliveira, C. Seither, H. Warzecha, S. C. Oliveira, G. H. Giambartolomei, and J. Cassataro, "An oral vaccine based on U-Omp19 induces protection against B. abortus mucosal challenge by inducing an adaptive IL-17 immune response in mice," PLoS One, Vol. 6, No. 1, 2011, pp. e16203.

[184] Pasquevich, K. A., C. Garcia Samartino, L. M. Coria, S. M. Estein, A. Zwerdling, A. E. Ibanez, P. Barrionuevo, F. S. Oliveira, N. B. Carvalho, J. Borkowski, S. C. Oliveira, H. Warzecha, G. H. Giambartolomei, and

J. Cassataro, "The protein moiety of Brucella abortus outer membrane protein 16 is a new bacterial pathogen-associated molecular pattern that activates dendritic cells in vivo, induces a Th1 immune response, and is a promising self-adjuvanting vaccine against systemic and oral acquired brucellosis," J Immunol, Vol. 184, No. 9, May 1 2010, pp. 5200–12.

[185] Fleischmann, R. D., M. D. Adams, O. White, R. A. Clayton, E. F. Kirkness, A. R. Kerlavage, C. J. Bult, J. F. Tomb, B. A. Dougherty, J. M. Merrick, and *et al.*, "Whole-genome random sequencing and assembly of Haemophilus influenzae Rd," Science, Vol. 269, No. 5223, Jul 28 1995, pp. 496–512.

[186] Giuliani, M. M., J. Adu-Bobie, M. Comanducci, B. Arico, S. Savino, L. Santini, B. Brunelli, S. Bambini, A. Biolchi, B. Capecchi, E. Cartocci, L. Ciucchi, F. Di Marcello, F. Ferlicca, B. Galli, E. Luzzi, V. Masignani, D. Serruto, D. Veggi, M. Contorni, M. Morandi, A. Bartalesi, V. Cinotti, D. Mannucci, F. Titta, E. Ovidi, J. A. Welsch, D. Granoff, R. Rappuoli, and M. Pizza, "A universal vaccine for serogroup B meningococcus," Proc Natl Acad Sci U S A, Vol. 103, No. 29, Jul 18 2006, pp. 10834–9.

[187] Pizza, M., V. Scarlato, V. Masignani, M. M. Giuliani, B. Arico, M. Comanducci, G. T. Jennings, L. Baldi, E. Bartolini, B. Capecchi, C. L. Galeotti, E. Luzzi, R. Manetti, E. Marchetti, M. Mora, S. Nuti, G. Ratti, L. Santini, S. Savino, M. Scarselli, E. Storni, P. Zuo, M. Broeker, E. Hundt, B. Knapp, E. Blair, T. Mason, H. Tettelin, D. W. Hood, A. C. Jeffries, N. J. Saunders, D. M. Granoff, J. C. Venter, E. R. Moxon, G. Grandi, and R. Rappuoli, "Identification of vaccine candidates against serogroup B meningococcus by whole-genome sequencing," Science, Vol. 287, No. 5459, Mar 10 2000, pp. 1816–20.

[188] Toneatto, D., S. Ismaili, E. Ypma, K. Vienken, P. Oster, and P. Dull, "The first use of an investigational multicomponent meningococcal serogroup B vaccine (4CMenB) in humans," Hum Vaccin, Vol. 7, No. 6, Jun 2011, pp. 646–53.

[189] Ross, B. C., L. Czajkowski, D. Hocking, M. Margetts, E. Webb, L. Rothel, M. Patterson, C. Agius, S. Camuglia, E. Reynolds, T. Littlejohn, B. Gaeta, A. Ng, E. S. Kuczek, J. S. Mattick, D. Gearing, and I. G. Barr, "Identification of vaccine candidate antigens from a genomic analysis of Porphyromonas gingivalis," Vaccine, Vol. 19, No. 30, Jul 20 2001, pp. 4135–42.

[190] Ross, B. C., L. Czajkowski, K. L. Vandenberg, S. Camuglia, J. Woods, C. Agius, R. Paolini, E. Reynolds, and I. G. Barr, "Characterization of two outer membrane protein antigens of Porphyromonas gingivalis that are protective in a murine lesion model," Oral Microbiol Immunol, Vol. 19, No. 1, Feb 2004, pp. 6–15.

[191] Brayton, K. A., L. S. Kappmeyer, D. R. Herndon, M. J. Dark, D. L. Tibbals, G. H. Palmer, T. C. McGuire, and D. P. Knowles, Jr., "Complete genome sequencing of Anaplasma marginale reveals that the surface is skewed to two superfamilies of outer membrane proteins," Proc Natl Acad Sci U S A, Vol. 102, No. 3, Jan 18 2005, pp. 844–9.

[192] Noh, S. M., K. A. Brayton, W. C. Brown, J. Norimine, G. R. Munske, C. M. Davitt, and G. H. Palmer, "Composition of the surface proteome of Anaplasma marginale and its role in protective immunity induced by outer membrane immunization," Infect Immun, Vol. 76, No. 5, May 2008, pp. 2219–26.

[193] Galletti, M. F., M. W. Ueti, D. P. Knowles, Jr., K. A. Brayton, and G. H. Palmer, "Independence of Anaplasma marginale strains with high and low transmission efficiencies in the tick vector following simultaneous acquisition by feeding on a superinfected mammalian reservoir host," Infect Immun, Vol. 77, No. 4, Apr 2009, pp. 1459–64.

[194] Gomez, G., J. Pei, W. Mwangi, L. G. Adams, A. Rice-Ficht, and T. A. Ficht, "Immunogenic and Invasive Properties of Brucella melitensis 16M Outer Membrane Protein Vaccine Candidates Identified via a Reverse Vaccinology Approach," PLoS One, Vol. 8, No. 3, 2013, pp. e59751.

[195] Bharti, A. R., J. E. Nally, J. N. Ricaldi, M. A. Matthias, M. M. Diaz, M. A. Lovett, P. N. Levett, R. H. Gilman, M. R. Willig, E. Gotuzzo, and J. M. Vinetz, "Leptospirosis: a zoonotic disease of global importance," Lancet Infect Dis, Vol. 3, No. 12, Dec 2003, pp. 757–71.

[196] McBride, A. J., D. A. Athanazio, M. G. Reis, and A. I. Ko, "Leptospirosis," Curr Opin Infect Dis, Vol. 18, No. 5, Oct 2005, pp. 376–86.

[197] Goure, J., W. A. Findlay, V. Deslandes, A. Bouevitch, S. J. Foote, J. I. MacInnes, J. W. Coulton, J. H. Nash, and M. Jacques, "Microarray-based comparative genomic profiling of reference strains and selected Canadian field isolates of Actinobacillus pleuropneumoniae," BMC Genomics, Vol. 10, 2009, pp. 88.

[198] Chen, X., Z. Xu, L. Li, H. Chen, and R. Zhou, "Identification of conserved surface proteins as novel antigenic vaccine candidates of

Actinobacillus pleuropneumoniae," J Microbiol, Vol. 50, No. 6, Dec 2012, pp. 978–86.

[199] Montigiani, S., F. Falugi, M. Scarselli, O. Finco, R. Petracca, G. Galli, M. Mariani, R. Manetti, M. Agnusdei, R. Cevenini, M. Donati, R. Nogarotto, N. Norais, I. Garaguso, S. Nuti, G. Saletti, D. Rosa, G. Ratti, and G. Grandi, "Genomic approach for analysis of surface proteins in Chlamydia pneumoniae," Infect Immun, Vol. 70, No. 1, Jan 2002, pp. 368–79.

12

Systems Biology Approaches to New Vaccine Development

Jing Sun[1], Guang Lan Zhang[1,2] and Vladimir Brusic[1,2]

[1]Cancer Vaccine Center,
Dana-Farber Cancer Institute,
Boston, MA, USA
[2]Department of Computer Science,
Metropolitan College, Boston University,
Boston, MA, USA

12.1 Introduction

Advances in instrumentation, informatics, and biotechnology have enabled us to measure biological processes and their components on a massive scale. We can rapidly and effectively identify genes, proteins, metabolites, and other components of living cells, their forms, expression, quantity, and change over time or under some stimulus. Understanding the components of living cells is important and these details can provide information about their presence or absence, function, interactions, and we can establish correlations between genes and proteins and certain physiological states. Properties of individual genes and proteins (and other components) alone do not provide insights into their collective function. To understand biological function at cellular, tissue, or organism level we need to understand the behavior and relationships of all components of the observed system, their interconnectivity, and dynamics of change of their relationships [1, 2]. To understand systemic behavior of biological systems we need to study four major properties: system structure, systems dynamics, control mechanisms, and the underlying model [2]. System structure focuses on gene and protein interactions, and resulting networks and pathways [3]. Such maps are largely static and it

Post-genomic Approaches in Drug and Vaccine Development, 323–346.

is essential to understand the systems dynamics, particularly the perturbations in these interactions that result from various stimuli and can result in pathological changes that link cells to tissues and organ systems [4]. The underlying molecular mechanisms that control the state of the cell and signaling between cells must be understood so that we can distinguish healthy from the pre-disease and disease states [5–7]. Information about network maps, their dynamics, and control mechanisms define the model that describes biological states, processes, and their perturbations. Such models enable *in silico* simulations that, combined with selected experimentation enable systemic approach in biomedicine [8]. Smaller models are integrated into large-scale simulation systems that transcend cellular, organ, and organism levels [9–11].

Infectious diseases are a main public health problem worldwide. Diseases such as tuberculosis, malaria, hepatitis C, and AIDS (HIV) are currently widespread. Vaccines against these diseases are non-existent or ineffective. In addition, vaccines against influenza are not effective against future strains and annual epidemics and occasional pandemics occur. Systems biology focuses on characterizing interactions between the host and pathogen at different stages of infectious cycle [12]. The examples of ongoing efforts include study of tuberculosis [13], HIV [14], influenza [15], flaviviruses[16, 17], fungi [18], as well as various other zoonotic pathogens [19]. Understanding the pathogen and host-pathogen interactions provides critical knowledge for further vaccine applications. Data analysis and integration of multiple sources along with targeted validation leads to the development of models that capture properties and behavior of the host-pathogen system. These models can be used to predict behavior of the system as well as the outcomes of infection (or vaccination). Validatory experiments based on prediction help refine the model and increase its accuracy, leading to more sophisticated tools for systems biology applications [12].

Vaccines offer protection against a range of infectious diseases and are considered among the most effective tools of public health medicine. The vast majority of successful vaccines were developed using empirical approach of trial-and-error but new technologies that support rational vaccine development are emerging. Developing and optimizing modern vaccines requires a multi-disciplinary approach involving basic science, immunology, molecular biology, epidemiology, clinical vaccinology, systems biology, and bioinformatics [20]. Rational vaccine design has several key questions: (a) what antigens are suitable or best targets, (b) what type of immunity is needed for protection, (c) what technologies need to be applied so that vaccines can fulfill the desired goals, and (d) what is needed to translate results from preclinical and ex vivo results into clinical vaccine applications [21]? The analysis of genotype–phenotype data and immune profiling is necessary for understanding the effects of genetic variation, molecular interactions, and the heterogeneiety of immune responses to vaccines [22]. Systems biology is an important component of this quest and will play a critical role in deciphering combinatorial complexity of immune responses to both infection and vaccination.

12.2 Systems Biology for Vaccine Development

Despite the highly successful development of vaccines that are able to elicit potent and protective immune responses, the majority of vaccines developed so far empirically and mechanistic events leading to protective immune responses are often poorly understood [23–25]. Immune-mediated protection is assessed by measuring biological correlates from standard readouts of antibody titres and ELISPOT assays [26]. Systems biology offers new approaches for vaccine design. These approaches seek the understanding of molecular networks activated or perturbed by vaccination [27]. Systems biology approaches in vaccinology include capturing and investigating large global sets of biological data and provide new methods for identification of early innate signatures. These early signatures are thought to predict immunogenicity of vaccines, and generate hypotheses for understanding the mechanisms that underpin successful immunogenicity [27–29].

Systems biology is the comprehensive and quantitative analysis of the interactions between large numbers of the components of biological systems measured over time. Systems biology involves an iterative cycle, in which emerging biological problems drive the development of new technologies and computational tools. These technologies and tools then open new frontiers that revolutionize biology [27].

12.2.1 Integration of Biological Data

Biological systems are more than simple collections of genes and proteins – they are complex, intricately interacting sets of functional and sometimes redundant pathways that collectively produce coherent behaviors[2, 30]. Both the innate and adaptive immune responses are examples of such behavior. With vaccine administration, the immune system will be stimulated to develop first innate immune responses followed by adaptive immune responses. The immune responses to vaccination involve coordinated induction of master transcription factors that leads to the development of a broad, poly-functional and persistent immune response integrating all effector cells of the immune systems [12]. Over the past decade or so, many high dimensional assays ("omics" technologies) become available. They can be organized according to the biological system or network within an organism that is measured, including molecular measurements of DNA sequences, RNA and protein expression levels, microRNAs, protein–protein and protein–DNA interactions and metabolite biology [29, 31, 32]. Systems biology approaches enable integration of large datasets from a variety of assays [33].

Every vaccine requires unique features based on the biology of the pathogen, the nature of the disease and the target population for vaccination [20]. Future vaccine development will be increasingly more difficult for more complex organisms. Large-scale analysis of related pathogens also requires advanced statistical, bioinformatics, and modeling tools. Of potential help for this effort, the discipline of systems biology is designed to take a holistic approach to understanding a biological system by integrating analyses of many measurements of the system under different perturbations [29]. Systems biology approaches enable the integration of large biological datasets from hierarchical levels and visualization of 'emergent properties' in immune responses, which are not demonstrated by individual parts from immune system and cannot be predicted from the parts alone [29], become possible.

12.2.2 Prediction of Vaccine Efficacy

The response of an individual to vaccination depends on a multitude of interacting genetic, molecular and environmental factors spanning numerous temporal and spatial scales. This is in stark contrast to the empirical "isolate, inactivate, inject" paradigm of vaccine development. Empirical approaches have limited capacity to take into account the pathogen and host variability [30]. One of the major applications of systems biology in vaccinology is prediction of vaccine efficacy. In cancer genomics, gene expression signatures have been used to predict the patient's clinical outcome and response to therapies.

In vaccinology, patterns of gene expression induced after vaccination (i.e., signatures) could be used to predict immunogenicity or efficacy [32].

One of the earliest systems biology applications in vaccinology is the yellow fever vaccine YF-17D. A machine-learning approach was adopted to identify early gene signatures that were capable of predicting the immunogenicity (to be specific, the responses of B-cells and CD8 + T-cells) of individuals vaccinated with the YF-17D or the inactivated influenza vaccine [16] This example demonstrates and validates the systems biology approach as a powerful tool for predicting vaccine immunogenicity [26].

Individual immune responses after vaccination might be largely diverse among different groups of people. The identification of signatures in individuals who respond sub-optimally to vaccination (e.g., the elderly, infants, and the immunocompromised) is crucial. For example, the immune system changes by age, with innate and adaptive cell components becoming increasingly dysfunctional. Vaccine responses in the elderly may be impaired in ways that differ depending on the type of vaccine (e.g., live attenuated, polysaccharide, conjugate, or subunit) and the mediators of protection (e.g., antibody and/or T-cell). Future application of systems biology to vaccination in the elderly may help to identify gene signatures that predict suboptimal responses and help to identify more accurate correlates of protection [34]. The identification of immune signatures can also provide utility for rapid screening of first responders during emergency outbreaks, vaccinees who respond sub-optimally, identify non-responders in partially effective vaccines (e.g., RTS, S malaria vaccine), rapidly assess vaccine immunogenicity and efficacy (e.g., new meningococcal vaccine) and identify novel correlates of immunity and/or protection [32].

By discovering molecular signatures and predicting how the immune system will behave, the researchers can address optimization of vaccines in the earliest stages of development, before it reaches expensive late-stage trials. Clinical development will be accelerated by performing more efficient vaccine trials [28]. These studies can be facilitated using predictive models based on available data. An ideal predictive model should be sufficiently accurate and detailed to allow predictions whether novel vaccines will lead to protective responses.

12.2.3 Identification of Novel Immune Regulation Mechanisms

The specific mechanisms of action of successful vaccines (empirically made) remain largely unknown. Designing a vaccine that achieves better protection, however, will require a more complete understanding of vaccine mechanisms of action and correlates of protection [33]. With integration of large datasets from a variety of assays and introduction of new approaches to identify vaccine-related immune response signatures, systems biology can provide insights into the molecular mechanisms in immune responses. One example of systems biology approaches is the systematic analysis of the yellow fever vaccine YF-17D, one of the most efficacious vaccines ever developed. In this case, systems biology has provided insight into its mechanism of action [24]. The recent modest success of the RV144 HIV vaccine trial in Thailand has also shown the power of systems biology approach [33].

Another field that will benefit from systems approaches is the deciphering of mechanisms by which adjuvants work [32]. Comprehensive analysis of immune responses to vaccines and immunotherapies at the system level (vaccinomics or systems vaccinology) will provide knowledge in adjuvant antigens for optimized vaccine development, inducing the sought cluster of genes and immune pathways leading to the required adaptive immune response [35].

Systems biology provides key technologies and approaches for deciphering complex relationships between pathogen, host, and vaccines and enables an insight into collective workings of the genetic, proteomic, and other components of the living cells in response to challenges – infection and vaccination. The interplay of these components and its temporal dynamics provide clues for vaccine design, related diagnostics, and optimization of vaccines and vaccination protocols.

12.3 Technologies

The current advancement of new high-throughput technologies enables the analysis of the genome, transcriptome, and proteome creating the opportunity for understanding mechanism of action and optimization of vaccine design. These technologies enable better understanding of pathogen biology, the host immune system, and host-pathogen interactions [31, 36]. Systems biology enables the rational design of new antimicrobials because their design requires understanding of drug action against microbial targets at the molecular level, of drug distribution and pharmacokinetics at the host level, and the effects of mutations on microbial fitness in immunocompetent and immunodeficient hosts at the population level [37].

12.3.1 Genomics

The completion of the first bacterial genome, that of *Haemophilusinfluenzae* in 1995 [38], represented a major milestone in both biological research and vaccine development. With an exponential growth of genome sequence information, at least one genome sequence is now available for each major human pathogen [39]. As of October 2013, 6851 bacterial genomes have been completed, sequencing of 18679 are ongoing and 1111 represent targeted genome projects (GOLD Genomes OnLine Database, http://www.genomesonline.org) [40].

The characterization of pathogens and host genomes provide huge amounts of data and new knowledge that will help modern vaccine design. The sequence-based "reverse vaccinology" approaches represent rational approach to vaccine design [41]. Reverse vaccinology starts with bioinformatics screening of entire pathogenic genome to find target genes. These genes are then screened for features that indicate antigenicity, such as protein coding properties, cellular localization, signal peptides, B- and T-cell cell epitopes. The selected targets are then scored for their potential as vaccine targets. Selected proteins then are verified in wet lab experiments for their potential to induce immune responses. The ability to sequence many pathogen genomes provides access to the entire antigenic repertoire of the studied pathogen helps identify and further study best vaccine targets [42].

Serogroup B *Neisseria meningitidis* (MenB), is the most common cause of meningococcal disease in the developed world. It was the first pathogen addressed by the reverse vaccinology approach. Several decades of conventional vaccine development failed to produce a comprehensive vaccine [39], yet the use of reverse vaccinology by Novartis Vaccines [43] identified more vaccine candidates in 18 months than had been discovered during the previous 40 years. This technological breakthrough has enabled the MenB vaccine into clinical development. After that, a series of vaccines have been developed against other bacterial pathogens with reverse vaccinology. *Streptococcus agalactiae* (GBS) is a multi-serotype bacterial pathogen that mostly transmits from mother to the newborn during delivery, causing sepsis, pneumonia and meningitis in the first month of life [42]. The genomes of eight GBS isolates were analyzed and 312 surface proteins were tested as vaccines. A vaccine composed of four proteins was able to protect against all serotypes of the GBS [44]. This example marks the transition of reverse vaccinology the classic one-genome approach to the study of pan-genome. Traditional genome-based approaches, such as comparative genomics, subtractive genomics, and incorporation of proteomics, were subsequently used for vaccine development of various pathogens, such as *Streptococcus pyogenes* (GAS) [45, 46], and *Streptococcus pneumonia* [47–49].

12.3.2 Proteomics

Proteomics is the large-scale study of the protein products of a cell,particularly their expression, structure and function. The availability of a growing number of complete genomes and related genomics data provide researchers with the information of their protein products. However, genomics alone is not sufficient to decipher protein products of a genome. The reason for that is proteome is dynamic, and there is no strict linear relationship between gene and protein expression, splice variation, and post-translational modifications [50]. With the advances in protein preparation, identification and analyses, the elucidation of total protein components of a given cellular population becomes a feasible task [51,52]. Technologies for proteomics-based approaches are rapidly advancing and are complementary to classical genomics-based approaches that focus on gene expression and characterization.

For most bacterial pathogens, proteins that are able to elicit protective immune responses are either secreted proteins or surface-exposed proteins, and these represent the most promising vaccine candidates. *In silico* analysis of the complete sequence of several bacterial genomes predicts that surface-associated proteins constitute 30%–40% of all bacterial proteins. 2D-PAGE coupled to mass spectrometry (MS), chromatographic techniques, and protein arrays are the principal proteomic methods used for analysis. They are usually in high-throughput mode targeting the complete protein profile of a microorganism, including protein localization, protein-protein interactions, post-translation modifications, and differential expression in specified conditions [39].

The proteomics-based approach was used in the study of *Streptococcus pyogenes* (GAS). Bacterial surface proteins were isolated [53] with proteolytic enzymes under conditions that preserve cell viability, and the peptides released were analyzed by the MS. The effectiveness of this approach is supported by the fact that 95% of the 72 proteins identified in this way from the completely sequenced M1-SF370 strain of GAS [54] were predicted to be cell wall proteins, lipoproteins, transmembrane-spanning proteins, and secreted proteins and included most of the protective antigens described to date plus a novel protective antigen. Surface exposure of most of these proteins was confirmed by FACS analysis [53].

12.3.3 Bioinformatics

The past decades have witnessed rapid development of biological technologies, along with computational methods that include databases, tools and computational models for data management and presentation, text mining and analysis. These methods enable the unprecedented descriptions of complex biological systems, as well as computational simulations and predictions. Bioinformatics methods are advancing along with the high-throughput technologies. Their iterative and combinatorial application furthers the development systems biology discipline [55].

12.3.3.1 Databases

Databases are essential sources of information for the systems biology. The most widely used databases are major repositories of nucleotide and protein sequences such as GenBank[56] and UniProt[57]. Specialized databases provide additional information such as mutations [58] or splice variation [59]. Additional relevant information includes molecular interactions databases NPIDB [60], STRING [61], IntAct[62], BioGRID[63], ChEMBL[64], iRefIndex[65] and MINT [66]. Databases of molecular pathways include KEGG [67], and Reactome[68], among others. The issues with the databases include the completeness, accuracy of entries, missing data, data interchange standards, inconsistent annotations between different sources. These issues can be addressed by evaluating data from different sources, and use data standards for description of datasets, the workflows used to generate these data and the scientific context for the work. These standards are defined in the Minimal Information for Biological and Biomedical Investigations (MIBBI) – a project focused on developing minimum information guidelines for description of experiments and reporting of biological and biomedical data, terminology artifacts, and exchange formats [69]. MIBBI has more than 30 subprojects covering standards in a diversity of bioscience and bioinformatics fields (http://mibbi.sourceforge.net). The BioSharing site lists more than 500 standards and more than 600 major databases that are standards–compliant [70]. Some of the related standards include various ontologies including Vaccine Ontology (VO) [71]. A rich set of links to vaccine resources can be found at the www.violinet.org web site. This site lists databases of immune epitopes (T- and B-cell epitopes), vaccine components (antigens, adjuvants, and vectors), vaccine responses, virulence factors, and vaccine related host-pathogen interactions [71]. The new generation of vaccine databases integrates data and analytical tools with the analysis workflows to enable data mining and automatic generation of reports. Examples of such databases include systems for knowledge discovery and vaccine applications such as FLAVIdB[72] and HPVdb[73] that enable vaccine applications for flaviviruses and human papilloma virus, respectively.

12.3.3.2 Tools

A variety of open sources tools have been developed for researches in systems biology and most have been integrated within several platforms. The Systems Biology Toolbox for MATLAB offers systems biologists an open and extensible environment, that enables them to explore ideas, perform prototyping and share new algorithms, and build applications for the analysis and simulation of biological and biochemical systems [74]. R and Bioconductor are also popular toolsets widely used in systems biology studies [75].

Rapid accumulation of biological data in the 'omics' era creates the problem of data representation. Visualization of large and complex datasets becomes increasingly important for data analysis and interpretation. The clear, meaningful and integrated visualization tools, which facilitate biological insights and prevent overwhelming by the intrinsic complexity of the data, are in high demand [76]. For example, Cytoscape[77], that enables interactive changes of the layout of the network, is one of the excellent protein network visualization tools. At the same time, a wide range of tools have been developed to aid the visualization of high-throughput expression profiling data [78], such as ExpressionProfile[79] and GenePattern[80].

12.3.3.3 Computational models

Accompanied with the development of systems biology databases and tools, scientific computing and mathematical modeling of biological processes have started to fundamentally impact the research [55]. A network perspective on cancer was strongly motivates the application of computational modeling approaches in all other fields [81, 82]. Methodologies for computational analysis can vary widely depending on the question being posed and the experimental data at hand, ranging from highly abstracted models using correlative regression to highly specified models using differential equations, with network component interaction and logic modeling techniques intermediate to these. The growth of the datasets and their complexity tests the limits of available statistical analyses, since the discovery of patterns and correlations within large-scale data provides clues to functional significance of individual components and their combinations. The data-rich environment exploration heavily depends on the use of computational models and in particular performing simulations that provide clues about the system behavior. Computational models are important for several reasons. They [83]:

- provide coherent analytical framework for data interpretation
- focus on central concepts that can be applied broadly

- can be used for *de novo* characterization of concepts and phenomena
- help identify key or components of a system
- enable linking of components described on multiple levels
- provide for the formalization of otherwise intuitive understanding
- help screening of a large space of possible hypotheses
- inform the design of experiments
- predict variables that are impossible to measure
- perform qualitative and quantitative predictions
- Systems biology creates an enormous search space, large number of possible binary relationships, and enormous number of possible relationships involving three or more features. This creates a combinatorial problem of enormous scale and statistical methods are insufficient for study of this space. Therefore, the need for mathematical models and their computational implementations is necessary for making sense of systems biology data. These models must also include well-defined concepts that capture biological principles, structures and processes that define studied systems.

12.4 Challenges

Systems biology enables rational vaccine design but application of systems biology is not straight forward. A number of issues arise both in general and in relation to vaccinology. First, there are technical and methodological issues related to the analysis of the large, noisy and multidimensional data generated by different high-throughput techniques. The reproducibility of results reported in the studies is generally low between different studies, and between different but similar patient cohorts; significant differences exist between different technology platforms [84–86]. Another major challenge concerns the biological complexity underlying the immune system [14, 29]. Constructing the bridges between molecular regulation processes and the innate and adaptive immunity to composition of vaccines and vaccination protocols is essential. The identified signatures that are proposed to be predictive of outcomes have not yielded any mechanistic insights. The disconnection between measurable features and the ability to describe the causality is a major gap in our knowledge. The data should provide insights into the mechanisms and processes that drive protection afforded by vaccination. Signatures that are both predictive of protection or other outcomes along with the explanation of causal relationships will lead to faster and cheaper vaccine trials and ultimately to broadly protective, optimized vaccines that serve the needs of population and individuals. Furthermore, high-throughput methods produce increasing quantities of data that, in turn, lead to increasing amounts of errors, biases, and artifacts. The fundamental property of high-throughput systems is a very low ratio of number of samples to number of measurable features, which alone creates the environment where statistically significant correlations arise more often due to sample biases, peculiarities of the datasets, and experimental conditions than the actual biological processes [87]. Some of these are preventable through proper design of experiments, careful checking of data, adequate statistics, and inclusion of constraints based on well-established knowledge about underlying biology. Finally, as a trans-disciplinary collaboration between immunologists and computational scientists, the application of systems biology on vaccinology requires that both sides – immunology/vaccinology and information science/biostatistics are utilized to the full capacity.

12.5 Examples

12.5.1 Yellow Fever Vaccine YF-17D

Yellow fever, a mosquito-borne viral hemorrhagic disease, is one of the most lethal diseases of humankind. The yellow fever vaccine YF-17D is one of the most successful vaccines ever developed [88, 89]. This vaccine stimulates polyvalent innate immune responses and adaptive immune responses that persist for decades after vaccination [90, 91]. A systems biology approach has been used in the vaccine efficacy analysis of YF-17D [91]. In this study, the patterns of early gene "signatures", including complement protein C1qB and eukaryotic translation initiation factor 2 alpha kinase 4 (an orchestrator of the integrated stress response), have been identified to correlate with YF-17D CD8 + T-cell responses. They proposed that these signatures can be used to predict immune responses in individuals vaccinated with the live attenuated vaccine YF-17D [16, 24]. This result is a demonstration of systems biology approaches in finding gene signatures and predicting vaccine efficacy, which marks the great potential utilization of systems biology in optimal vaccine design. Further studies are needed to elucidate the mechanisms of protection.

12.5.2 Seasonal Influenza Vaccine

Another promising application of systems biology is the signature discovery in individuals vaccinated against seasonal influenza. The study of innate and adaptive responses to vaccination against influenza in humans during three consecutive influenza seasons was performed [15]. Unlike the situation with YF-17D, majority of the individuals vaccinated against seasonal influenza had already been exposed to the virus, through previous infections or vaccinations. Thus, an additional complexity of this study was that the induced immune

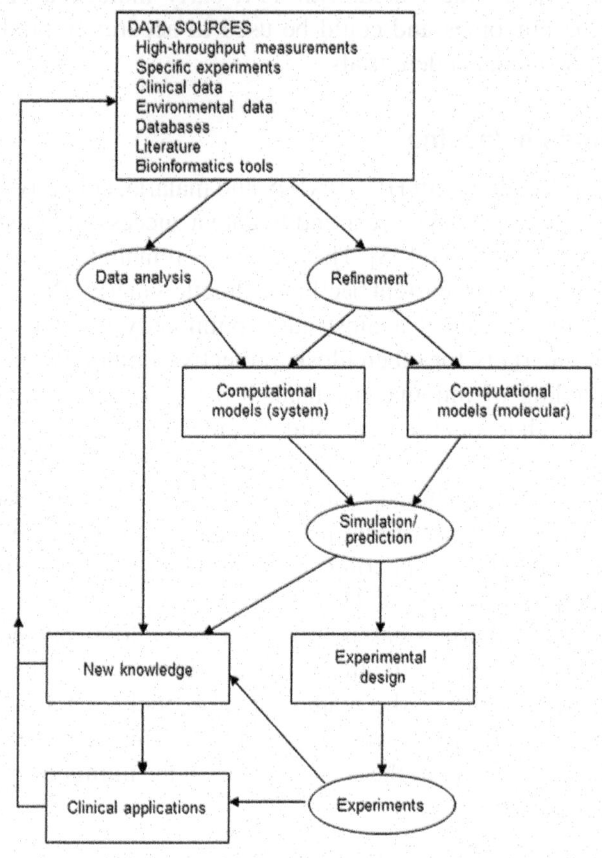

Figure 12.1 The Immunomics paradigm [96, 97]. The existing data are analyzed and used to build and subsequently refine computational models for simulation of studied systems and processes. In silico simulations generate new testable hypotheses or produce new knowledge outright. Subsequent experimental design is targeted to specifically test the newly generated hypotheses. The experiments can be either large scale measurements or focused experiments that test individual hypotheses. New knowledge produced by combination of data analysis, computational simulations, and experiments together produce is returned to enrich data sources, or streamlined into clinical applications, in the areas of diagnosis, therapy or prevention. Vaccine development, being a combinatorial problem, will benefit significantly from the Immunomics paradigm.

responses could result from newly developed immunogenicity or recall of memory responses. This study focused on predicting the immunogenicity of recall immune responses [14]. Healthy adults were vaccinated with trivalent inactivated influenza vaccine (TIV) or live attenuated influenza vaccine (LAIV). TIV induced higher antibody titers and more plasmablasts than LAIV vaccinations. In subjects vaccinated with TIV, early molecular signatures correlated with the outcomes and could be used to accurately predict later antibody titers in two independent trials.

12.5.3 Tuberculosis Vaccine

Tuberculosis (TB), together with HIV-1/AIDS and malaria, collectively cause more than five million deaths per year and to date a successful conventional vaccine has not been developed [28]. These diseases remain the major global public health challenges. The current bacillus Calmette-Guérin (BCG) vaccine has limited capacity to protect against adult pulmonary tuberculosis. The lack of defined biomarkers for tuberculosis protective immunity and lack of detailed understanding of immune mechanism between host and pathogen constitute the critical barriers for successful vaccine development [13].

Systems biology approaches have been proven useful in furthering the understanding of the host-pathogen interactions as well as elucidating the related patterns of activation of various metabolic pathways [13]. Tuberculosis was used as a case study by deciphering complex network interactions which focused on the temporal analysis of TB interaction with the host [37]. Mathematical models and techniques based upon systems biology data have been used for the development novel treatment for tuberculosis and other clinical disease states [92, 93]. One of the mathematical models of the early lung immune response to tuberculosis was proposed [94]. This model effectively illustrated and brought into focus the consequences of the immunosuppressive nature of the lung environment.

12.6 Conclusions

Vaccination is one of the great triumphs in modern medicine by preventing or protecting humans from infectious diseases. In the past decades, advances in systems biology and related technologies have largely influenced vaccinology and made remarkable insights into vaccine development. We are currently undergoing a paradigm change in vaccinology thinking. The empiric trial-and-error approach in the discovery and development of vaccines produces a low number of suitable vaccine candidates and vaccines of low efficiency. Systems biology, reverse vaccinology, vaccine modeling, and immunoinformatics [22, 26, 27, 29, 30, 32, 35, 41, 89, 95, 96] together with experimental and clinical vaccinology are leading the change in the vaccine development paradigm. With the accumulation of data and deepening the knowledge of key concepts, mechanisms and processes, we are getting closer to rational design of vaccines. The "Vaccinomics paradigm" [22, 35, 95] fits within the "Immunomics paradigm" [97] that combines data, models, experimentation and clinical immunology/vaccinology for generation of both new knowledge and clinical applications. The immunomics paradigm is explained in Figure 12.1. Systems biology is a key driver of modern approaches that help develop deep understanding of mechanisms of how vaccines stimulate immune responses, discovery of biomarkers that predict vaccine response, and understanding of factors that lead to vaccine failure or to undesirable side effects of vaccination. Integrating hierarchical levels of immunological data, predicting vaccine efficacy, and understanding the fine workings of human immune responses stimulated by vaccination are the goals of systems biology in vaccinology. With the development of genomics and post-genomics technologies, systems biology approaches will become more powerful in providing insights to complex systems issues.

References

[1] Ideker T, Galitski T, Hood L: A new approach to decoding life: systems biology. Annual review of genomics and human genetics 2001, 2: 343–372.

[2] Kitano H: Systems biology: a brief overview. Science 2002, 295(5560): 1662–1664.

[3] Kelley R, Ideker T: Systematic interpretation of genetic interactions using protein networks. Nature biotechnology 2005, 23(5): 561–566.

[4] Barabasi AL, Gulbahce N, Loscalzo J: Network medicine: a network-based approach to human disease. Nature reviews Genetics 2011, 12(1): 56–68.

[5] Hahn WC, Weinberg RA: Modelling the molecular circuitry of cancer. Nature reviews Cancer 2002, 2(5): 331–341.

[6] Calvano SE, Xiao W, Richards DR, Felciano RM, Baker HV, Cho RJ, Chen RO, Brownstein BH, Cobb JP, Tschoeke SK, Miller-Graziano C, Moldawer LL, Mindrinos MN, Davis RW, Tompkins RG, Lowry SF: A network-based analysis of systemic inflammation in humans. Nature 2005, 437(7061): 1032–1037.

[7] Shapira SD, Gat-Viks I, Shum BO, Dricot A, de Grace MM, Wu L, Gupta PB, Hao T, Silver SJ, Root DE, Hill DE, Regev A, Hacohen N: A physical and regulatory map of host-influenza interactions reveals pathways in H1N1 infection. Cell 2009, 139(7): 1255–1267.
[8] Meng TC, Somani S, Dhar P: Modeling and simulation of biological systems with stochasticity. In silico biology 2004, 4(3): 293–309.
[9] Salwinski L, Eisenberg D: In silico simulation of biological network dynamics. Nature biotechnology 2004, 22(8): 1017–1019.
[10] Pappalardo F, Halling-Brown MD, Rapin N, Zhang P, Alemani D, Emerson A, Paci P, Duroux P, Pennisi M, Palladini A, Miotto O, Churchill D, Rossi E, Shepherd AJ, Moss DS, Castiglione F, Bernaschi M, Lefranc MP, Brunak S, Motta S, Lollini PL, Basford KE, Brusic V: ImmunoGrid, an integrative environment for large-scale simulation of the immune system for vaccine discovery, design and optimization. Briefings in bioinformatics 2009, 10(3): 330–340.
[11] Hofmann-Apitius M, Fluck J, Furlong L, Fornes O, Kolarik C, Hanser S, Boeker M, Schulz S, Sanz F, Klinger R, Mevissen T, Gattermayer T, Oliva B, Friedrich CM: Knowledge environments representing molecular entities for the virtual physiological human. Philosophical transactions Series A, Mathematical, physical, and engineering sciences 2008, 366(1878): 3091–3110.
[12] Aderem A, Adkins JN, Ansong C, Galagan J, Kaiser S, Korth MJ, Law GL, McDermott JG, Proll SC, Rosenberger C, Schoolnik G, Katze MG: A systems biology approach to infectious disease research: innovating the pathogen-host research paradigm. mBio 2011, 2(1): e00325–00310.
[13] Wang CC, Zhu B, Fan X, Gicquel B, Zhang Y: Systems approach to tuberculosis vaccine development. Respirology 2013, 18(3): 412–420.
[14] Nakaya HI, Pulendran B: Systems vaccinology: its promise and challenge for HIV vaccine development. Current opinion in HIV and AIDS 2012, 7(1): 24–31.
[15] Nakaya HI, Wrammert J, Lee EK, Racioppi L, Marie-Kunze S, Haining WN, Means AR, Kasturi SP, Khan N, Li GM, McCausland M, Kanchan V, Kokko KE, Li S, Elbein R, Mehta AK, Aderem A, Subbarao K, Ahmed R, Pulendran B: Systems biology of vaccination for seasonal influenza in humans. Nature immunology 2011, 12(8): 786–795.
[16] Querec TD, Akondy RS, Lee EK, Cao W, Nakaya HI, Teuwen D, Pirani A, Gernert K, Deng J, Marzolf B, Kennedy K, Wu H, Bennouna S,

Oluoch H, Miller J, Vencio RZ, Mulligan M, Aderem A, Ahmed R, Pulendran B: Systems biology approach predicts immunogenicity of the yellow fever vaccine in humans. Nature immunology 2009, 10(1): 116–125.

[17] Suthar MS, Brassil MM, Blahnik G, McMillan A, Ramos HJ, Proll SC, Belisle SE, Katze MG, Gale M, Jr.: A systems biology approach reveals that tissue tropism to West Nile virus is regulated by antiviral genes and innate immune cellular processes. PLoS pathogens 2013, 9(2): e1003168.

[18] Carvalho A, Cunha C, Iannitti RG, Casagrande A, Bistoni F, Aversa F, Romani L: Host defense pathways against fungi: the basis for vaccines and immunotherapy. Frontiers in microbiology 2012, 3: 176.

[19] Adams LG, Khare S, Lawhon SD, Rossetti CA, Lewin HA, Lipton MS, Turse JE, Wylie DC, Bai Y, Drake KL: Enhancing the role of veterinary vaccines reducing zoonotic diseases of humans: linking systems biology with vaccine development. Vaccine 2011, 29(41): 7197–7206.

[20] Hoft DF, Brusic V, Sakala IG: Optimizing vaccine development. Cellular microbiology 2011, 13(7): 934–942.

[21] Ulmer JB, Sztein MB: Promising cutting-edge technologies and tools to accelerate the discovery and development of new vaccines. Current opinion in immunology 2011, 23(3): 374–376.

[22] Poland GA, Oberg AL: Vaccinomics and bioinformatics: accelerants for the next golden age of vaccinology. Vaccine 2010, 28(20): 3509–3510.

[23] Ahmed N, Gottschalk S: How to design effective vaccines: lessons from an old success story. Expert review of vaccines 2009, 8(5): 543–546.

[24] Gaucher D, Therrien R, Kettaf N, Angermann BR, Boucher G, Filali-Mouhim A, Moser JM, Mehta RS, Drake DR, 3rd, Castro E, Akondy R, Rinfret A, Yassine-Diab B, Said EA, Chouikh Y, Cameron MJ, Clum R, Kelvin D, Somogyi R, Greller LD, Balderas RS, Wilkinson P, Pantaleo G, Tartaglia J, Haddad EK, Sekaly RP: Yellow fever vaccine induces integrated multilineage and polyfunctional immune responses. The Journal of experimental medicine 2008, 205(13): 3119–3131.

[25] Pulendran B, Ahmed R: Immunological mechanisms of vaccination. Nature immunology 2011, 12(6): 509–517.

[26] Trautmann L, Sekaly RP: Solving vaccine mysteries: a systems biology perspective. Nature immunology 2011, 12(8): 729–731.

[27] Six A, Bellier B, Thomas-Vaslin V, Klatzmann D: Systems biology in vaccine design. Microbial biotechnology 2012, 5(2): 295–304.

[28] Rappuoli R, Aderem A: A 2020 vision for vaccines against HIV, tuberculosis and malaria. Nature 2011, 473(7348): 463–469.
[29] Zak DE, Aderem A: Systems biology of innate immunity. Immunological reviews 2009, 227(1): 264–282.
[30] Oberg AL, Kennedy RB, Li P, Ovsyannikova IG, Poland GA: Systems biology approaches to new vaccine development. Current opinion in immunology 2011, 23(3): 436–443.
[31] Joyce AR, Palsson BO: The model organism as a system: integrating 'omics' datasets. Nature reviews Molecular cell biology 2006, 7(3): 198–210.
[32] Pulendran B, Li S, Nakaya HI: Systems vaccinology. Immunity 2010, 33(4): 516–529.
[33] Andersen-Nissen E, Heit A, McElrath MJ: Profiling immunity to HIV vaccines with systems biology. Current opinion in HIV and AIDS 2012, 7(1): 32–37.
[34] Duraisingham SS, Rouphael N, Cavanagh MM, Nakaya HI, Goronzy JJ, Pulendran B: Systems biology of vaccination in the elderly. Current topics in microbiology and immunology 2013, 363: 117–142.
[35] Buonaguro L, Wang E, Tornesello ML, Buonaguro FM, Marincola FM: Systems biology applied to vaccine and immunotherapy development. BMC systems biology 2011, 5: 146.
[36] Gay CG, Zuerner R, Bannantine JP, Lillehoj HS, Zhu JJ, Green R, Pastoret PP: Genomics and vaccine development. Rev Sci Tech 2007, 26(1): 49–67.
[37] Young D, Stark J, Kirschner D: Systems biology of persistent infection: tuberculosis as a case study. Nature reviews Microbiology 2008, 6(7): 520–528.
[38] Fleischmann RD, Adams MD, White O, Clayton RA, Kirkness EF, Kerlavage AR, Bult CJ, Tomb JF, Dougherty BA, Merrick JM, et al.: Whole-genome random sequencing and assembly of Haemophilus influenzae Rd. Science 1995, 269(5223): 496–512.
[39] Rinaudo CD, Telford JL, Rappuoli R, Seib KL: Vaccinology in the genome era. The Journal of clinical investigation 2009, 119(9): 2515–2525.
[40] Pagani I, Liolios K, Jansson J, Chen IM, Smirnova T, Nosrat B, Markowitz VM, Kyrpides NC: The Genomes OnLine Database (GOLD) v.4: status of genomic and metagenomic projects and their associated metadata. Nucleic acids research 2012, 40(Database issue): D571–579.

[41] Rappuoli R: Reverse vaccinology. Current opinion in microbiology 2000, 3(5): 445–450.
[42] Seib KL, Zhao X, Rappuoli R: Developing vaccines in the era of genomics: a decade of reverse vaccinology. Clinical microbiology and infection : the official publication of the European Society of Clinical Microbiology and Infectious Diseases 2012, 18 Suppl 5: 109–116.
[43] Pizza M, Scarlato V, Masignani V, Giuliani MM, Arico B, Comanducci M, Jennings GT, Baldi L, Bartolini E, Capecchi B, Galeotti CL, Luzzi E, Manetti R, Marchetti E, Mora M, Nuti S, Ratti G, Santini L, Savino S, Scarselli M, Storni E, Zuo P, Broeker M, Hundt E, Knapp B, Blair E, Mason T, Tettelin H, Hood DW, Jeffries AC, Saunders NJ, Granoff DM, Venter JC, Moxon ER, Grandi G, Rappuoli R: Identification of vaccine candidates against serogroup B meningococcus by whole-genome sequencing. Science 2000, 287(5459): 1816–1820.
[44] Maione D, Margarit I, Rinaudo CD, Masignani V, Mora M, Scarselli M, Tettelin H, Brettoni C, Iacobini ET, Rosini R, D'Agostino N, Miorin L, Buccato S, Mariani M, Galli G, Nogarotto R, Nardi-Dei V, Vegni F, Fraser C, Mancuso G, Teti G, Madoff LC, Paoletti LC, Rappuoli R, Kasper DL, Telford JL, Grandi G: Identification of a universal Group B streptococcus vaccine by multiple genome screen. Science 2005, 309(5731): 148–150.
[45] Mora M, Bensi G, Capo S, Falugi F, Zingaretti C, Manetti AG, Maggi T, Taddei AR, Grandi G, Telford JL: Group A Streptococcus produce pilus-like structures containing protective antigens and Lancefield T antigens. Proceedings of the National Academy of Sciences of the United States of America 2005, 102(43): 15641–15646.
[46] Falugi F, Zingaretti C, Pinto V, Mariani M, Amodeo L, Manetti AG, Capo S, Musser JM, Orefici G, Margarit I, Telford JL, Grandi G, Mora M: Sequence variation in group A Streptococcus pili and association of pilus backbone types with lancefield T serotypes. The Journal of infectious diseases 2008, 198(12): 1834–1841.
[47] Hava DL, Camilli A: Large-scale identification of serotype 4 Streptococcus pneumoniae virulence factors. Molecular microbiology 2002, 45(5): 1389–1406.
[48] Paton JC, Giammarinaro P: Genome-based analysis of pneumococcal virulence factors: the quest for novel vaccine antigens and drug targets. Trends in microbiology 2001, 9(11): 515–518.

[49] Wizemann TM, Heinrichs JH, Adamou JE, Erwin AL, Kunsch C, Choi GH, Barash SC, Rosen CA, Masure HR, Tuomanen E, Gayle A, Brewah YA, Walsh W, Barren P, Lathigra R, Hanson M, Langermann S, Johnson S, Koenig S: Use of a whole genome approach to identify vaccine molecules affording protection against Streptococcus pneumoniae infection. Infection and immunity 2001, 69(3): 1593–1598.
[50] Pandey A, Mann M: Proteomics to study genes and genomes. Nature 2000, 405(6788): 837–846.
[51] Grandi G: Antibacterial vaccine design using genomics and proteomics. Trends in biotechnology 2001, 19(5): 181–188.
[52] Serruto D, Rappuoli R: Post-genomic vaccine development. FEBS letters 2006, 580(12): 2985–2992.
[53] Rodriguez-Ortega MJ, Norais N, Bensi G, Liberatori S, Capo S, Mora M, Scarselli M, Doro F, Ferrari G, Garaguso I, Maggi T, Neumann A, Covre A, Telford JL, Grandi G: Characterization and identification of vaccine candidate proteins through analysis of the group A Streptococcus surface proteome. Nature biotechnology 2006, 24(2): 191–197.
[54] Ferretti JJ, McShan WM, Ajdic D, Savic DJ, Savic G, Lyon K, Primeaux C, Sezate S, Suvorov AN, Kenton S, Lai HS, Lin SP, Qian Y, Jia HG, Najar FZ, Ren Q, Zhu H, Song L, White J, Yuan X, Clifton SW, Roe BA, McLaughlin R: Complete genome sequence of an M1 strain of Streptococcus pyogenes. Proceedings of the National Academy of Sciences of the United States of America 2001, 98(8): 4658–4663.
[55] Cho CR, Labow M, Reinhardt M, van Oostrum J, Peitsch MC: The application of systems biology to drug discovery. Current opinion in chemical biology 2006, 10(4): 294–302.
[56] Benson DA, Cavanaugh M, Clark K, Karsch-Mizrachi I, Lipman DJ, Ostell J, Sayers EW: GenBank. Nucleic acids research 2013, 41(Database issue): D36–42.
[57] Update on activities at the Universal Protein Resource (UniProt) in 2013. Nucleic acids research 2013, 41(Database issue): D43–47.
[58] Stenson PD, Ball EV, Mort M, Phillips AD, Shaw K, Cooper DN: The Human Gene Mutation Database (HGMD) and its exploitation in the fields of personalized genomics and molecular evolution. Current protocols in bioinformatics / editoral board, Andreas D Baxevanis [et al] 2012, Chapter 1: Unit1 13.
[59] Martelli PL, D'Antonio M, Bonizzoni P, Castrignano T, D'Erchia AM, D'Onorio De Meo P, Fariselli P, Finelli M, Licciulli F, Mangiulli M, Mignone F, Pavesi G, Picardi E, Rizzi R, Rossi I, Valletti A, Zauli A,

Zambelli F, Casadio R, Pesole G: ASPicDB: a database of annotated transcript and protein variants generated by alternative splicing. Nucleic acids research 2011, 39(Database issue): D80–85.

[60] Kirsanov DD, Zanegina ON, Aksianov EA, Spirin SA, Karyagina AS, Alexeevski AV: NPIDB: Nucleic acid-Protein Interaction DataBase. Nucleic acids research 2013, 41(Database issue): D517–523.

[61] Franceschini A, Szklarczyk D, Frankild S, Kuhn M, Simonovic M, Roth A, Lin J, Minguez P, Bork P, von Mering C, Jensen LJ: STRING v9.1: protein-protein interaction networks, with increased coverage and integration. Nucleic acids research 2013, 41(Database issue): D808–815.

[62] Kerrien S, Aranda B, Breuza L, Bridge A, Broackes-Carter F, Chen C, Duesbury M, Dumousseau M, Feuermann M, Hinz U, Jandrasits C, Jimenez RC, Khadake J, Mahadevan U, Masson P, Pedruzzi I, Pfeiffenberger E, Porras P, Raghunath A, Roechert B, Orchard S, Hermjakob H: The IntAct molecular interaction database in 2012. Nucleic acids research 2012, 40(Database issue): D841–846.

[63] Chatr-Aryamontri A, Breitkreutz BJ, Heinicke S, Boucher L, Winter A, Stark C, Nixon J, Ramage L, Kolas N, O'Donnell L, Reguly T, Breitkreutz A, Sellam A, Chen D, Chang C, Rust J, Livstone M, Oughtred R, Dolinski K, Tyers M: The BioGRID interaction database: 2013 update. Nucleic acids research 2013, 41(Database issue): D816–823.

[64] Willighagen EL, Waagmeester A, Spjuth O, Ansell P, Williams AJ, Tkachenko V, Hastings J, Chen B, Wild DJ: The ChEMBL database as linked open data. Journal of cheminformatics 2013, 5(1): 23.

[65] Razick S, Magklaras G, Donaldson IM: iRefIndex: a consolidated protein interaction database with provenance. BMC bioinformatics 2008, 9: 405.

[66] Licata L, Briganti L, Peluso D, Perfetto L, Iannuccelli M, Galeota E, Sacco F, Palma A, Nardozza AP, Santonico E, Castagnoli L, Cesareni G: MINT, the molecular interaction database: 2012 update. Nucleic acids research 2012, 40(Database issue): D857–861.

[67] Kanehisa M, Goto S, Sato Y, Furumichi M, Tanabe M: KEGG for integration and interpretation of large-scale molecular datasets. Nucleic acids research 2012, 40(Database issue): D109–114.

[68] Croft D, O'Kelly G, Wu G, Haw R, Gillespie M, Matthews L, Caudy M, Garapati P, Gopinath G, Jassal B, Jupe S, Kalatskaya I, Mahajan S, May B, Ndegwa N, Schmidt E, Shamovsky V, Yung C, Birney E, Hermjakob H, D'Eustachio P, Stein L: Reactome: a database of reactions, pathways and

biological processes. Nucleic acids research 2011, 39(Database issue): D691–697.

[69] Kettner C, Field D, Sansone SA, Taylor C, Aerts J, Binns N, Blake A, Britten CM, de Marco A, Fostel J, Gaudet P, Gonzalez-Beltran A, Hardy N, Hellemans J, Hermjakob H, Juty N, Leebens-Mack J, Maguire E, Neumann S, Orchard S, Parkinson H, Piel W, Ranganathan S, Rocca-Serra P, Santarsiero A, Shotton D, Sterk P, Untergasser A, Whetzel PL: Meeting Report from the Second "Minimum Information for Biological and Biomedical Investigations" (MIBBI) workshop. Standards in genomic sciences 2010, 3(3): 259–266.

[70] Field D, Sansone S, Delong EF, Sterk P, Friedberg I, Gaudet P, Lewis S, Kottmann R, Hirschman L, Garrity G, Cochrane G, Wooley J, Meyer F, Hunter S, White O, Bramlett B, Gregurick S, Lapp H, Orchard S, Rocca-Serra P, Ruttenberg A, Shah N, Taylor C, Thessen A: Meeting Report: BioSharing at ISMB 2010. Standards in genomic sciences 2010, 3(3): 254–258.

[71] He Y, Xiang Z: Databases and in silico tools for vaccine design. Methods Mol Biol 2013, 993: 115–127.

[72] Olsen LR, Zhang G, Reinherz E, Brusic V: FLAVIdB: a data mining system for knowledge discovery in flaviviruses with direct applications in immunology and vaccinology. Immunome Research 2011, 7(3).

[73] Zhang GL, Riemer A, Keskin D, Chitkushev L, Reinherz E, Brusic V: HPVdb: A data mining system for knowledge discovery in Human Papillomavirus with direct applications in T-cell immunology and vaccinology. Proceedings of ACM-BCB 2013, Washington DC 2013.

[74] Schmidt H, Jirstrand M: Systems Biology Toolbox for MATLAB: a computational platform for research in systems biology. Bioinformatics 2006, 22(4): 514–515.

[75] Gentleman R, Carey V, Huber W, Irizarry R, Dudoit S: Bioinformatics and computational biology solutions using R and Bioconductor, vol. 746718470: Springer New York; 2005.

[76] Gehlenborg N, O'Donoghue SI, Baliga NS, Goesmann A, Hibbs MA, Kitano H, Kohlbacher O, Neuweger H, Schneider R, Tenenbaum D, Gavin AC: Visualization of omics data for systems biology. Nature methods 2010, 7(3 Suppl): S56–68.

[77] Shannon P, Markiel A, Ozier O, Baliga NS, Wang JT, Ramage D, Amin N, Schwikowski B, Ideker T: Cytoscape: a software environment for integrated models of biomolecular interaction networks. Genome research 2003, 13(11): 2498–2504.

[78] Quackenbush J: Computational analysis of microarray data. Nature reviews Genetics 2001, 2(6): 418–427.

[79] Kapushesky M, Kemmeren P, Culhane AC, Durinck S, Ihmels J, Korner C, Kull M, Torrente A, Sarkans U, Vilo J, Brazma A: Expression Profiler: next generation–an online platform for analysis of microarray data. Nucleic acids research 2004, 32(Web Server issue): W465–470.

[80] Reich M, Liefeld T, Gould J, Lerner J, Tamayo P, Mesirov JP: GenePattern 2.0. Nature genetics 2006, 38(5): 500–501.

[81] Hornberg JJ, Bruggeman FJ, Westerhoff HV, Lankelma J: Cancer: a Systems Biology disease. Bio Systems 2006, 83(2–3): 81–90.

[82] Anderson AR, Quaranta V: Integrative mathematical oncology. Nature reviews Cancer 2008, 8(3): 227–234.

[83] Wooley JC, Lin HS. Catalyzing Inquiry at the Interface of Computing and Biology. National Research Council (US) Committee on Frontiers at the Interface of Computing and Biology. Washington (DC): National Academies Press (US); 2005.

[84] Zhang M, Yao C, Guo Z, Zou J, Zhang L, Xiao H, Wang D, Yang D, Gong X, Zhu J, Li Y, Li X. Apparently low reproducibility of true differential expression discoveries in microarray studies.Bioinformatics. 2008, 24(18): 2057–63.

[85] Begley CG, Ellis LM. Drug development: Raise standards for preclinical cancer research. Nature. 2012, 483(7391): 531–3.

[86] Leek JT, Scharpf RB, Bravo HC, Simcha D, Langmead B, Johnson WE, Geman D, Baggerly K, Irizarry RA. Tackling the widespread and critical impact of batch effects in high-throughput data. Nat Rev Genet. 2010, 11(10): 733–9.

[87] Macarthur D. Methods: Face up to false positives. Nature. 2012, 487(7408): 427–8.

[88] Monath TP: Yellow fever vaccine. Expert review of vaccines 2005, 4(4): 553–574.

[89] Pulendran B: Learning immunology from the yellow fever vaccine: innate immunity to systems vaccinology. Nature reviews Immunology 2009, 9(10): 741–747.

[90] Querec T, Bennouna S, Alkan S, Laouar Y, Gorden K, Flavell R, Akira S, Ahmed R, Pulendran B: Yellow fever vaccine YF-17D activates multiple dendritic cell subsets via TLR2, 7, 8, and 9 to stimulate polyvalent immunity. The Journal of experimental medicine 2006, 203(2): 413–424.

[91] Barrett AD, Teuwen DE: Yellow fever vaccine - how does it work and why do rare cases of serious adverse events take place?Current opinion in immunology 2009, 21(3): 308–313.

[92] An G, Hunt CA, Clermont G, Neugebauer E, Vodovotz Y: Challenges and rewards on the road to translational systems biology in acute illness: four case reports from interdisciplinary teams. Journal of critical care 2007, 22(2): 169–175.

[93] Day J, Schlesinger LS, Friedman A: Tuberculosis research: going forward with a powerful "translational systems biology" approach. Tuberculosis (Edinb) 2010, 90(1): 7–8.

[94] Day J, Friedman A, Schlesinger LS: Modeling the immune rheostat of macrophages in the lung in response to infection. Proceedings of the National Academy of Sciences of the United States of America 2009, 106(27): 11246–11251.

[95] Poland GA, Kennedy RB, McKinney BA, Ovsyannikova IG, Lambert ND, Jacobson RM, Oberg AL. Vaccinomics, adversomics, and the immune response network theory: individualized vaccinology in the 21st century.Semin Immunol. 2013, 5(2): 89–103.

[96] Brusic V, Petrovsky N. Immunoinformatics and its relevance to understanding human immune disease. Expert Rev Clin Immunol. 200, 1(1): 145–57.

[97] Brusic V, August JT, Petrovsky N. Information technologies for vaccine research. Expert Rev Vaccines. 2005, 4(3): 407–17.

13

Inhibition of Virulence Potential of *Vibrio Cholerae* by Herbal Compounds

Sinosh Skariyachan

Department of Biotechnology Engg, Dayananda Sagar Institutions, Bangalore, India

13.1 Introduction

Vibrio cholerae is notorious pathogenic bacteria responsible for enormous public health burden, especially in developing countries all over the world. Most common impact of this human pathogen is cholera, a self-limiting illness. The outbreak of cholera is responsible for approximately 120, 000 deaths every year and has a major impact on the health of young children between the ages of 1 and 5 years [1]. At present, the treatments against cholera and related outbreaks are very critical issue worldwide as most of the strains have developed multiple drug resistance to all classes of conventional chemotherapeutic agents. These circumstances underline the necessity for alternative and promising strategies and development of novel therapeutic solutions. The present chapter is mainly focused on the utility of natural remedies against *Vibrio cholerae* infection when all conventional drugs seem to have failed. Initial section of the chapter emphasizes the basic microbiology, pathogenicity, virulent factors and their mode of action for *Vibrio cholerae.*Later sections explain the current global scenario of the evolution of multidrug resistant (MDR) strains and development of alternative natural remedies against MDR strains of *Vibrio cholerae*.

13.2 Microbiology of *Vibrio Cheolrae*

Vibrio cholerae is one of the most notorious pathogenic bacterium and belongs to the *Vibrionaceae* family. It is Gram-negative, aerobic or facultatively anaerobic, non-sporulating, curved rod shaped organism of 1.4–2.6 mm

Post-genomic Approaches in Drug and Vaccine Development, 347–374.

length and undergoes respiratory and fermentative mode of metabolism. This organism is non–capsulated and motile with single, sheathed, polar flagellum. Vibrios are natural inhabitants in seawater and freshwater ecosystem [2].

Vibrio cholerae shows wide varieties of metabolism that can be analyzed by various biochemical reactions. It forms indole, catalase, oxidase but does not produce methyl red and urease. It reduces nitrate (NO^{3-}) to nitrite (NO^{2-}) and ferments sugars such as glucose, sucrose, mannitol, mannose without gas formation. Similarly, it is a late lactose fermenter and does not ferment arabinaose, inisitol and dulcitol. It liquefies gelatin and decarboxylateslysin and Ornithine but not arginine [3–5].

On solid medium, nutrient agar *Vibrio cholerae* produces moist, translucent colonies with light bluish/greenish tinge. Since it is late lactose fermenter, it develops pink coloured colonies on MacConkey's agar after 24 hours of incubation. Most vibrios are toxigenic and produce hemolytic colonies on Blood agar. The optimum temperature for the growth of this bacterium is 37°C but it prefers high pH. On selective medium, TCBS (thiosulfate citrate bile salt sucrose) agar *Vibrio cholerae* produces large yellow (Figure 13.1), moist and convex colonies [6].

13.2.1 Mechanism of Infection

The common disease caused by *Vibrio cholerae* is cholera, most feared epidemic diarrheal disease. It is an infection of the small intestine that causes a large amount of watery diarrhea. The bacterium releases a toxin that causes increased release of water from cells in the intestines, which produces severe diarrhoea. Cholera remains a global threat and a big issue where safe

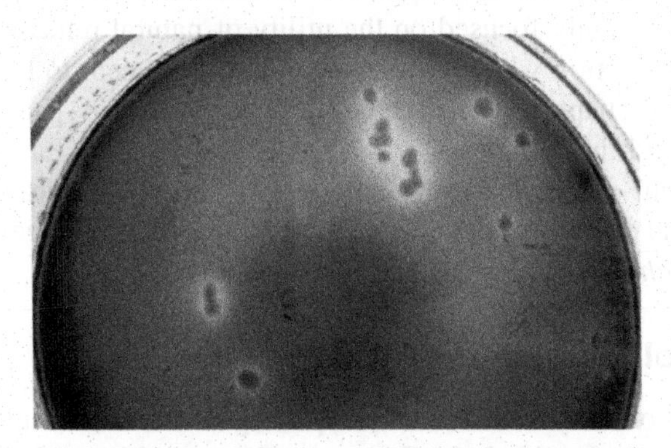

Figure 13.1 Growth of Vibrio cholerae on TCBS agar (Source: Skariyachan *et al.*, unpublished data).

drinking water and adequate sanitation facilities are limited [7–8]. Hence, the transmission is very common in areas with poor sewage systems and unclean drinking water.

Cholera normally begins with the oral ingestion of food or water contaminated with *V. cholerae*. After ingestion, the bacteria pass through gastric acid barrier of the stomach and penetrate the intestinal mucosa. It has been reported that the infectious dose was determined to be high, varied from 10^6 to 10^{11}CFU (colony-forming units) [9]. Subsequently, the bacteria survive and colonize the intestinal epithelial cells and produce the cholera toxin (CT) and causing cholera symptoms [10]. The primary site of bacterialcolonization is the small intestine. The bacterial cells are exposed to many environmental changes such as temperature, acidity and osmolarity during the entry from the aquatic environment to the intestinal environment. The intestinal mucosa contains growth inhibitory substances such as bile salts and organic acids, factors of the innate immune system, mainly complement secreted by intestinal epithelial cells [11]. Hence, the organism developed the ability to survive, colonize, and express virulence factors such as CT even in the extreme environmental conditions. After the secretion of the toxin the symptoms of cholera will occur [12].

The symptoms of cholera begins within 6 hours to 5 days of exposure, range from mild or asymptomatic to severe infection, characterized by large volumes of watery diarrhea, vomiting and leg cramps. This results in severe dehydration and shock in these individuals. Loss of skin plasticity, sunken eyes, fast heart beat, low blood pressure and rapid weight loss are the major signs of this infection. Shock may result due to the collapse of circulatory system. Huge lose of protein-free fluid and associated electrolytes, bicarbonates and ions within one or two days are common in those patients are suffering from cholera. The major reason is the activity of the cholera toxin, an enterotoxin, which activates adenylatecyclase enzyme in the intestinal cells, converting them into pumps which extract water and electrolytes from blood and tissues and pump it into the lumen of the intestine [13]. The loss of fluid leads to dehydration, anuria, acidosis and shock. The watery diarrhea is speckled with flakes of mucus and epithelial cells and contains enormous numbers of vibrios. The loss of potassium ions may result in cardiac complications and circulatory failure [14].

13.3 Global Status of Cholera

Transmission of the disease generally results from consuming water that is contaminated with feces from an infected person or by eating or drinking contaminated food or water. Food borne transmission may also occur when an individual eats raw or undercooked shellfish. For example, people in the US have occasionally contracted cholera from eating undercooked shellfish from the Gulf of Mexico [15]. Globally, every developing country faces cholera outbreaks which include Africa, Asia, Mexico, South and Central America [16–17].

As per the recent reports, there are many serogroups of *Vibriocholerae* that have been identified worldwide based on the antigenic variations. About 200 such serogroups have identified, however, only, serogroup-O1 and O139 have been associated with severe infections and cholera pandemics [17–24]. These strains mainly exist as facultative human pathogens in natural aquatic ecosystems. Many reports revealed that by the starting of the 20th century, there had been six major cholera pandemics. Subsequently, the world is now fighting its seventh pandemic, caused mainly by a new strain of *Vibrio cholerae*, El T first reported in Haiti, North America [25]. Out of 535,000 cases, over 7,000 deaths have been reported so far in Haiti as a result of various cholera outbreaks. Over 100 serotypes of *V. cholerae* have been identified. The strain responsible for Haiti epidemic has been characterized as toxigenic *Vibrio cholerae*, serogroup O1, serotype Ogawa, biotype El T and serotype Inaba [25–26]. Similar outbreaks have also been reported in various parts of the world which includes Indonesia, India and Bangladesh, USSR, Iran and Iraq [17–25].

13.4 Introduction

The common virulent factors of *V. cholerae* are explained below:

- Cholera toxin (Hemolysins)
- Toxin coregulated plus (TCP)
- Adhesin factor (ACF)
- Hemagglutination-protease (Hap, Mucinase)
- Neuramindase
- Siderophores
- Outer membrane proteins and Lipoproteins

13.4.1 Cholera Toxin

13.4.1.1 Structure and function

Cholera toxin is an enterotoxin produced by most strains of *Vibrio cholerae* and which act as the major virulent factor. The toxin is a heat labile protein of molecular weight 856000kD (Figure 13.2). It is a hetero-oligomer consisting of a single A subunit of molecular weight 27215kDa and five B units each of molecular weight 11677 kDa. [27–28].3D crystal structure of the toxin revealed that it has 7 chains namely A, C (A subunit), D, E, F, G and H (B subunit) [27]. The chain A of the toxin has 194 amino acids and consist of 21% of helical structures (10 helices; 42 residues) and 21% of beta sheet (9 strands, 41 residues). The C chain has 46 residues which includes 89% of helices (2 helices; 41 residues). The chains D, E, F, G and H have 103 amino acids each. The D chain constitutes 23% helices (2 helices; 24 residues) and 29% beta sheets (7 strands; 30 residues). The E chain includes 22% helical structures

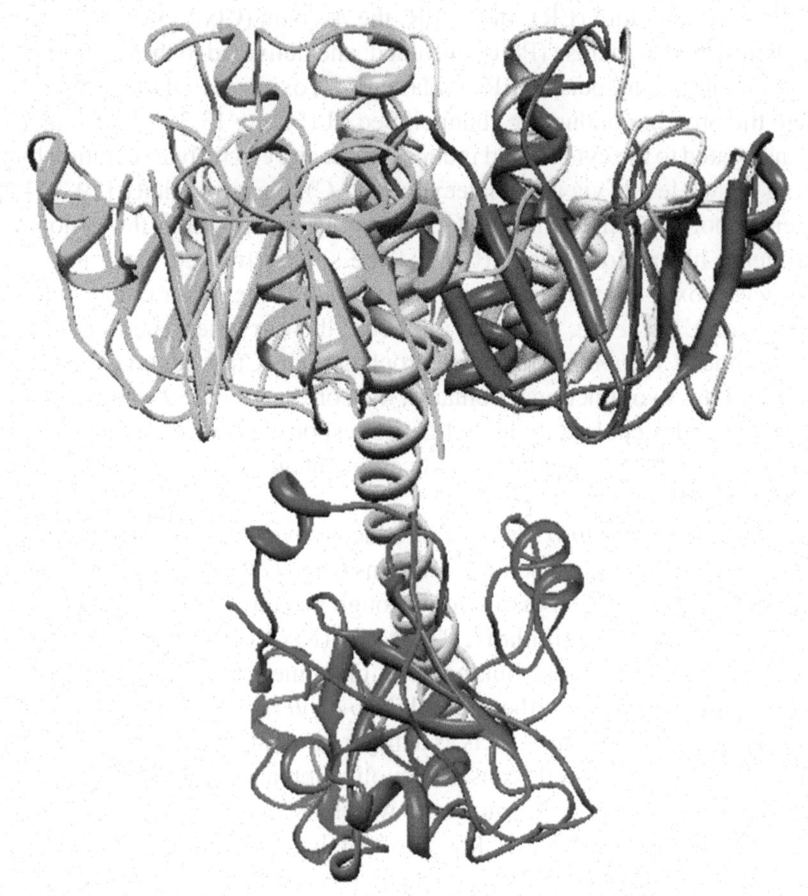

Figure 13.2 The 3D structure of cholera toxin showing its two functional subunits (PDB ID: 1XTC) [27].

(2 helices; 23 residues) and 26% of sheets (9 strands; 27 residues). The chain F of the toxin consists of 21% helical structures (3 helices; 22 residues) and 37% sheets (8 strands; 39 residues). The chain G has 20% helices (3 helices; 21 residues) and 33% sheets (6 strands; 35 residues). Similarly, Chain H constitutes 22% helices (4 helices; 23 residues) and 34% sheets (6 strands; 36 residues) [27].

The subunit B is responsible for specific binding to the GM1 ganglioside receptor of epithelial cells. CtxA and CtxB are the two main genes responsible for the expression of the A and B subunit of the toxin. Once secreted, CTX B-chain (CtxB) binds to ganglioside GM1 on the surface of the host's cells. After binding, the entire CTX complex is carried from plasma membrane (PM) to endoplasmic reticulum (ER). In the ER, the A-chain (CtxA) is recognized by protein disulfideisomerase (PDI), unfolded, and delivered to the membrane where the membrane-associated ER-oxidase, Ero1, oxidizes PDI to release the CtxA into the protein-conducting channel, Sec 61 (Figure 13.3). CtxA is then retro-translocated to the cytosol and induces water and electrolyte secretion by increasing cAMP levels via adenylatecyclase (AC) to exert toxicity [29–35].

Recent reports revealed that the Ctx genes are lying within the genome of a lysogenic filamentous phage (CtxΦ). The single-stranded CtxΦ infects *V. choleraevia* toxin-coregulatedpilus (TCP) which is the major colonization determinant of *V. cholerae*. After the infection, the CtxΦ genome integrates into the genome of *V. cholerae* to form a prophage. The CtxΦ genome is of 6.9 kb in size and has two functional domains, the core and the RS2 regions. The core region encodes cholera toxin and genes responsible for phage morphogenesis. The RS2 gene is essential for replication, integration, and regulation of CtxΦ [34].

In addition to Ctx, *Vibrio cholerae* produces several toxins that are also act as major virulent factors. Zonulaoccludens toxin (ZOT) is one such gene product causes tight junction disruption through protein kinase C-dependent actin polymerization [35]. Similarly, RTX toxin (RtxA) causes actin depolymerization by covalently cross-linking actin monomers into dimers, trimers, and higher multimers [36]. Furthermore, *Vibrio cholerae* cytolysin (VCC), an important pore-forming toxin, so acts as a major virulent factor. The assembly of VCC anion channels in cells causes vacuolization and lysis [37].

The cholera toxin is a haemolysin, which is an exotoxin that lyses erythrocyte membranes with the liberation of haemoglobin. Haemolysins lyse erythrocytes membranes and create pores, which removes the iron-binding proteins such as haemoglobin, transferrin and lactoferrin. This iron can be picked up by various siderophores and taken up through receptors in

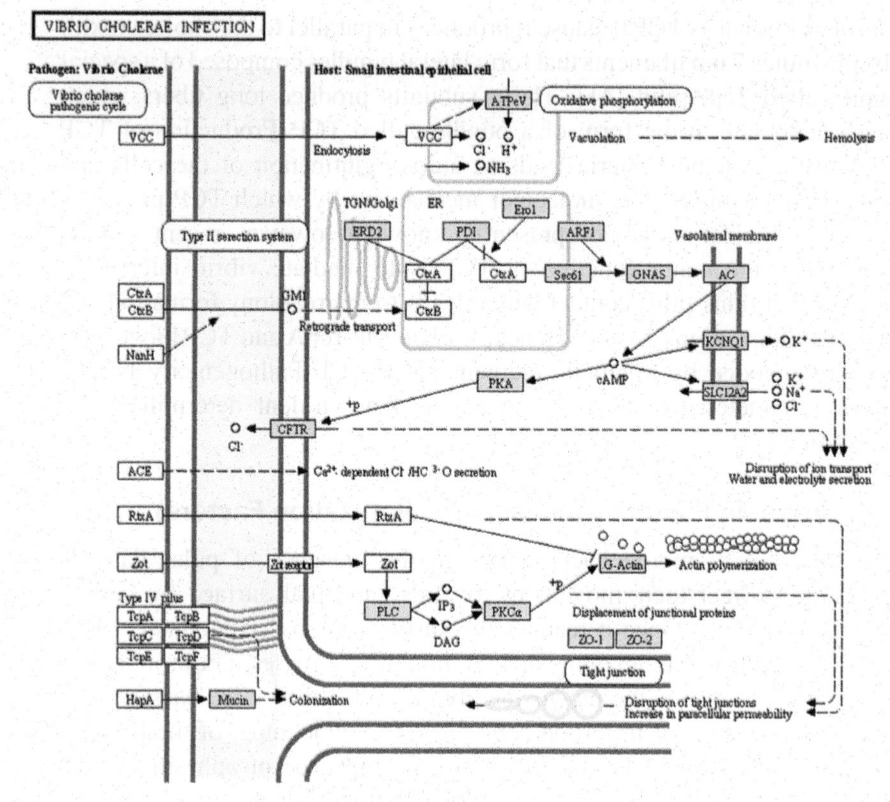

Figure 13.3 Pathway showing the molecular mechanism of toxin involved in *Vibrio cholerae* infection retrieved from KEGG database [34–41].

the cell membrane. In addition to erythrocytes, the pore-forming activity of haemolysin is extended to other cell types such as mast cells, neutrophils and polymorphonuclear cells and enhances virulence by causing tissue damage. There are four representative haemolysin families present in *Vibrio spp.*, including the TDH (thermostable direct haemolysin) family, the HlyA (E1 Tor haemolysin) family, the TLH (thermolabile haemolysin) family and the d-VPH (thermostable haemolysin) family. The TDH of *Vibrio parahaemolyticus* and HlyA of *Vibrio cholerae* have been reported as the most common hemolysins [38–39].

13.4.2 Toxin coregulated pilus (TCP)

Toxin coregulatedpilus (TCP) acts as another important virulent factor for many vibrio-mediated outbreaks [40–44]. The type IV pilus encoded by *V.*

cholerae is known as TCP because it produces in parallel to the cholera toxin. TCP constitutes 7 nm filaments that form lateral bundles composed of a special subunit called TcpApilin [41]. These subunits produce long fibers which laterally associate in the form of a bundle called TCP. Production of TCP on the surface of the bacteria leads to auto-agglutination of the cells and intestinal colonization. The molecular mechanism by which TCP promotes intestinal colonization is not presently known. However, recent reports revealed that the major function of TCP is to mediate vibrio interaction through direct pilus-pilus contact which leads to microcolony formation and intestinal colonization [43–44]. Hence, *V. cholerae* TcpA and TCP biogenesis genes are encoded by a genetic element (TCP-ACF Pathogenicity Island) with the characteristics of a phage act as key virulent determinants of *V. cholerae* [44].

13.4.3 Adhesin Factors (Accessory Colonization Factors)

Vibrio cholerae is motile bacteria due to the presence of polar flagella. The bacteria direct their movement towards intestinal surface by certain chemotactic signals. Once it reaches the surface, the bacteria adhere to the cell surface and secrete enzyme such as mucinase, lipases and proteinases. These enzymes degrade the mucus membrane and facilitate the bacterium to penetrate the surface of intestinal cells [45]. The attachment of the bacteria to the intestinal mucosa is mainly due to the presence of adhesin factors. The adhesion factors (accessory colonization factor) code for Acf genes [46]. The Tcp and Acf gene clusters are linked on the *V. cholerae* chromosome and are flanked by bacteriophage attachment (att) sites. These two regions constitute around 25 Kb of chromosomal DNA code for the pathogenicity island. Genomic analysis of *V. cholerae* strains indicate that only vibrios capable of causing epidemic cholera possess this "pathogenicity island" [47]. Recent reports revealed that the pathogenicity island carrying the Tcp/Acf genes is a prophage (*Vibrio* pathogenicity island -VPI) [48].

13.4.4 Hemagglutinin

Vibrio cholerae of serogroups O1 and O139, causative agents of epidemic cholera, adhere and colonize the small intestine and secrete cholera toxin (CT), which causes the major clinical symptoms of the disease. *V. cholerae* produces a soluble Zn-dependent metalloprotease called hemagglutinin (HA-protease) encoded by hapA [49–50]. HA-protease can proteolytically activate cholera toxin A subunit and can hydrolyze several important proteins such as mucin, fibronectin, and lactoferrin [51–52]. HA-protease perturbs the paracellular barrier of cultured intestinal epithelial cells by acting on tight-junction-associated proteins and promotes the detachment of vibrios from monolayers and mucin [53]. HA-protease spilts mucus and fibronectin and induces intestinal inflammation and releases free vibrios from mucosa to the intestinal lumen. Hence, HA-protease, hemagglutinins also act as major virulent factors for vibrio infections.

13.4.5 Neuraminidase

Vibrio cholerae neuraminidase (NANase) acts synergistically with cholera toxin (CT) and increases the binding and penetration of CT to intestinal absorptive cells. NANase catalyzes the conversion of higher-order gangliosides to mono-sialo tetra hexosylganglioside (GM1) [53]. It has been regarded that NANase may produce high concentrations of the GM1 receptor for CT. This enhances the binding and internalization of CT and results in greater efflux of water and electrolytes [54]. The role of *V. cholerae* NANase in the binding and penetration of CT has been reported by *in vitro* studies using purified NANase and CT with various substrates [55]. The 3D structure of the *Vibrio cholerae* neuraminidase (NANase) is shown in Figure 13.4.

13.4.6 Siderophore

Iron is an essential requirement for *Vibrio cholerae* in the human host as well as its surroundings medium. It has multiple systems for iron acquisition; including the TonB-dependent transport of heme [56], the endogenous siderophore vibriobactin and several siderophores that are produced by other microorganisms. There is also a Feo system for the transport of ferrous iron and

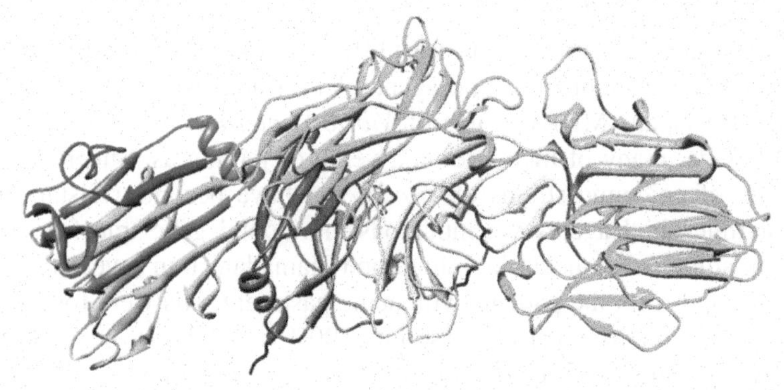

Figure 13.4 Crystal structure of *Vibrio choleare* neuramidase (PDB ID: 1 KIT) [96].

an ABC transporter, Fbp, which transports ferric iron. The genes associated with multiple iron transport systems (siderophores) play a critical role in the virulent potential of *Vibrio cholerae* [57].

13.4.7 Outer Membrane Proteins and Lipopolysacchides

Lipopolysacchrides (LPS) are the most exposed molecules in the outer membrane and contributes a major virulent function [58] in many Gram-negative bacteria including *Vibrio cholerae*. During infection, *V. cholerae* cells are exposed to a series of changes, such as temperature, acidity, osmolarity and antibacterial agents, and innate immune system components. The outer membrane prevents the entry of noxious compounds into the bacterial cell, help them to recognize the host and facilitate colonization. LPS and capsule structures represent important virulent determinants of pathogenic *V. cholerae* strains. Lipopolysacchride consists of Lipid-A, O-antigen and a core polysaccharide. Recent studies revealed that the core-OS-encoding gene clusters (wav genes) are mainly associated with many pathogenic isolates of *Vibrio cholerae*. These studies indicate that the core-OS have a critical role in the virulent function of *Vibrio cholerae* [59]. In addition to LPS, two porins (outer membrane proteins) named OmpU and OmpT are also contribute to *V. cholerae* virulence. These proteins are involved in virulent mechanism due to their expression strictly regulated by ToxR which activates the transcription of ompU and represses the transcription of ompT. ToxR also regulates CT and TCP, other key virulent factors of *Vibrio cholerae* [60].

13.5 Emergence of Multidrug Resistant *Vibrio Cholerae*: "The Superbug"

The outbreaks caused by *Vibrio sps* are treated by various classes of antibiotics, most popularly by chemotherapeutic agents. Unfortunately, the emergence of multidrug resistant (MDR) strains of *Vibrio cholera* e are very critical issues worldwide, many of them are on a verge of joining as "superbugs". The majority of epidemic strains of *V. cholerae* have become resistant to multiple antimicrobial agents via mutations, horizontal gene transfer, etc. [61]. Recent studies conducted in Tehran and Hamadan, Iran revealed that samples collected from the patients with acute gastroenteritis were found to be multidrug resistant. Among isolated strains, 100% were found to be resistant to nalidixic acid, furazolidone and amoxicillin, 95.7% identified as resistant to trimethoprim-sulfamethoxazole and erythromycin 77.4% to erythromycin and few strains were found to be resistant to ciprofloxacin [62–63]. Similar studies conducted in various regions of India [64–65] and Bangladesh [66] also indicated that many strains responsible for epidemic cholera have become resistant to conventionally used antibiotics such as nalidixic acid, polymyxin B, streptomycin, sulfamethoxazole, trimethoprim and rifampicin. Recent studies reported that there is a high emergence of MDR *Vibrio cholerae* in northern Vietnam [67], Zimbabwe [68], Nepal [69], Namibia [70], Cameroon [71], Madagascar [72], Indonesia [73] and Pakistan [74].

Since most strains of *Vibrio cholerae* developed multidrug resistance to conventional antibiotics the treatment against the infections are a major issue worldwide. The spread of the multidrug resistant superbugs urges the need for an alternative and promising therapy. Screening of bioactive compounds from natural sources, including compounds that can specifically target bacterial virulence cascade without affecting their growth, is one such approach that could be used as novel therapeutic interventions [75]. Computer-aided approach is a novel platform to screen and select better therapeutic substances from wide varieties of lead molecules. Currently, plant products are considered to be important alternative sources of new antimicrobial drugs against antibiotic-resistant microorganisms [76]. Many herbal-derived compounds have significant inhibitory and antimicrobial properties against broad range of pathogenic microorganisms [77]. We have recently reported the computer-aided virtual screening and selection of novel lead molecules against the probable drug targets of multidrug resistant *Clostridium perfringens* [78], *Staphylococcus aureus* [79], *Salmonella typhi*, *Shigella dysenteriae* and *Vibrio cholerae* [80]. Since the virulent factors of *Vibrio choleare* are well known these factors could be used as probable drug targets. The inhibitory properties of many natural compounds especially, herbal compounds against the virulent potential of *Vibrio cholerae* have been studied [80].

13.6 Inhibition of Virulent Factors of *Vibrio Cholerae* by Natural Compounds

13.6.1 Cholera Toxin Inhibitors

As most strains of *Vibrio cholerae* have emerged as superbugs due to multiple antibiotic resistances against present generation drugs the treatments against the outbreaks are a critical issue worldwide. Hence, there is a high scope and emergency to screen alternative therapeutic solutions to treat cholera and related out breaks. In historical times, traditional therapeutics were prepared for the prevention of infections caused by *Vibrio cholerae* from various medicinal plants. Since most of the infections are mainly due to certain virulent factors, the best remedial measure is inhibiting the virulent function of such virulent genes or genes products [81]. The most common method to administer such natural remedies is via infusions or decoctions by the formulation of various natural products. The active compounds present in varieties of herbal extracts can treat cholera by various pharmacological mechanisms such as direct antimicrobial activity against *V. cholerae* or prevention of adhesion of cholera toxin to the GM1 receptors at epithelial cell surface and inhibit ADP-ribosylation of active unit of CT. The better knowledge about the molecular mechanism of CT toxin and improved biological assays has allowed the identification of variety of pharmacologically active compounds from natural origin. The popular phytoligands showing pronouncing inhibitory properties against cholera toxins are explained below and their structures are shown in Figure. 13.5

Recent reports indicated that many kinds of herbal-derived products are polyphenols isolated from green and black tea, green apples, hops bract and the Chinese rhubarb rhizome. Garlic extract is another example of a promising therapy against cholera. Recent researches have recognized galactan polysaccharide as a major anti-choleric component of garlic [82–87].

Toda *et al.* (1992) reported that polyphenol catechins such as EGC (epigallocatechin), ECG (epicatechingallate) and EGCG (epigallocatechingallate) isolated from green tea (*Camellia sinensis*) have good pharmacological activity against *V. cholerae* O1. EGCG and ECG also provide better protection against hemolysin in a high concentration. Animal studies indicated that these catechins reduced the fluid accumulation, primary cause of cholera fatality,

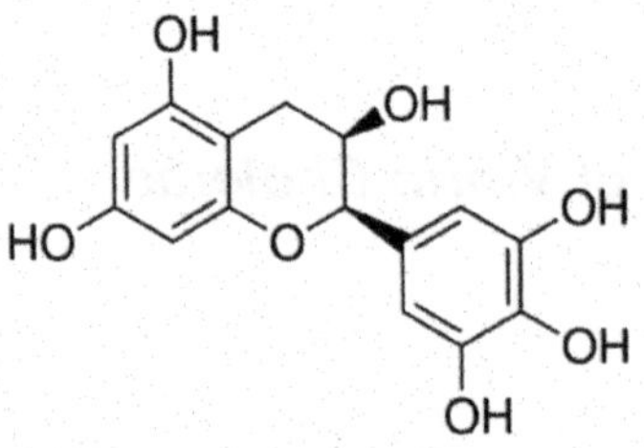

(−)-Epigallocatechin (ECG)

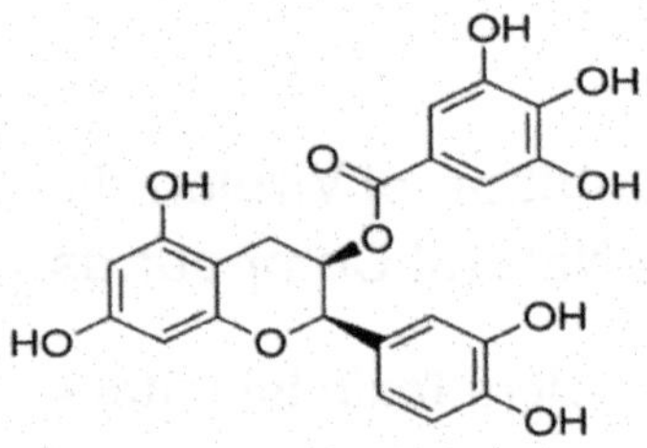

(−)-Epicatechingallate (ECG)

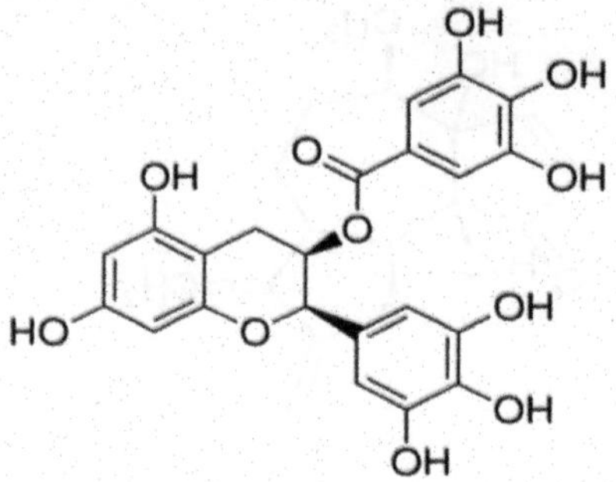

(−)-Epigallocatechingallate (EGCG)

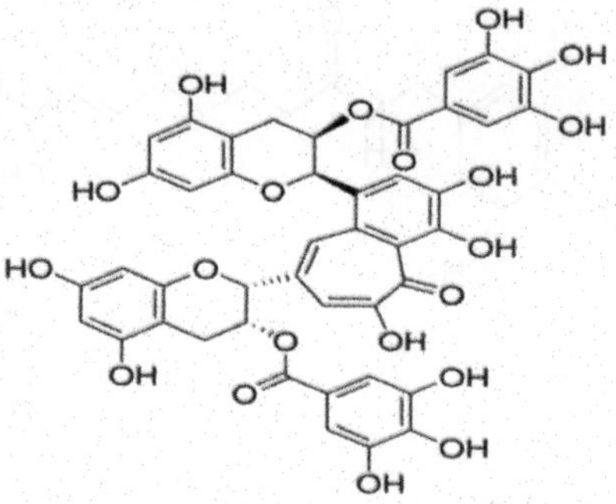

Theaflavin 3,3′-digallate

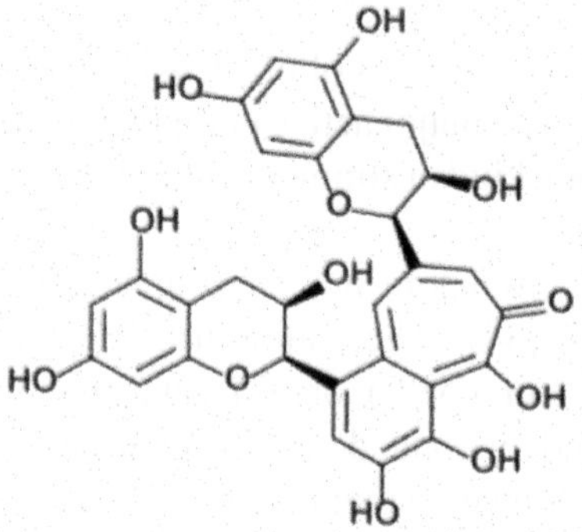

Thearubigin

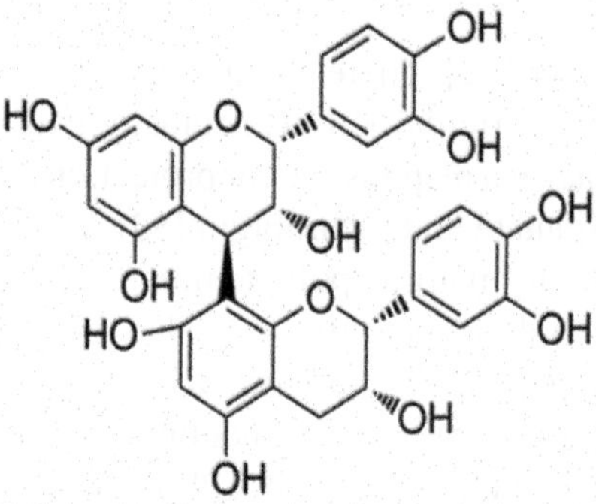

Procyanidine B1

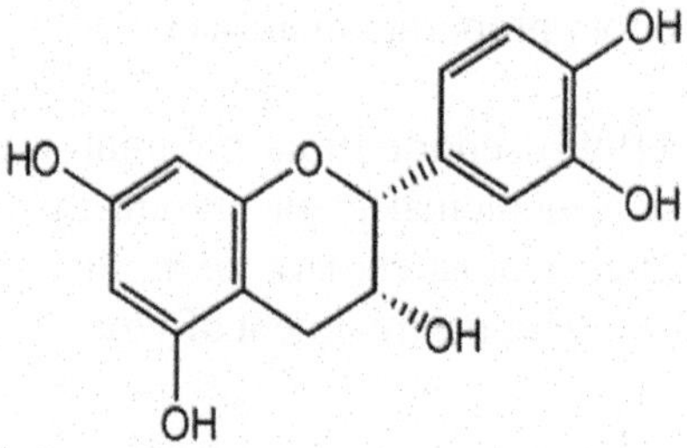

Proanthocyanidin

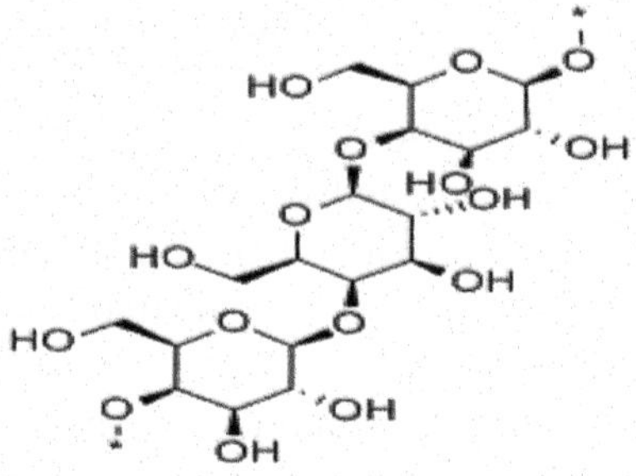

Galactan

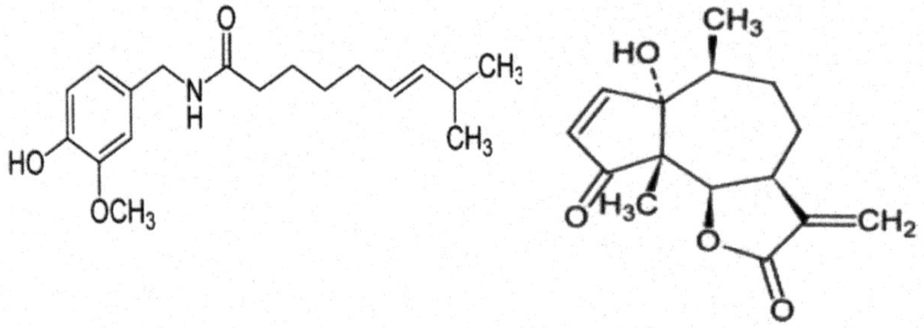

Capsaicin **Parthenin**

Figure 13.5 Chemical structures of common phytoligands used against *Vibrio cholerae* infections.

from CT [82]. There are also reports indicating that extracts of black tea have anti-bactericidal activity against *V. cholerae* O1. The major active components of black tea responsible for protection against such kind of strains are aflavin-3, 3'-digallate and thearubigin [83].

Many natural polyphenols extracted from immature apples with good anti-choleric activity was also been reported [84]. The inhibitory effect of apple (*Malus domestica*) polyphenols extract (APE) on CT-catalyzed ADP-ribosylation of agmantine is mainly dependent on its dose. The pharmacological mechanism of APE is the inhibition of the enzymatic activity of the A subunit of CT. Apple extracts consist of a highly polymerized catechin, procyanidine, key inhibitory components. Other therapeutic constituents present in apple extracts are non-catechin-type polyphenols including chlorogenic acid, phloridzin, phloretin, caffeic acid and p-coumaric acid, monomeric catechins, which have shown slightly less inhibitory properties against cholera toxin [84].

Similar studies conducted by Hor *et al.* (1995) reported that proanthocyanidines extracted from *Guazumaulimfolia*, a medicinal plant present in Mexico, can provide *in vivo* inhibitory properties against cholera toxin, this therapy is very promising and traditional in Mexico for the treatment of *Vibrio* related diarrhea [85].

Oi *et al.* (2002) reported the pharmacological properties of rhubarb galloyl tannin (RG-tannin), an active compound isolated from Rhei rhizome (*Rheum palmatum*), against CT including ADP-ribosylation and fluid accumulation. Studies conducted in mouse and rabbit indicated that the heterologuos polyphenol gallate inhibits fluid accumulation induced by CT, especially cholera toxin A subunit. However, RG-tannin had no effect on the binding of cholera toxin B to the ganglioside GM1 or an endogenous ADP-ribosylation of membrane proteins [86].

The inhibitory properties of galactan [high molecular weight polysaccharide, isolated from garlic (*Allium sativum*) extracts)] against the B subunit of CT were reported by Politi *et al.* (2006). The interaction of cholera toxin and galactan was confirmed by NMR methods (Saturation Transfer Difference methods); this technique is used to measure interaction between ligands and target receptor. This study revealed that galactan polymer could effectively bind with the CTB subunits. Upon binding, galactan forms high molecular weight aggregates with CTB and prevent adhesion of the toxin to cell-surface [87].

Chatterjee *et al.* (2010) reviewed that the bioactive compound present in red chilli (*Capsicum annuum*), a common pungent spice used for many purposes has good inhibitory properties against *Vibrio cholerae* toxin [88]. One of the active ingredients in red chilli is capsaicin (N-anillyl- 8-methyl-nonenamide), which can also act as an antimicrobial agent against many bacterial pathogens such as *Vibrio cholerae*, *Bacillus sps.*, *Helicobacter pylori*, etc. [89–91]. Capsaicin is one of the phytoligand present in red chilli that can suppress CT production. The inhibitory property of CT production by capsaicin is due to the enhancement of transcription of the hns gene. In the presence of red chilli methanol extract and capsaicin, the transcription of the CtxA gene was repressed in many *V. cholerae* strains (Figure 13.6). The transcription of the CtxAB gene is regulated with that of tcpA by a regulator protein called ToxT [90]. The working principle of capsaicin is ToxT-dependent process which occurs in the presence of the transcription of tcpA and toxT genes. Enhancement of hns gene transcription in the presence of capsaicin supports that this gene play a vital role in the reduction of transcriptions of CtxA and tcpA. A hypothesis revealed that capsaicin might directly or indirectly activate the hns transcription; resulting in the down-regulation of the transcription of toxT, ctxA and tcpA genes [89]. The mechanism of virulent gene regulation by capsaicin can now be studied by real-time quantitative reverse transcription-PCR (qRTPCR) assay.

13.6.2 Antimicrobial Potentials of Other Phyto Extracts Against *Vibrio Cholerae*

Hannan *et al.* (2012) reported the antimicrobial activities of onion (*Allium cepa*) against clinical isolates of *Vibrio cholerae*. The antibacterial property

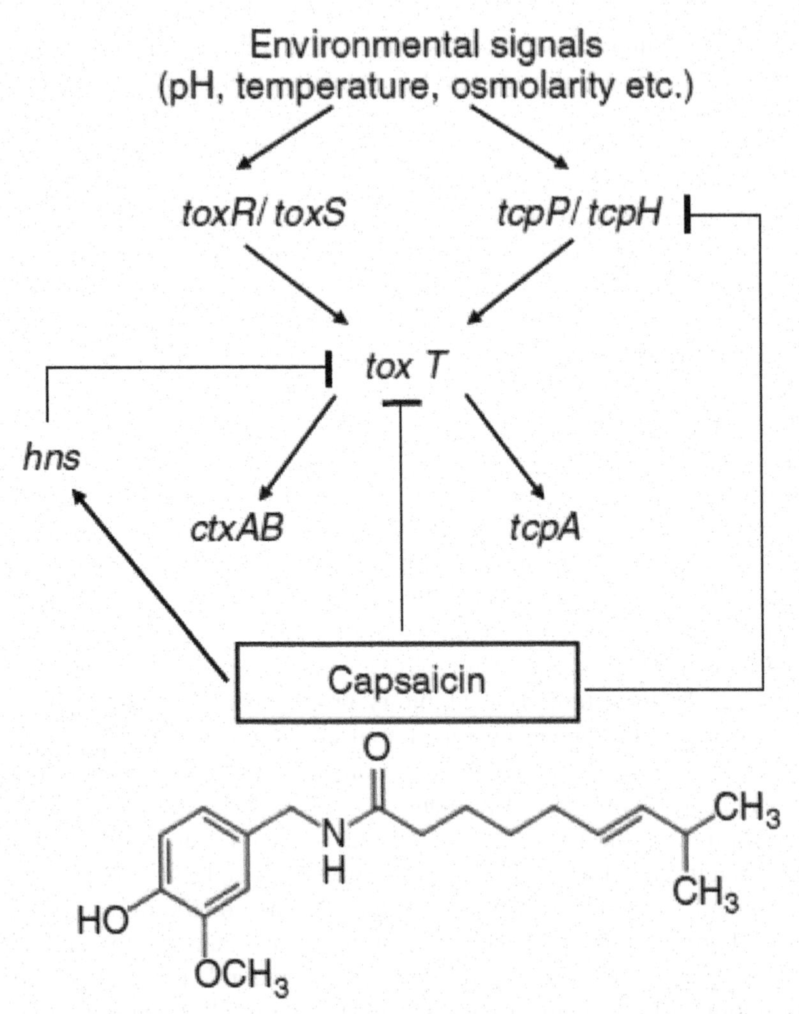

Figure 13.6 Regulatory mechanism for the transcriptions of CtxAB and tcpA genes in *Vibrio cholerae* in the presence of capsaicin [88].

of onion extracts, prepared by reflux extraction method, was tested against various clinical isolates of *V. cholerae* by agar well diffusion techniques. The MIC (minimum inhibitory concentration) was further determined. The results revealed that all tested strains of *V. cholerae* were sensitive to onion extracts of two types (purple and yellow). Purple type of extract had MIC range of 19.2–21.6 mg/ml and yellow type had MIC of 66– 68.4 mg/ml [92]. However, the active compound responsible for inhibition should be identified and the anti-bacterial properties of onion should be tested in animal model in order to make conformation of the inhibitory action.

Recent reports indicate that the active substance present in a type of carrot weed (*Partheniumhysterophorus*) has good antimicrobial properties against many diarrheal pathogens including *V. cholerae* [93]. This medicinal plant is abundantly present in Machillipatnam region, Andhra Pradesh, India and many other parts of the country. Phytochemical analysis reported that parthenin (Figure 13.5), a sesquiterpene lactone, is the major phtyoligand identified from such carrot weeds. Subsequent investigations with parthenin have shown significant allelopathic and cytotoxic effects towards many strains of *V. cholerae* [93].

Oi *et al.* (2002) reported that an instant mix, named as Kampo formulations, can provide good cure against *Vibrio cholerae* outbreaks [86]. Kampo formulations are traditional herbal medications used in China and Japan for many centuries to treat diarrheal diseases such as cholera. The Kampo formulation includes Daio-kanzoto (a mixture of rhubarb and licorice) and Daio (*Rheirhizoma*). This formulations is known to have inhibitory properties against cholera toxin especially ADP-ribosylation and cell elongation

Similarly, the herbal extracts of basil, nopal cactus, sweet acacia and white sagebrush have good antimicrobial properties against various strains of *V. cholerae*. The methanolic extracts of basil (*Ocimum basilicum* L *.),*nopal cactus (*Opuntiaficus-indicavar Villanueva* L *.),* sweet acacia (*Acacia farnesiana* L.) and white sagebrush (*Artemisia ludoviciana* Nutt.) are the most effective extracts against *V. cholerae* with minimum inhibitory concentrations ranging from 0.5 to 3.0 mg/ml. Studies reveled that extracts from these plants were able to cause many detrimental impacts to the bacteria including disruption of cell membranes, increased membrane permeability, decreased cytoplasmic pH, hyperpolarization of cell membrane and decreased cellular ATP concentrations [94–95]. Hence, these four plant extracts could be used as therapeutic alternatives in future against *V. cholerae* contaminations and associated diseases.

13.7 Conclusions

Vibrio cholerae is a notorious pathogen causing many health issues among which cholera outbreaks are a very serious concern. It has spread all over the world, seven cholera pandemics have already been reported. At present, been all strains of *Vibrio cholerae* have become multi-drug resistant (MDR) to most of the conventional drugs and emerged as "superbugs". Hence, there is a high demand for alternative therapeutic solutions. The bacteria posses many virulent factors include cholera toxin (hemolysins), toxin coregulated plus (TCP), adhesin factor (ACF), hemagglutination-protease (hap, mucinase), neuramindase, siderophores and outer membrane proteins and lipopolysac-chrides. Cholera toxins (CT) are the most important virulent factor in many vibrio-mediated outbreaks and it is regarded as the probable drug target. The inhibition of the virulent function of CT by various approaches is going to be a promising therapy against cholera outbreaks. Many herbal-derived substances have good antimicrobial potential against wide varieties of bacterial pathogens including *Vibrio cholerae*. The natural herbal remedies are always welcomed because of their safety and lack of side effects, main drawbacks of conventional chemotherapeutic agents. Many phytoligands have shown good inhibitory activities against cholera toxin. The polyphenols and similar substances isolated from green and black tea, carrot weed, red chillies, green apples, hops bract and rhubarb rhizome have been reported as CT inhibitors. Similarly, the extracts of onion, basil, nopal cactus, sweet acacia and white sagebrush, and instant mixes such as kampo formulations are good herbal remedies against vibrio-mediated infections. Hence, these phytocompounds can be used as better pharmacological solutions than antibiotics, which have failed to cure cholera and other outbreaks due to the emergence of MDR strains, to treat *Vibrio cholerae* infections.

References

[1] WHO. Meeting on the Potential Role of New Cholera Vaccines in the Prevention and Control of Cholera outbreaks during Acute Emergencies. 1995. Document CDR/GPV/95.1. Geneva: World Health Organization.

[2] Baumann, P., Furniss A. L and J. V. Lee.Genus I. Vibrio. In Krieg and Holt (Editors), Bergey'sManual of Systematic Bacteriology, 1st Ed., 1984. Vol1, The Williams & Wilkins Co., Baltimore. pp. 518–538.

[3] Jorgensen, J. H., J. D. Turnidge and J. A. Washington. Antibacterial susceptibility tests: dilution and disk diffusion methods. In: Murray PR, Pfaller MA, Tenover FC, Baron EJ, Yolken RH, ed. Manual of clinical microbiology, 7th ed., Washington, DC: ASM Press; 1999: pp. 1526–1543

[4] Choopun, N., V. Louis, A. Huq and R. R. Colwell, "Simple Procedure for Rapid Identification of Vibrio cholerae from the Aquatic Environment". Appl Environ Microbiol. Vol. 68, No. 2, February 2002, pp. 995–998.

[5] Jesudason, M. V., V. Balaji, U. Mukundan and C. J. Thomson, "Ecological study of Vibrio cholerae in Vellore", Epidemiol Infect. Vol. 124, No. 2, April 2000, pp. 201–206.

[6] Parija, S. C. Bacteriology: Vibrio, Aeromonas and Plesiomonas, Text book of Microbiology and immunology, Elsevier. 2009, pp. 307–316.

[7] Richardson S. H., D. G. Evans and J. C. Feeley, "Biochemistry of *Vibrio cholerae* Virulence I. Purification and Biochemical Properties of PF/Cholera Enterotoxin", Infect Immun. Vol. 1, No. 6, Aug 1976, pp. 546–554.

[8] Bennish, M. L., Cholera: pathophysiology, clinical features and treatment. In: *Vibrio cholerae* and Cholera: Molecular to Global Perspectives (Wachsmuth, K.I., Blake, P.A. and Olsik, O., Eds.), pp.229–255. American Society for Microbiology, Washington, DC.

[9] Holmes, R. K., M. L. Vasil and R. A. Finkelstein RA, "Studies on toxinogenesis in *Vibrio cholerae*. III. Characterization of nontoxinogenic mutants in vitro and in experimental animals",.J Clin Invest. Vol. 55, No. 3, March 1975, pp. 551–560.

[10] Holmgren, J and A.M., "Svennerholm, Mechanisms of disease and immunity in cholera: a review", J. Infect. Dis. Vol. 136 Suppl: pp. S105–12.

[11] Andoh, A., Y. Fujiyama, H. Sakumoto, H. Uchihara, T. Kimura, S. Koyama and T. Bamba, "Detection of complement C3 and factor B gene expression in normal colorectal mucosa, adenomas and carcinomas", Clin. Exp. Immunol. Vol 111, No. 3, March 1998, pp. 477–483.

[12] Steinberg, E. B., K. D. Greene, C. A. Bopp, D. N. Cameron, J. G. Wells and E. D Mintz, "Cholera in the United States, 1995–2000: trends at the end of the twentieth century," J Infect Dis, Vol. 184, No. 6, September 2001, pp. 799–802.

[13] Cassel, D and Z. Selinger. "Mechanism of adenylate cyclase activation by cholera toxin: inhibition of GTP hydrolysis at the regulatory site", Proc Natl Acad Sci U S A. Vol. 74, No. 8, August 1977, pp. 3307–3311.

[14] Reimer, A. R., G. Van Domselaar, S. Stroika, M. Walker, H. Kent, C. Tarr, D. Talkington et al, "Comparative genomics of *Vibrio cholerae* from Haiti, Asia, and Africa. Emerg Infect Dis. Vol. 17, No. 11, November 2011, pp. 2113–2121.

[15] Adagbada, A. O., S. A Adesida, F. O Nwaokorie, M. T. Niemogha and A.O Coker, "Cholera epidemiology in Nigeria: an overview. Pan Afr Med J. Vol. 12, July 2012, pp 59.

[16] Zuckerman, J. N., L. Rombo and A. Fisch A, "The true burden and risk of cholera: implications for prevention and control", Lancet Infect Dis, Vol. 7, No. 8, August 2007, pp. 521–30.

[17] Rodrigue, D.C., T. Popovic and K. Wachsmuth. "Nontoxigenic *Vibrio cholerae* O1 infections in the United States. In: Vibrio cholerae and Cholera: Molecular to Global Perspectives (Wachsmuth, K.I., Blake, P.A. and Olsvik, O., Eds.), pp. 69–76. American Society for Microbiology, Washington, DC.

[18] Morris, J.G., "Non-O group1 Vibrio cholerae strains not associated with epidemic disease. *In: Vibrio cholerae* and Cholera: Molecular to Global Perspectives (Wachsmuth, I.K., Blake, P.A., Olsvik, O., Eds.), pp. 103–116. American Society for Microbiology, Washington.

[19] Saha, P. K., H. Koley, A. K. Mukhopadhyay, S. K. Bhattacharya, G. B. Nair, B. S. Ramakrishnan, S. Krishnan, T. Takeda and Y. Takeda, "Non-toxigenic *Vibrio cholerae* 01 serotype Inaba biotype El Tor associated with a cluster of cases of cholera in southern India", J. Clin. Microbiol. Vol. 34, No. 5, May 1996, pp. 1114–1117.

[20] Sharma C, M. Thungapathra, A. Ghosh, A. K. Mukhopadhyay, A. Basu and R. Mitra, "Molecular analysis of non-O1, non-O139 *Vibrio cholerae* associated with an unusual upsurge in the incidence of cholera-like disease in Calcutta, India", J. Clin. Microbiol. Vol. 36, No. 3, March 1998, pp. 756–763.

[21] Colwell, R.R., J. Kaper and S. W. Joseph, S.W, "Vibrio cholerae, Vibrio parahaemolyticus and other Vibrios: occurrence and distribution in Chesapeake Bay, Science Vol. 198, No. 4315, October 1977, pp. 394–396.

[22] Colwell, R.R, "Global climate and infectious disease: The cholera paradigm", Science, Vol. 274, No. 5295, December 1996, pp. 2025–2031.

[23] Garay, E., A. Arnau and C. Amaro, "Incidence of Vibrio cholerae and related Vibrios in a coastal lagoon and seawater influenced by lake discharges along an annual cycle", Appl. Environ. Microbiol, Vol. 50, No. 2, August 1985, pp. 426–430.

[24] Islam, M.S., B. S. Drasar and R. B. Sack, "The aquatic fora and fauna as reservoirs of *Vibrio cholerae:* a review", J. Diarrh. Dis. Vol. 12, No. 2, June 1994, pp. 87–96.

[25] Poirier, M. J., R. Izurieta, S. S. Malavade and M. D. McDonald, " Re-emergence of Cholera in the Americas: Risks, Susceptibility and Ecology", J Glob Infect Dis. Vol. 4, No. 3, July 2012, pp. 162–171.

[26] Ryan, E. T., "Haiti in the context of the current global cholera pandemic", Emerg Infect Dis. Vol. 17, No. 11, November 2011, pp. 2175–2176.

[27] Zhang, R. G.,D. L. Scott, M. L. Westbrook, S. Nance, B. D. Spangler, G. G. Shipley GG and E. M. Westbrook EM, "The three-dimensional crystal structure of cholera toxin", J Mol Biol. Vol. 251, No. 4, August 1995, pp. 563–573.

[28] King, C.A. and W.A. van Heyningen, "Deactivation of cholera toxin by a sialidase-resistant monosialosylganglioside", J. Infect. Dis. Vol. 127, No. 6, June 1973, pp. 639–647.

[29] Pierce, N.F, "Differential inhibitory effects of cholera toxoids and ganglioside on the enterotoxin of *Vibrio cholerae* and *Escherichia coli", J. Exp. Med.*, Vol. 137, No. 4. April 1973, 1009–1023.

[30] Lencer, W. I., "Microbes and microbial Toxins: paradigms for microbial-mucosal toxins. *V. Cholerae*: invasion of the intestinal epithelial barrier by a stably folded protein toxin", Am J Physiol Gastrointest Liver Physiol, Vol. 280, No. 5, May 2001, pp. G781-G786.

[31] Tsai, B and T. A. Rapoport, "Unfolded cholera toxin is transferred to the ER membrane and released from protein disulfide isomerase upon oxidation by Ero1", J Cell Biol, Vol.159, No. 2, pp. October 2002, pp. 207–216

[32] Fujinaga, Y and A. A Wolf, C. Rodighiero, H. Wheeler, B. Tsai, L. Allen, M. G, Jobling, T. Rapoport, R. K. Holmes and W. I. Lencer, "Gangliosides that associate with lipid rafts mediate transport of cholera and related toxins from the plasma membrane to endoplasmic reticulm", Mol Biol Cell. Vol. 14, No. 12, December 2003, pp. 4783–4793.

[33] Lencer, W. I. and Tsai B, "The intracellular voyage of cholera toxin: going retro", Trends Biochem Sci. Vol. 28, No. 12, December 2003, pp. 639–645.

[34] Olson, R and E. Gouaux E, "Vibrio cholerae cytolysin is composed of an alpha-hemolysin-like core", Protein Sci. Vol. 12, No. 2, February 2003, pp. 379–83.

[35] Chinnapen, D. J., H. Chinnape, D. Saslowsky and W. I. Lencer, "Rafting with cholera toxin: endocytosis and trafficking from plasma membrane to ER", FEMS Microbiol Lett, Vol. 266, No. 2, January 2007, 129–137.

[36] Uzzau, S, P. Cappuccinelli and A. Fasano A, "Expression of *Vibrio cholerae* zonula occludens toxin and analysis of its subcellular localization", Microb Pathog. Vol. 27, No. 6, December 1999, pp. 377–385.

[37] Lin, W, K. J. Fullner, R. Clayton, J. A. Sexton, M. B. Rogers, K. E. Calia, S. B. Calderwood, C. Fraser, J. J. Mekalanos, "Identification of a *Vibrio*

cholerae RTX toxin gene cluster that is tightly linked to the cholera toxin prophage", Proc Natl Acad Sci U S A, Vol. 96, No. 3, February 1999, pp. 1071–1076.
[38] Peterson, K. M., "Expression of *Vibrio cholerae* virulence genes in response to environmental signals", Curr Issues Intest Microbiol. Vol. 3, No. 2, September 2002, pp. 29–38.
[39] Yamamoto, K, Y. Ichinose, N. Nakasone, M. Tanabe, M. Nagahama, J. Sakurai and M. Iwanaga, "Identity of hemolysins produced by *Vibrio cholerae* non-O1 and *V. cholerae* O1, biotype El Tor", Infect Immun, Vol. 1, No. 3, March 1986, pp. 927–931.
[40] Hall, R. H and B. S. Drasar, "*Vibrio cholerae* HlyA hemolysin is processed by proteolysis", Infect Immun, Vol.58, No. 10, October 1990, pp. 3375–3379.
[41] Li, J, M. S. Lim, S. Li, M. Brock, M. E. Pique, V. L. Jr, "Woods and L. Craig L, "*Vibrio cholerae* toxin-coregulated pilus structure analyzed by hydrogen/deuterium exchange mass spectrometry", Structure. Vol. 16, No. 1, January 2008, pp. 137–148.
[42] Kirn, T. J., M. J. Lafferty, C. M. Sandoe, R. K. Taylor RK, "Delineation of pilin domains required for bacterial association into microcolonies and intestinal colonization by *Vibrio cholerae*. Mol Microbiol. Vol. 35, No. 4, February 2000, pp. 896–910.
[43] Manning, P. A, "The tcp gene cluster of *Vibrio cholerae*", Gene, Vol. 192, No. 1, June 1997, 63–70.
[44] Freter, R and P. C. O'Brien, "Role of chemotaxis in the association of motile bacteria with intestinal mucosa: chemotactic responses of *Vibrio cholerae* and description of motile nonchemotactic mutants", Infect Immun. Vol. 34, No. 1, October 1981, pp. 215–221.
[45] Everiss, K. D., K. J. Hughes, K. M Peterson, "The accessory colonization factor and toxin-coregulated pilus gene clusters are physically linked on the *Vibrio cholerae* 0395 chromosome", DNA Seq. Vol. 5, No., 1, June 1994, pp. 51–55.
[46] Kovach, M. E., M. D. Shaffer and K. M. Peterson, "A putative integrase gene defines the distal end of a large cluster of ToxR-regulated colonization genes in *Vibrio cholerae*", Microbiology. Vol. 142 No. Pt 8, August 1996, pp. 2165–2174.
[47] Rajanna, C, J. Wang J, D. Zhang, Z. Xu, A. Ali, Y. M. Hou, D. K. Karaoli, "The Vibrio pathogenicity island of epidemic *Vibrio cholerae* forms precise extrachromosomal circular excision products. J Bacteriol, Vol. 185, No. 23, December 2003, pp. 6893–6901.

[48] Booth, B. A., M. Boesman-Finkelstein and R. A. Finkelstein, "*Vibrio cholerae* soluble hemagglutinin/protease is a metalloenzyme" Infect. Immun. Vol. 42, No. 2, November 1983, pp. 639–644.

[49] Finkelstein, R. A., and L. F. Hanne, "Purification and characterization of the soluble hemagglutinin (cholera lectin) produced by *Vibrio cholerae*", Infect. Immun, Vol. 36, No. 3, June 1982, pp. 1199–1208.

[50] Nagamune, K., K. Yamamoto, A. Naka, J. Matsuyama, T. Miwatani, and T. Honda, "In vitro proteolytic processing and activation of the recombinant precursor of E1 Tor cytolysin/hemolysin (Pro-HlyA) of *Vibrio cholerae* by soluble hemagglutinin/protease of V. cholerae, trypsin, and other proteases", Infect. Immun, Vol. 64, No. 11, November 1996, pp. 4655–4658.

[51] Finkelstein, R. A., M. Boesman-Finkelstein, and P. Holt, "*Vibrio cholerae* hemagglutinin/lectin/protease hydrolyzes fibronectin and ovomucin: F. M. Burnet revisited", Proc. Natl. Acad. Sci. USA, Vol. 80, No. 4, February 1983, pp. 1092–1095.

[52] Wu, Z., P. Nybom, and K. E. Magnusson, "Distinct effects of *Vibrio cholerae* hemagglutinin/protease on the structure and localization of the tight junction-associated proteins occludin and ZO-1" CellMicrobiol, Vol. 2, No. 1, February 2000, pp. 11–17.

[53] Galen, J. E, J. M. Ketley, A. Fasano, S. H. Richardson, S. S. Wasserman and J. B. Kaper, "Role of *Vibrio cholerae* neuraminidase in the function of cholera toxin" Infect Immun, Vol.60, No. 2, February 1992, pp. 406–415.

[54] Holmgren, J., I. Lonnroth, J.E. Mansson, and L. Svennerholm, "Interaction of cholera toxin and membrane GM, ganglioside of small intestine", Proc. Natl. Acad. Sci. USA, Vol. 72, No. 7, July 1975, pp. 2520–2524.

[55] Wyckoff, E. E., A. R. Mey and S. M. Payne, "Iron acquisition in Vibrio cholerae", Biometals, Vol. 20, No. 3–4, June 2007, pp. 405–416.

[56] Mey, A. R. and S. M. Payne, "Haem utilization in *Vibrio cholerae* involves multiple TonB-dependent haem receptors", Mol Microbiol, Vol. 42, No. 3, November 2001, pp. 835–849.

[57] Henderson, D. P. and S. M. Payne, "*Vibrio cholerae* iron transport systems: roles of heme and siderophore iron transport in virulence and identification of a gene associated with multiple iron transport systems", Infect Immun. Vol. 62, No. 11, November 1994, pp. 5120–5125.

[58] Fishmann, P.H, "Mechanism of action of cholera toxin. In:ADP-ribosylating Toxins and G Proteins" (Moss, J. and Vaughan, M., Eds.), pp. 127–137. American Society for Microbiology, Washington, DC.

[59] Nesper, J, A. Kraiss, S. Schild, J. Blass, K. E. Klose, J. Bockemühl, J. Reidl, "Comparative and genetic analyses of the putative *Vibrio cholerae* lipopolysaccharide core oligosaccharide biosynthesis (wav) gene cluster", Infect Immun, Vol. 70, No. 5, May 2002, pp. 2419–2433.

[60] Reidl, J and K. E. Klose, "*Vibrio cholerae* and cholera: out of the water and into the host", FEMS Microbiol Rev, Vol. 26, No. 2, June 2002, pp. 125–139.

[61] Mwansa, J. C., J. Mwaba, C. Lukwesa, N. A. Bhuiyan, M. Ansaruzzaman, T. Ramamurthy, M. Alam and G. B. Nair, "Multiply antibiotic-resistant *Vibrio cholerae* O1 biotype El Tor strains emerge during cholera outbreaks in Zambia", Epidemiol Infect, Vol. 135, No. 5, July 2007, pp. 847–853.

[62] Ranjbar, M, E. Rahmani, A. Nooriamiri, H. Gholami, A. Golmohamadi and H. Barati, "High prevalence of multidrug-resistant strains of *Vibrio cholerae,* in a cholera outbreak in Tehran-Iran, during June-September 2008", Trop Doct, Vol. 40, No. 4, October 2010,pp. 214–216.

[63] Keramat, F, S. H. Hashemi, M. Mamani, M. Ranjba and H. Erfan, "Survey of antibiogram tests in cholera patients in the 2005 epidemic in Hamadan, Islamic Republic of Iran. East Mediterr Health J, Vol. 14, No. 4, July 2008, pp. 768–775.

[64] Waldor, M. K and J. J. Mekalanos, "Emergence of a new cholera pandemic: molecular analysis of virulence determinants in *Vibrio cholerae* O139 and development of a live vaccine prototype", J Infect Dis, Vol. 170, No. 2, August 1994, pp. 278–283.

[65] Jain, M, Goel, A. K, P. Bhattacharya, M. Ghatole and D. V. Kamboj, "Multidrug resistant *Vibrio cholerae* O1 El Tor carrying classical ctxB allele involved in a cholera outbreak in South Western India", Acta Trop, Vol. 117, No. 2, February 2011, pp.152–156.

[66] Kitaoka, M, S. T. Miyata, D. Unterweger, S and Pukatzki S, "Antibiotic resistance mechanisms of *Vibrio cholerae*. J Med Microbiol, Vol60, No. Pt 4, April 2011, pp. 397–407.

[67] Tran, H.D., M. Alam, N. V. Trung, N. V. Kinh, H. H. Nguyen, V. C. Pham, M. Ansaruzzaman, S. M. Rashed, N. A. Bhuiyan, T. T. Dao, H. P. Endtz and H. F. Wertheim HF, "Multi-drug resistant *Vibrio cholerae* O1 variant El Tor isolated in northern Vietnam between 2007 and 2010", J Med Microbiol, Vol. 61, No. Pt 3, March 2012, pp. 431–437.

[68] Islam, M. S, Z. H. Mahmud, M. Ansaruzzaman, S. M. Faruque, K. A. Talukder, F. Qadri F et al, "Phenotypic, genotypic, and antibiotic

sensitivity patterns of strains isolated from the cholera epidemic in Zimbabwe", J Clin Microbiol, Vol.49, No. 6, March 2011, pp. 2325–2327.

[69] Karki, R, D. R. Bhatta, S. Malla, S. P. Dumre, B. P. Upadhyay, S. Dahal and D. Acharya, "Resistotypes of *Vibrio cholerae* 01 Ogawa Biotype El Tor in Kathmandu, Nepal', Nepal Med Coll J, Vol. 13, No. 2,June 2011, pp. 84–87.

[70] Smith, A. M., K. H. Keddy and L. De Wee L, "Characterization of cholera outbreak isolates from Namibia, December 2006 to February 2007", Epidemiol Infect. September 2008 Vol. 136, No. 9, pp. 1207–1209.

[71] Ngandjio, A., M. Tejiokem, M. Wouafo, I. Ndome, M. Yonga, A. Guenole, L. Lemee, M. L. Quilici and M. C. Fonkoua, "Antimicrobial resistance and molecular characterization of *Vibrio cholerae* O1 during the 2004 and 2005 outbreak of cholera in Cameroon", Foodborne Pathog Dis, Vol. 6. No. 1, February 2009, pp. 49–56.

[72] Dromigny, J. A., O. Rakoto-Alson, D. Rajaonatahina, R. Migliani, J. Ranjalahy and P. Mauclére, "Emergence and rapid spread of tetracycline-resistant *Vibrio cholerae* strains, Madagascar", Emerg Infect Dis, Vol. 8, No. 3, March 2002,pp. 336–338.

[73] Tjaniadi, P., M. Lesmana, D. Subekti, N. Machpud, S. Komalarini, W. Santoso et al, "Antimicrobial resistance of bacterial pathogens associated with diarrheal patients in Indonesia", Am J Trop Med Hyg, Vol. 68, No. 6, June 2003, pp. 666–670.

[74] Jabeen, K., J. Siddiqui, A. Zafar, S. Shakoor, N. Ali and A. K. Zaidi, "*Vibrio cholerae* O1 bacteremia in Pakistan: analysis of eight cases. Trans R Soc Trop Med Hyg, Vol. 104, No. 8, August 2010, pp. 563–565.

[75] Chatterjee, S., M. Asakura, V. Chowdhury, S. B. Neogi, N. Sugimoto, S. Haldar, S. P. Awasthi, A. Hinenoya, S. Aoki and S. Yamasaki S, "Capsaicin, a potential inhibitor of cholera toxin production in *Vibrio cholerae*", FEMS Microbiol Lett, Vol. 306, No. 1, May 2010, pp. 54–60.

[76] Sittiwet, C., and D. Puangpronpitag, "Antimicrobial properties of Derris scandens aqueous extract", J Biol. Sci, Vol. 9, No. 6, June 2009, pp. 607–611.

[77] Palombo, E. A., "Traditional medicinal plant extracts and natural products with activity against oral bacteria: Potential application in the prevention and treatment of oral diseases", Evid Based Complement Alternat Med. January 2012, 680354.

[78] Skariyachan, S., R. S. Krishnan, S. B. Siddapa, C. Salian, P. Bora and D. Sebastian, "Computer aided screening and evaluation of herbal

therapeutics against MRSA infections", Bioinformation, Vol. 7, No. 5, October 2011, pp. 222–233.
[79] Skariyachan, S., A. B. Mahajanakatti, N. Sharma and M. Sevanan M, "Selection of herbal therapeutics against deltatoxin mediated Clostridial infections", Bioinformation, Vol. 6, No. 10, August 2011, 375–379.
[80] Skariyachan, S, A. B. Mahajanakatti, N. Sharma, S. Karanth, S. Rao and N. Rajeswari, "Structure based virtual screening of novel inhibitors against multidrug resistant superbugs", Bioinformation, Vol. 8, No. 9, May 2012, pp. 420–425.
[81] Fan E, C. J, O'Neal, D. D. Mitchell, M. A. Robien, Z. Zhang and J. C. Pickens, "Structural biology and structure-based inhibitor design of cholera toxin and heat-labile enterotoxin." Int J Med Microbiol, Vol. 294, No. 4, October 2004, pp. 217–223.
[82] Toda, M., S. Okubo, H. Ikigai, T. Suzuki, Y. Suzuki, Y. Hara and T. Shimamura, "The Protective Activity of Tea Catechins against Experimental-Infection by *Vibrio cholerae* O1." Microbiol Immunol, Vol. 36, No. 9, April 1992, pp. 999–1001.
[83] Toda, M., S. Okubo, H. Ikigai, T. Suzuki, Y. Suzuki and T. Shimamura, "The protective activity of tea against infection by *Vibrio cholerae* O1," J Appl Bacteriol, Vol. 70, No. 2, February 1991, pp. 109–112.
[84] Saito, T., M. Miyake, M. Toba, H. Okamatsu, S. Shimizu and M. Noda, "Inhibition by apple polyphenols of ADPribosyltransferase activity of cholera toxin and toxin-induced fluid accumulation in mice," Microbiol Immunol, Vol. 46, No. 4, may, 2002, pp. 249–255.
[85] Hör, M., H. Rimpler and M. Heinrich, "Inhibition of intestinal chloride secretion by proanthocyanidins from Guazuma ulmifolia," Planta Med Vol. 61, No. 3, June 1995, pp. 208–212.
[86] Oi, H., D. Matsuura, M. Miyake, M. Ueno, I. Takai, T. Yamamoto, M. Kubo, J. Moss and M. Noda "Identification in traditional herbal medications and confirmation by synthesis of factors that inhibit cholera toxin-induced fluid accumulation," Proc Natl Acad Sci U S A, Vol. 99, No. 5, March 2002, pp. 3042–3046.
[87] Politi, M, J. Alvaro-Blanco, P. Groves, A. Prieto, J. A. Leal, F. J. Cañada and J. Jiménez-Barbero, "Screening of garlic water extract for binding activity with cholera toxin B pentamer by NMR spectroscopy - An old remedy giving a new surprise," European Journal of Organic Chemistry, Vol. 2006, No. 9, may 2006, pp. 2067–2073.
[88] Chatterjee, S., M. Asakura, N. Chowdhury, S. B. Neogi, N. Sugimoto, S. Haldar, et al, "Capsaicin, a potential inhibitor of cholera toxin production

in *Vibrio cholerae*, FEMS Microbiol Lett, Vol. 306, No. 1, May 2010, pp. 54–60.

[89] Cichewicz, R. H and P. A. Thorpe PA, "The antimicrobial properties of chile peppers (Capsicum species) and their use in Mayan medicine", J Ethnopharmacol, Vol. 52, No. 2, June 1996, pp. 61–70.

[90] Jones, N. L, S. Shabib, P. M. Sherman, "Capsaicin as an inhibitor of the growth of the gastric pathogen *Helicobacter pylori*," FEMS Microbiol Lett, Vol.146, No. 2, January 1997, pp. 223–227.

[91] DiRita, V. J., C. Parsot, G. Jander and J. J Mekalanos, "Regulatory cascade controls virulence in *Vibrio cholerae*", P Natl Acad Sci USA, Vol. 88, No. 12, June 1991, pp. 5403–5407.

[92] Hannan, A., T. Humayun, M. K. Hussain, M. Yasir and S. Sikandar, "In vitro antibacterial activity of onion (Allium cepa) against clinical isolates of *Vibrio cholerae*", J Ayub Med Coll Abbottabad, Vol. 22, No. 2, June 2012, pp. 160–163.

[93] Siddhardha, B., G. Ramakrishna and M. V. B. Rao, "In vitro antibacterial efficacy of a sesquiterpene lactone, parthenin from Parthenium hysterophorus L (compositae) against enteric bacterial pathogens, International journal of pharmaceutical, chemical and biological sciences. Vol. 2, No. 3, March 2012, pp. 206–209.

[94] Sittiwet, C., and D. Puangpronpitag, "Antimicrobial properties of Derris scandens aqueous extract", J Biol. Sci, Vol. 9, No. 6, June 2009, pp. 607–611.

[95] Sánchez E, S. García, and N. Heredia, "Extracts of edible and medicinal plants damage membranes of *Vibrio cholerae*", Appl Environ Microbiol, Vol. 76, No. 20, October 2010, pp. 6888–94

[96] Crennell, S., E. Garman, G. Laver, E. Vimr, G. Taylor, "Crystal structure of *Vibrio cholerae* neuraminidase reveals dual lectin-like domains in addition to the catalytic domain", Structure. Vol. 2, No. 6, June 1994, 535–544.

14

In silico Approaches for Dealing With Gene Regulatory Networks to Understand Molecular Mechanisms of Immunity in Infectious Disease

Milsee Mol and Shailza Singh

National Centre for Cell Science,
NCCS Complex, University of Pune Campus,
Ganeshkhind, Pune - 411007

14.1 Introduction

Jacob and Monad laid the concept of mechanism of gene regulation by studying the lac operon in *E coli* in 1961and since then several traditional approaches have been applied to analyze this phenomenon considering natural systems as a model or by analyzing abstract models to obtain general results. Post-human genome project and advances in high-throughput measuring techniques has resulted in volumes of data, identifying the gene and protein components making up an organism. In an organism, these components interact to form a complex network to achieve their biological function. Qualitative and quantitative understanding of these networks in an individual organism and in combination (host and parasite) would help uncover novel biological pathways and interactions between cells in an infectious disease state. This is an important step in finding disease markers and deciding the mode of treatment, or in the near future towards building synthetic components to intervene the prognosis of a disease.

When one talks of biological networks it encompasses varieties of interactions that coexist in synchrony with each other in the cell. Signal transduction and gene regulatory networks (GRNs) in a cell are marked by transient protein – protein and protein – DNA interactions, which make an indispensible part of important decision – making mechanism in response to cues that could

Post-genomic Approaches in Drug and Vaccine Development, 375–390.

arise from the intracellular milieu or from extracellular environment. One such extracellular stimulus is a causative agent involved in infectious disease such as leishmaniasis, sleeping sickness and malaria.

In general, the complexity of data generated from global and high-throughput technologies, compels one to use computational methods for comprehensive data analysis. Moreover, the non-linearity and stochasticity which is an integral part of biological system makes it difficult to predict biological behavior without extensive modeling. Using computational models one can easily describe the GRN model and also refined analysis and visualization can reveal emergent properties of the system, making interpretation of the model easier. Also, large number of '*in silico*' experiments can be performed at low cost, giving insight of roles of different regulatory interactions in the system and guide '*in vitro*' experimental planning. They can also highlight irregularities in the assumptions made, when a GRN model does not reproduce the expected experimental observations.

Alterations in signaling pathways and the connected GRNs can change cellular decisions and thereby trigger the onset of infection. Understanding the functional role of changes during an infection and thereby predicting novel targets for efficient intervention can be achieved by the identification of novel signaling components by using genome wide RNAi and protein – protein interaction screens with generation of quantitative information and dynamic pathway modeling. The generation of data suitable for mathematical modeling can be achieved by large-scale screening of cellular networks, quantitative proteomics and multi-parameter imaging; the generated models can be inferred, analyzed, mathematically modeled, numerically simulated and optimized for translational research applications. Thus, integration of large amounts of experimental data and previous knowledge can enhance novel biological pathway discovery. *In silico* approaches describing mathematical model, design implementation and testing gene regulatory networks with respect to immune signal transduction and infectious disease are innumerable; this chapter describes the recent developments in this regard giving representative models, starting with the general introduction to signal transduction and GRNs.

14.1.1 Signal Transduction, Gene Regulatory Networks and Infectious Disease

In general, signal transduction is a mechanism of cellular communication, where the cells perceive and convey information (signal) from outside the cell to the nucleus or other sites within the cell, leading to a cellular response, mediated by receptors found on the cell membrane.This cellular response could involve molecular mechanisms like activation or repression of a gene product or just a change in cytoskeletal configuration, vesicle fusion, which finally leads to cellular responses like cell differentiation, cell growth and cell death. Restricting our discussion to transcription factors, these pathways often involve the phosphorylation (post-translational modifications) of the receptor protein which in turn phosphorylates other cellular proteins, in this way the signal is passed through several cellular proteins where eventually a transcription factor is modified, which is trans-located into the nuclear compartment from the cytoplasmic compartment for activating or repressing transcription of a related gene (Figure 14.1).

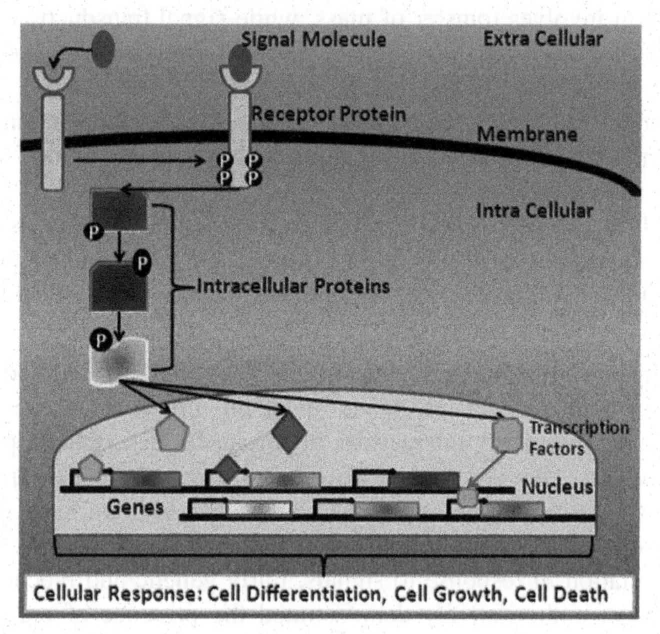

Figure 14.1 The general mechanism of signal transduction and cellular response (Gene activation → Gene repression →).

14.1.1.1 Characters of a signal transduction pathway [1]

A signal transduction pathway has high **specificity**, which is achieved by precise molecular complementarity between the signal and receptor molecule, mediated by non-covalent forces that mediate enzyme–substrate and antigen–antibody interactions. These interactions are **cooperative** in nature i.e. the binding of ligand to the receptor results in large changes in receptor activation with a very small change in the stimulus. As the signal is passed from the receptor to the interior of the cell in the cascade the signal is **amplified**, which is due to the activation of larger enzyme molecules in the subsequent stages of activation. But when the signal molecule is present continuously, **desensitization/adaptation** of the receptor occurs and when the stimulus falls below the threshold level the system again becomes sensitive. The signal transduction pathway has the ability to **integrate** the system i.e. to receive multiple signals and produce unified response appropriate to the needs of the cell.

At this juncture, it should be noted that a signaling pathway differs from a metabolic pathway in ways outlined below:

- Metabolism involves transfer of mass, while signal transduction is the transfer of information, quantified by material converted (μM or mM) and molecules (10 to 10^4) per cell respectively.
- A metabolic pathway is determined by the state (active or inactive) of present set of enzymes and signaling pathway is determined by the dynamic assembly of signaling proteins.
- The catalyst to substrate ratio is held low (explained by quasi-steady-state assumption in Michaelis-Menten kinetics) in a metabolic pathway, whereas in a signaling pathway the amount of catalyst and substrate are held in the same order of magnitude.

A signaling network interacts with the GRN by way of modified transcription factors (TFs) which are made up of all molecular species (TFs and non-TFs) and regulatory interactions important for describing a pattern of gene expression. This is closely associated with controlling the protein levels with respect to time, in a cell which are involved in various cellular processes. A GRN is concerned with the control of gene expression i.e. it's up and down regulation in response to signals. Early genetic and biochemical experiments have authoritatively demonstrated the presence of regulatory elements (sequences) in close proximity of genes to be transcribed and the existence of proteins that are able to bind to these elements and control the activity of genes either by activation or repression of transcription. These regulatory proteins are themselves encoded by genes, as exemplified by the *lac* operon in bacteria and an imaginary transcription factor network (***Figure 14.2***), forming a complex regulatory network, which further includes positive and negative feedback loops. These principles of gene regulation apply to prokaryotes as well as to eukaryotes [2].

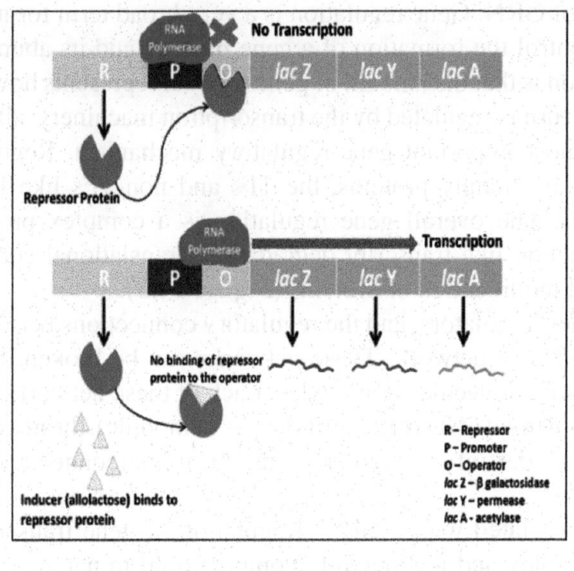

(a)

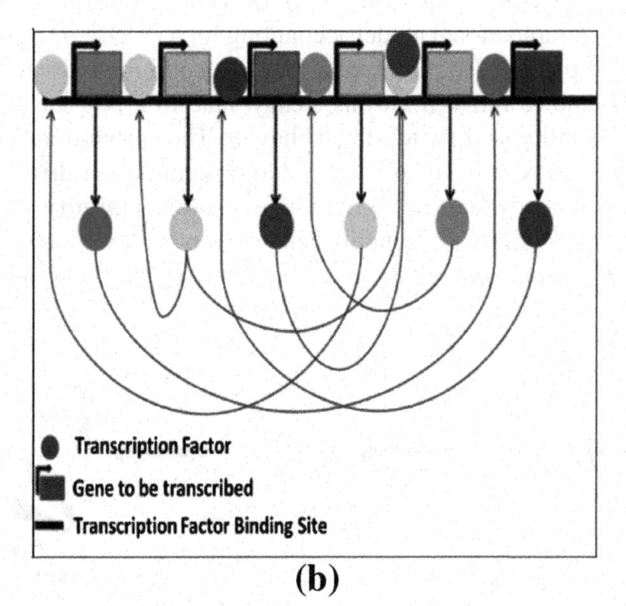

(b)

Figure 14.2 'Gene network' exemplified by (a) Lac Operon and (b) an imaginary transcription factor network.

Architecture of GRN: Gene regulation is a very broad term for molecular processes that control the formation of a gene product and its abundance. A general assumption is that the amount of gene product represents how active a gene is, which in turn is regulated by the transcription machinery, all together making up the most important gene-regulatory mechanism. Regulators of the transcription are mainly proteins, the TFs and non-TFs like RNA and small metabolites. The overall gene regulation is a complex process and includes phenomenon like transcript degradation, translational control, and post-translational modification of proteins (***Figure 14.3***).

Thus, the genes, regulators, and the regulatory connections between them form a gene regulatory network. These networks can be broken down into groups of interacting molecules or modules. Each of these networks are characterized by their own set of distinct motifs (types of module), the identification of such motifs provides information about the local interconnection patterns in the network.

Characters associated with a GRN: Regulation of gene transcription in eukaryotes is complex and is inherently combinatorial in nature. Transcriptional synergy is a key element of such combinatorial control in gene regulation networks. It requires **cooperative** binding of multiple TFs and is intrinsically non-linear in nature. Considering these synergistic functioning is an important part of building a computational model accounting for a precise interpretation of the underlying biology [3]. Gene products in equilibrium (steady state) when perturbed tend to move from this steady state to a new steady state, implying the possibility of a switch-like behavior. The system remains near one of its steady states until a sufficiently large perturbation drives to the vicinity of another steady state and giving rise to **multi-stability**. Negative feedback increases stability in generic gene regulatory systems, whereas positive feedback decreases stability, the action of these feedback loops

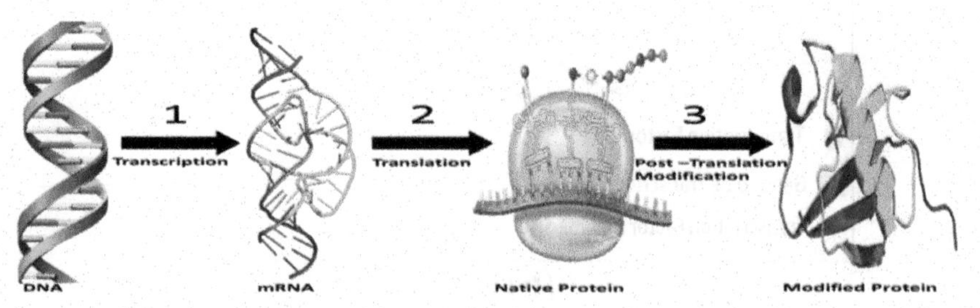

Figure 14.3 Gene Regulation: 1. Regulation by RNA processing and Transport 2. Regulation of Protein Synthesis 3. Regulation of Protein Modification.

introduces a feature; that of oscillatory behavior in the system [4]. Another characteristic feature of GRN is of robustness and evolvability, which seem to be very contradictory as the former requires that genetic alterations leave the phenotype intact, while the latter requires these alterations to be used for the exploration of new phenotypes. Several empirical analysis have documented that robustness often facilitates evolvability [5, 6].

The interaction between the signaling network and the gene network is important in deciding the fate of an agent causing an infectious disease. Appropriate and timely response by the immune signaling against the causative agent can eliminate it. But, through evolutionary changes in the causative agent better survival mechanisms have evolved, which can evade the host immune response, making the host a suitable home for its wellbeing. Most of these evading strategies are such that the pathogen can directly interact with the host signaling mechanism and modulate the immune response, leading to disease progression. Most of these modulations are by the transcription of immunosuppressive chemokines like IL10, TFGβ and also by inducing natural negative regulators of the inflammatory signaling pathway and phosphatases. Several clinically important human pathogens like *Mycobacterium tuberculosis (MtB), Helicobacter pylori* and HIV-1 induce crosstalk between toll like receptors (TLRs) and the C-type lectin DC-SIGn leading to high levels of IL-10 production by dendritic cell (DC) s [7]. Similarly, *Plasmodium falciparum* selectively inhibits IL-12 production and suppresses DC maturation and T-cell activation through interactions with the scavenger receptor CD36 and TLR2 through the erythrocyte membrane protein 1 (PfEMP1), expressed on infected erythrocytes [8]. *Leishmania major*, an intracellular parasite of macrophages, seems to benefit from complement activation and C5aR-induced inhibition of Th1 cell-mediated immunity [9]. Many bacterial pathogens use effector proteins to target the kinase signaling cascades like the nuclear factor κB (NF-κB), mitogen-activated protein kinase (MAPK), phosphatidylinositol 3-kinase (PI3K)/Akt, and p21-activated kinase (PAK) pathways and manipulating the cross talk points [10]. cAMPsignaling pathways have been described in *Trypanosoma cruzi* and *Trypanosoma brucei* that regulates PKA with striking effects on cell survival and cell proliferation, through the inhibition of DNA, RNA and protein synthesis [11].

The host-pathogen interactions are not well understood, but computational modeling of the affected signaling pathway and the corresponding gene network, can help us interrogate these interactions for the better understanding of the dynamic relationship shared by the two partners, for the development of improved therapeutic strategies.

14.1.2 Modeling Signaling Network and Gene Regulatory Network in Infectious Disease

Typically, a signaling network and a GRN is modeled as a network composed of nodes (**V**) (representing genes, proteins and/or metabolites) and edges (**E**) (representing molecular/physical interactions such as protein–protein bindings or protein), depicted in a graph, called the interaction graph. This type of graph has allowed biologists to visually examine the graph and map out information, such as functional annotation, cellular localization and expression level. Also, it allows definition of functional context, from which the biological role of individual proteins or genes can be inferred. To depict disease-gene network – a bipartite graph with two sets of vertices are reconstructed, one set represents diseases of a given organism (e.g. Humans) and the other set represents the organism's genes. A gene and a disease are connected by an edge if the gene is involved in the disease (e.g. Causal Gene). Though parameters of the models are usually not obvious, since the inputs and outputs of the molecular processes are unclear and the underlying principles are unknown or too complex, therefore, the idea is to reconstruct the regulatory interactions of the system from a given amount of experimental data (e.g. sequence of proteins interaction, gene expression profiles, DNA sequence information). The general procedure involved in modeling are depicted in (Figure 14.4) [12].

In general, two ways of modeling can be distinguished: the physical approach and the influence approach [13]. The physical approach seeks to identify the true molecular interactions that are involved in the biological process to be modeled, e.g. identifying adaptor proteins, effector proteins, the TF and the DNA motifs to which they bind, for modeling a signaling network and GRN. The other approach is the influence approach, which aims at identifying regulatory influences between molecular species involved in the regulatory process without describing the molecular interactions, to capture implicitly the regulatory events at the proteomic and metabolomic level. The ability to capture hidden (implicit) regulatory mechanisms that are not experimentally (explicit) measured is the major advantage of the influence approach. The choice between physical and influence modeling strategy depends on several factors, like the availability of experimental data, sequence data, and the interaction data. Another consideration to be made while modeling is whether the model should depict a dynamic or static network. A static model does not have a time-component in them, i.e. they only yield the network topology. Dynamic models can be seen as more informative,

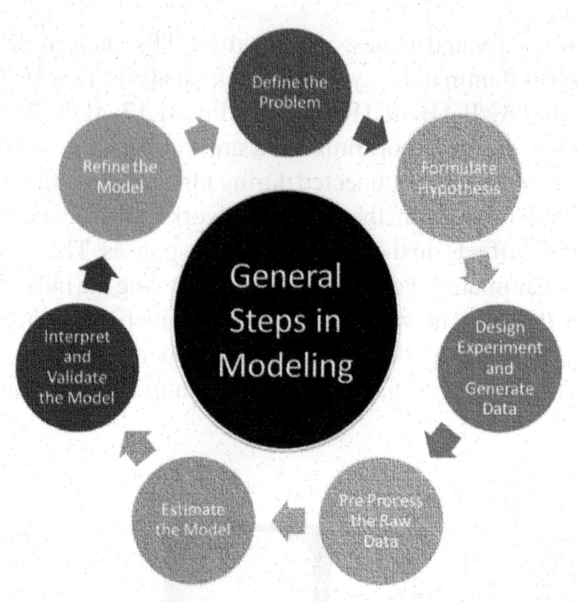

Figure 14.4 General steps involved in modeling – Defining the problem and hypothesis formulation are the essential parts of reconstruction and thereafter modeling of a signaling network and GRN.

since they don't only provide a hypothetical linking between the signaling and GRN component, but are also able to capture the dynamic behavior of the network, e.g. the change of gene expression with time, stimulus strength and space, which requires the time series data.

As stated earlier, a graph is the basic modeling scheme for network reconstruction. These graphs can be described mathematically as a system of (linearized) equations, as a Boolean network, as a Bayesian network, or as an association network. These graphs can be inferred by using different algorithms, depending upon the conclusion to be derived from the model. Few of these algorithms will be described very briefly in application of combined modeling of signaling networks and GRNs in infectious disease, in the following paragraphs (Figure 14.5).

As inflammation is the hallmark of any human disease, we begin by citing the work of Bor-Sen Chen *et al.* (2008) where they constructed a gene regulatory network of inflammation using data extracted from the Ensemble and JASPAR (2005) databases. They used the algorithms like cross correlation threshold, maximum likelihood estimation method and Akaike Information Criterion (AIC) on time-lapsed microarray data to refine the genome-wide transcriptional regulatory network in response to bacterial endotoxins in the context of dynamic activated genes, regulated by TFs such as NF-κB (an important modular inflammatory system). Their analysis revealed that the important genes (IL1A, IL1B, IL1R, IL6, TNFα, IL17, IL8, IL1R, TLR4 and TNFR) detected by the algorithms used are vital for the inflammatory response because they are more connected during inflammation than in normal conditions. In inflamed condition, they appear to work in accordance with each other to enhance their effects on the inflammatory responses. They inferred the robustness of the inflammatory gene network, by analyzing the hubs and "weak ties" structures of the gene network. Based on the kinetic parameters of the dynamic gene regulatory network, they identified properties like susceptibility to infection of the immune system, which is a factor important for translational research [14].

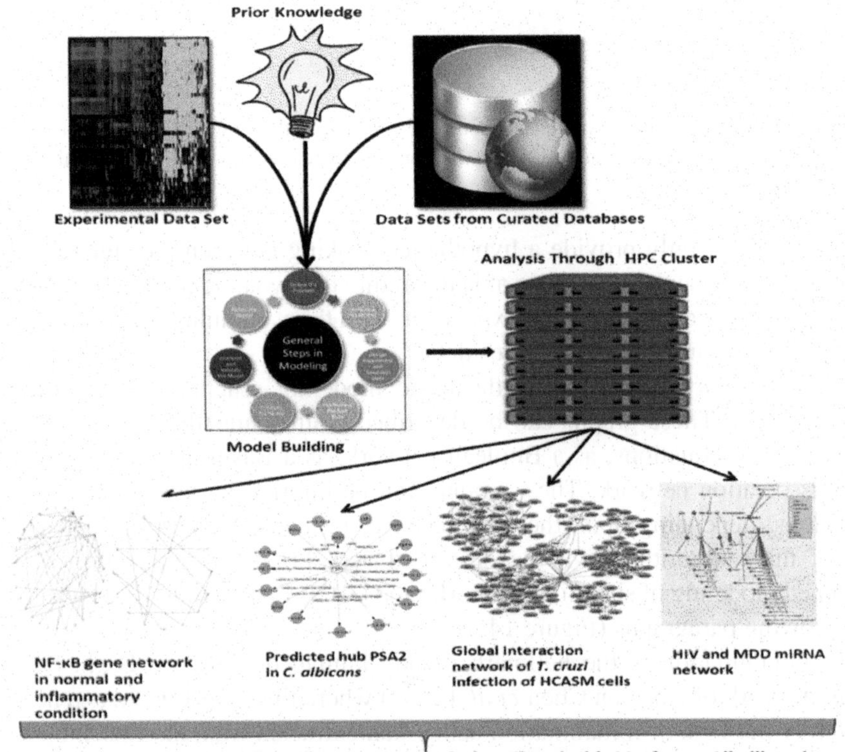

Figure 14.5 GRN and its inference exemplified in infectious disease.

Many pathogens such as *MtB*, *P. falciparum*, *Trypanosoma cruzi* etc. bring about a modulation in the inflammatory mechanism and reside silently in the human host. This modulation is thought to be brought about by the interaction of the parasite with host genome, at different stages of infection. Pius N. Nde *et al.*(2010) constructed a GRN of *T. cruzi* during an early infective stage of the Chagas' disease, which infects cardiac muscles leading to cardiac arrest followed by death. The genetic architectures in the early *T. cruzi* infection process of human cells are unknown. To understand the genetic architectures of the early invasion process of *T. cruzi*, they conducted gene transcription microarray analysis followed by gene network construction of the host cell response in primary human coronary artery smooth muscle cells infected with *T. cruzi* or exposed to *T. cruzi* gp83, a ligand used by the trypanosome to bind to host cells. Using 7 RT-PCR verified up-regulated genes (FOSB, ATF5, INPP1, CCND2, THBS1, LAMC1, and APLP2) as the seed for network construction, and they built an interaction network of the early *T. cruzi* infection process containing 165 genes, connected by 598 biological interactions, using the MiMICytoscape plugin. This interactome network is centered on the BCL6 gene as a hub and silencing the expression of two seed genes (THBS1 and LAMC1) by RNAi reduced *T. cruzi* infection, elucidating the significant and complex process involved in *T. cruzi* infection of HCASM cells at the transcriptome level [15].

Another example to cite is, when a commensal turns into a pathogen, such as *Candida albicans*. It causes systemic infection that may lead to the death of the host if the immune system has deteriorated. Despite a growing number of genome data available, little is known about it. In many organisms, central nodes in the interaction network (hubs) play a crucial role for information and energy transport. Knockouts of such hubs often lead to lethal phenotypes making them interesting drug targets. These central genes were identified (Robert Altwasser *et al.* 2012) via topological analysis, by using a gene regulatory network that is sparse and scale-free. Information from various sources (prior knowledge sources: BIND, TRANS, and PPI) were collected to complement the limited expression data available. A linear regression (LASSO) algorithm was used to infer the genome-wide gene regulatory interaction networks and they have identified 16 hubs (Yor353c, ASR1, Taf19, OPT8, AGP2, PSA2 to name a few) in the genome of the fungus which is susceptible to antifungal treatment [16].

Treatment options for bacterial and fungal disease are many, but for HIV, the struggle still continues to find an ideal druggable target, due to the very high frequency of mutation of the retrovirus. It also has been observed that a very small proportion of human population remains negative for HIV infection after repeated HIV-1 viral exposure, which is called HIV-1 resistance. Understanding this mechanism of HIV-1 resistance could be an important turning point in the development of HIV-1 vaccines and Acquired Immune Deficiency Syndrome (AIDS) therapies. Huang *et al.* (2011) analyzed the gene expression profiles of CD4 + T-cells from HIV-1-resistant individuals and HIV-susceptible individuals. One hundred eighty-five, (significantly enriched in the response to stress, defense response, immune response, cell communication and signal transduction) discriminative HIV-1 resistance genes were identified using the Minimum Redundancy-Maximum Relevance (mRMR) and Incremental Feature Selection (IFS) methods. The virus protein target enrichment analysis of the 185 HIV-1 resistance genes suggested that the HIV-1 protein nef might play an important role in HIV-1 infection. They also identified 29 infection information exchanger genes from the 185 HIV-1 resistance genes based on a virus-host interaction network analysis. The infection information exchanger genes are located on the shortest paths between virus-targeted proteins and are important for the coordination of virus infection. These proteins may be useful targets for AIDS prevention or therapy, as intervention in these pathways could disrupt communication with virus-targeted proteins and HIV-1 infection [17].

Secondary infections are many in a person suffering from AIDS, one of them being tuberculous meningitis (TBM) a fatal form of *MtB* infection of the central nervous system(CNS). The similarities in the clinical and radiological findings in TBM cases with or without HIV make the diagnosis very challenging. Identification of genes, which are differentially expressed in brain tissues of HIV positive and HIV negative TBM patients, would enable better understanding of the molecular aspects of the infection and would also serve as an initial platform to evaluate potential biomarkers. Sameer *et al.* (2012) reported the identification of 796 differentially regulated genes in brain tissues of TBM patients co-infected with HIV using oligonucleotide DNA microarrays and the GeneSpring GX v11.0.2 pathway analysis software. They also performed immunohistochemical validation and confirmed the abundance of four gene products-glial fibrillaryacidic protein (GFAP), serpin peptidase inhibitor, clade A member 3 (SERPINA3), thymidine phosphorylase (TYMP) and heat shock 70 kDa protein 8 (HSPA8). This study paves the way for understanding the mechanism of TBM in HIV positive patients and for further validation of potential disease biomarkers and hence are possible targets for drug discovery in MtB HIV coinfection [18].

HIV not only brings about changes in the immune response, causing secondary infections, but also disturbs the functioning of the central nervous system (CNS) activating inflammation. To find out the regulatory mechanism for such untoward inflammatory activation, Erick T. Tatro *et al.* (2010) conducted a retrospective study to construct a GRN using HIV and the glial cell system. MicroRNA plays a major role in host defense and neuronal homeostasis, but its role in HIV CNS infection is lesser known which was explored in this study. They studied the RNA alterations in the frontal cortex (FC) of HIV-infected individuals and those concurrently infected and diagnosed with major depressive disorder (MDD). They tried to identify changes in miRNA expression that occurred in the FC of HIV individuals, determine whether miRNA expression profiles of the FC could differentiate HIV from HIV/MDD, and adapt a method to meaningfully integrate gene expression data and miRNA expression data in clinical samples. The isolated RNA from the study groups were used for large scale miRNA profiling and these experimental data sets were used for a target bias analysis. The study pointed out three types of miRNAs: (a.) those with many dysregulated mRNA targets of less stringent statistical significance, (b.) fewer dysregulated target-genes of highly stringent statistical significance, and (c.) unclear bias. In HIV/MDD, more miRNAs were down-regulated than in HIV alone, providing evidence that certain miRNAs serve as key elements in gene regulatory networks in HIV-infected FC and may be implicated in neurobehavioral disorder [19].

14.2 Abbreviations

AIC	- Akaike Information Criterion
AIDS	- Acquired Immune Deficiency Syndrome
APLP2	- Amyloid beta (A4) Precursor-like Protein 2
ATF5	- Activating Transcription Factor (ATF/CREB) family
BIND	- Biomolecular Interaction Network Database
CCND2	- Cyclin Dependent Kinases 2
CD	- Cluster Determinant
CNS	- Central Nervous System
DCs	- Dendritic Cells
FC	- Frontal Cortex
FOSB	- FBJ murine osteosarcoma viral oncogene homolog B (FOS family)
GFAP	- Glial Fibrillary Acidic Protein

GRNs	- Gene Regulatory Networks
HCASM	- Human Coronary Artery Smooth Muscle
HIV	- Human Immuno-deficiency Virus
HSPA8	- Heat Shock 70 Kda Protein 8
IFS	- Incremental Feature Selection
IL	- Interleukin
JASPAR	- collection of transcription factor DNA-binding preferences, modeled as matrices
LAMC1	- Laminin γ-1
LASSO	-Least Absolute Shrinkage and Selection Operator
MAPK	- Mitogen-Activated Protein Kinase
MDD	- Major Depressive Disorder
MRMR	- Minimum Redundancy-Maximum Relevance
MtB	- *Mycobacterium tuberculosis*
NF- κ B	- Nuclear Factor κ B
PAK	- p21-activated kinase
PI3K	- Phosphatidylinositol 3-Kinase
PKA	- Protein Kinase A
PPI	- Protein Protein Interaction
RT-PCR	- Real Time Polymerase Chain Reaction
SERPINA 3	- Serpin Peptidase Inhibitor, Clade A Member 3
TBM	- Tuberculous Meningitis
TFGβ	- Transforming growth Factor β
TFs	- Transcription Factors
Th1/2	- T Helper1/2
THBS1	- Thrombospondin 1
TLRs	- Toll like Receptors
TNF	- Tumor Necrosis Factor
TNFR	- Tumor Necrosis Factor Receptor
TRANS	-Transcription Factor database
TYMP	- Thymidine Phosphorylase

14.3 Conclusions

The modeling of gene regulatory networks relies on characterization of the behavior of small subsystems, formation of hypotheses about how these subsystems interconnect translation of these hypotheses into a mathematical model and experimentation to yield results that indicate necessary changes to the original hypotheses. Using the natural system to build artificial (synthetic) network represents the important advances in the recent years of research in biology through an engineering perspective. Current progress in the study of both naturally occurring and synthetic genetic networks indicates that computational modeling should have an important role in hastening the bench top research to moving into the clinics in treating infectious diseases and also a better understanding of the host-parasite interaction.

Acknowledgement

The authors are thankful to Dr. Shekhar C. Mande, Director National Centre for Cell Science (NCCS) for supporting the Bioinformatics and High Performance Computing Facility. Milsee Mol acknowledges fellowship from CSIR, Government of India.

References

[1] D. L. Nelson, Lehninger Principles of Biochemistry, IV Edition –Chapter 12 Biosignaling.

[2] T. Schlitt, A. Brazma, Current approaches to gene regulatory network modeling. BMC Bioinformatics 2007, 8(Suppl 6):S9 doi:10.1186/1471-2105-8-S6–S9.

[3] D. Das, N. Banerjee, M. Q. Zhang, Interacting models of cooperative gene regulation. PNAS 2004 101(46):16234–16239.

[4] J. Hasty, D. Mac Millen, F. Isaacs, J. J. Collins, Computational studies of gene regulatory networks: in numero molecular biology Nature Review 2001 2:268–279.

[5] J. L. Payne, J. H. Moore, A. Wagner. Robustness, Evolvability, and the Logic of Genetic Regulation. Artificial life Early Access 2013:1–16. doi:10.1162/ARTL_a_00099.

[6] E. Ferrada, A. Wagner, Protein robustness promotes evolutionary innovations on large evolutionary time-scales. Proceedings of the Royal Society London B, 2008 275:1595–1602.

[7] S. I. Gringhuis,, J. den Dunnen, M. Litjens, M. van der Vlist, T. B. Geijtenbeek, Carbohydrate specific signaling through the DC-SIGN signalosome tailors immunity to Mycobacterium tuberculosis, HIV-1 and Helicobacter pylori. Nature Immunol. 2009 10:1081–1088.

[8] S. N. Patel, et al., Disruption of CD36 impairs cytokine response to Plasmodium falciparumglycosylphosphatidylinositol and confers susceptibility to severe and fatal malaria in vivo. J. Immunol. 2007 178:3954–3961.

[9] H. Hawlisch, et al., C5a negatively regulates Toll-like receptor 4-induced immune responses. Immunity 2005 22:415–426.

[10] A. M. Krachler, A. R. Woolery, K. Orth, Manipulation of kinase signaling by bacterial pathogens J. Cell Biol. 2011 195(7):1083–1092.

[11] K. A. McDonough, A. Rodriguez, The myriad roles of cyclic AMP in microbial pathogens: from signal to sword, Nature Reviews Microbiology.201210:27–38.

[12] V. Cherkassky, F. M. Mulier, Learning from Data: Concepts, Theory, and Methods. John Wiley & Sons, 2007.

[13] T.S. Gardner, J.J. Faith, Reverse-engineering transcription control networks. Physics of Life Reviews 2005 2:65–88.

[14] C. Bor-Sen, Y. Shih-Kuang, L. Chung-Yu, C. Yung-Jen, A systems biology approach to construct the gene regulatory network of systemic inflammation via microarray and databases mining. BMC Medical Genomics 2008 1:46.

[15] N. N. Pius, C. A. Johnson, S. Pratap, T. C. Cardenas, Y. Y. Kleshchenko, V. A. Furtak, K. J. Simmons, M. F. Lima, F. Villalta, Gene Network Analysis During Early Infection of Human Coronary Artery Smooth Muscle Cells by Trypanosomacruzi and Its gp83 Ligand ChemBiodivers. 2010 7(5):1051–1064. doi:10.1002/cbdv.200900320.

[16] R. Altwasser, J. Linde, E. Buyko, U. Hahn, R. Guthke, Genome wide scale-free network inference for Candida albicans. Front. Microbio. 20123:51. doi: 10.3389/fmicb.2012.00051.

[17] T. Huang, Z. Xu, L. Chen, Y-DCai, X. Kong, Computational Analysis of HIV-1 Resistance Based on Gene Expression Profiles and the Virus-Host Interaction Network. PLoS ONE 2011 6(3): e17291. doi:10.1371/journal.pone.0017291.

[18] G. S. Sameer Kumar, A. K. Venugopal, M. K. Kashyap, R. Raju, A. Marimuthu, et al., Gene Expression Profiling of Tuberculous Meningitis Co-infected with HIV. J Proteomics Bioinform 2012 5: 235–244. doi:10.4172/jpb.

[19] E. T. Tatro, E. R. Scott, T. B. Nguyen, S. Salaria, S. Banerjee, et al., Evidence for Alteration of Gene Regulatory Networks through MicroRNAs of the HIV Infected Brain: Novel Analysis of Retrospective Cases. PLoS ONE 2010 5(4):e10337 doi:10.1371/journal.pone.0010337.

15

Statistical Methods to Predict Drug Side-Effects

Yoshihiro Yamanishi

Division of System Cohort, Multi-scale Research Center
for Medical Science, Medical Institute of Bioregulation, Kyushu
University and Institute for Advanced Study, Kyushu University

15.1 Abstract

Drug side-effects, or adverse drug reactions, are a serious problem for public health. In this chapter, we review several recently proposed statistical methods to predict unknown side-effect profiles of drug candidate molecules from chemical structure data and target protein information. We show the usefulness of the methods on the simultaneous prediction of about one thousand side-effects for approved drugs and compare the performance between different methods. We also discuss the advantages and limitations of each method and provide some perspectives for future work. These statistical methods are expected to be useful at many stages of the drug development.

Keywords: side-effect prediction, adverse drug reactions, data integration, machine learning, statistical models.

15.2 Introduction

Drug side-effects, or adverse drug reactions, are a serious problem for public health. It is the main cause of failure in the process of drug development, and of withdrawal of developed drugs from the market. For example, serious drug side-effects are estimated to be the fourth leading cause of death in the United States, resulting in 100,000 deaths per year [6]. From the viewpoint of chemical

Post-genomic Approaches in Drug and Vaccine Development, 391–406.

genomics and systems biology, phenotypic effects of drugs can be caused by perturbations to the biological system [10, 18]. Actually, the body's response to a drug reflects not only expected favorable effects due to the interaction with its main target protein, but also unexpected adverse side-effects due to the interaction with additional proteins (off-targets). It was reported that drugs with similar side-effects tend to share similar protein targets, and the use of side-effect similarity was proposed to identify missing protein targets of known drugs [3].

The identification of potential adverse side-effects of drugs is a challenging issue in recent pharmaceutical research. A useful experimental approach for predicting side-effects is preclinical *in vitro* safety profiling that tests compounds with biochemical and cellular assays, but experimental detection of many side-effects is very difficult in terms of cost, efficiency and scalability [21]. Therefore, *in silico* prediction of potential side-effects is expected to improve this long and expensive drug development process and to provide safe therapies for patients. So far, a variety of computational methods for predicting drug side-effects have been developed, and the previous methods can be categorized into (i) chemical structure-based approach, (ii) biological target-based approach, and (iii) the integrative approach. Figure 15.1 shows an illustration of the three approaches.

The strategy of chemical structure-based approach is to relate drug chemical structures with drug side-effects, following the spirit of QSAR (quantitative structure-activity relationship) and QSPR (quantitative structure-property relationship) [12, 20, 19]. A link between chemical substructures and side-effects was suggested by a graph-based method [16]. A statistical method for predicting drug side-effect similarity information from chemical structures was proposed, but the method cannot be applied to predict high-dimensional side-effect profiles directly [26]. An algorithmic framework for predicting side-effect profiles from chemical structure data was established using canonical correlation analysis (CCA), which is a pioneering work in terms of simultaneous prediction of many side-effects [1]. A chemical fragment-based method was proposed to relate drug chemical substructures with side-effects using the sparse version of canonical correlation analysis (SCCA), and the extracted chemical substructures were used for predicting side-effect profiles [14].

The strategy of biological target-based approach is to relate drug targeted proteins or biological pathways with side-effects. A correlation between drug side-effects and biological pathways was confirmed by comparing biological pathways affected by toxic compounds and those affected by non-toxic

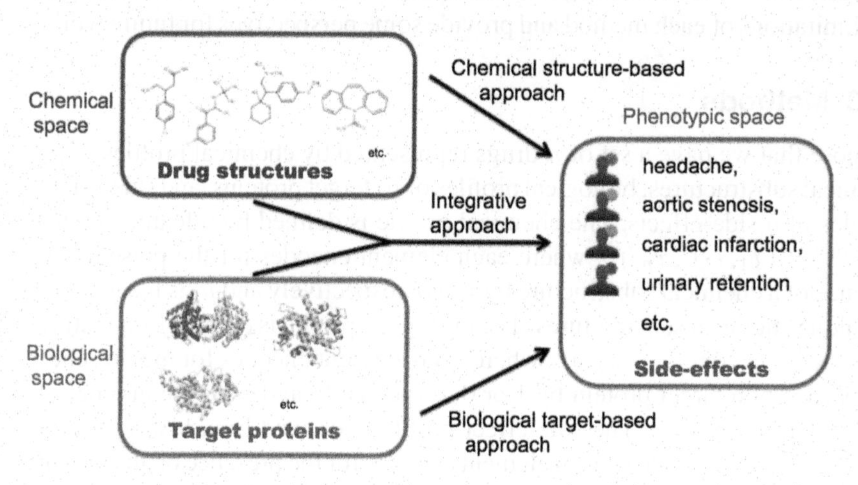

Figure 15.1 An illustration of the three approaches.

compounds [15]. A docking based-method for predicting side-effects was proposed by identifying off-targets of drugs, where a drug was docked into a protein's binding pocket similar to that of its primary target, and to map them onto biological pathways known to involve side-effects [24]. However, the method depends on the availability of protein 3D structures and the accuracy of biological pathway maps. Drug-protein interaction network was related with drug side-effects using sparse canonical correlation analysis (SCCA) [13]. However, the method depends on the availability of target protein information.

The strategy of the integrative approach is to use both chemical structure-based approach and biological target-based approach in an integrated framework. Recently, the joint use of chemical structure information and biological target information was proposed to predict drug side-effect profiles using an extension of kernel regression (KR), and it was shown that heterogeneous data integration improved the prediction accuracy [27].

In this chapter, we review several of recently proposed statistical methods to predict unknown side-effect profiles of drug candidate molecules from chemical structure data and target protein information. In this review, we focus on canonical correlation analysis (CCA), sparse canonical correlation analysis (SCCA), kernel regression (KR) and an extension of KR. In the results, we demonstrate the usefulness of the methods on the simultaneous prediction of about one thousand side-effects for approved drugs and compare the performance between different methods. We also discuss the advantages and limitations of each method and provide some perspectives for future work.

15.3 Methods

Suppose that we have a set of n drugs represented by chemical profiles of p_1 chemical substructures, biological profiles of p_2 target proteins, and side-effect profiles of q side-effects. The chemical profile is defined by a feature vector $\mathbf{x}^{(chem)} = (x_1, \cdots, x_{p_1})^T$, where each element encodes for the presence or absence of a chemical substructure by 1 or 0, respectively and p_1 is the number of unique chemical substructures. The biological profile is defined by a feature vector $\mathbf{x}^{(bio)} = (x_1, \cdots, x_{p_1})^T$, where each element encodes for the presence or absence of a target protein by 1 or 0, respectively and p_2 is the number of unique target proteins. The side-effect profile is defined by a feature vector $\mathbf{y} = (y_1, \cdots, y_q)^T$, where each element encodes for the presence or absence of a side-effect by 1 or 0, respectively and q is the number of unique side-effect terms. We consider the situation where we are given a new drug candidate molecule with chemical profile $\mathbf{x}_{new}^{(chem)}$ and biological profile $\mathbf{x}_{new}^{(bio)}$, and we want to predict its potential side-effect profile $\mathbf{y}_{new}$ based on the chemical and biological profiles.

15.3.1 Canonical correlation analysis (CCA)

The use of canonical correlation analysis (CCA) [8] was proposed to predict drug side-effect profiles from chemical profile [1]. We make a brief review of the CCA-based method.

Suppose that we have a set of n drugs $\{\mathbf{x}_i\}_{i=1}^n$ represented by data profiles (chemical profiles or biological profiles in this study) and $\{\mathbf{y}_i\}_{i=1}^n$ represented by side-effect profiles. Consider two linear combinations as $u_i = \alpha^T \mathbf{x}_i$ and $v_i = \beta^T \mathbf{y}_i$ $(i = 1, 2, \cdots, n)$, where $\alpha = (\alpha_1, \cdots, \alpha_p)^T$ and $\beta = (\beta_1, \cdots, \beta_q)^T$ are weight vectors. The goal of ordinary CCA is to find weight vectors α and β which maximize the following canonical correlation coefficient:

$$\rho = corr(u, v) = \frac{\sum_{i=1}^n \alpha^T \mathbf{x}_i \cdot \boldsymbol{\beta}^T \mathbf{y}_i}{\sqrt{\sum_{i=1}^n (\alpha^T \mathbf{x}_i)^2} \sqrt{\sum_{i=1}^n (\boldsymbol{\beta}^T \mathbf{y}_i)^2}}, \tag{15.1}$$

where u and v are assumed to be centered.

Multiple canonical components (m canonical components) can be obtained by the following manipulation: $(\alpha_k, \boldsymbol{\beta}_k) = \arg\max \rho = corr(u, v)$ under the following orthogonality constraints: $\alpha \perp \alpha_1, \cdots, \alpha_{k-1}$ and $\boldsymbol{\beta} \perp \boldsymbol{\beta}_1, \cdots, \boldsymbol{\beta}_{k-1}$ $(k = 1, 2, ..., m)$.

For the prediction of the side-effect profile of a new drug $\mathbf{x}_{new}$, we compute the following prediction score:

$$\mathbf{y}_{new} = B^{-T} A^T \mathbf{x}_{new}, \tag{15.2}$$

where $A = [\alpha_1, \cdots, \alpha_m]$, $B = [\boldsymbol{\beta}_1, \cdots, \boldsymbol{\beta}_m]$, m is the number of components, and B^{-T} is the peudo-inverse matrix of B^T. If the j-th element in $\mathbf{y}_{new}$ has a high score, the new molecule is predicted to have the j-th side-effect $(j = 1, 2, \cdots, q)$.

In practice, we consider three possible predictive models for a new drug candidate molecule with chemical profile $\mathbf{x}_{new}^{(chem)}$ and biological profile $\mathbf{x}_{new}^{(bio)}$, as follows:

- **CCA with chemical information (CCA chem):** The prediction is performed based only on $\mathbf{x}_{new}^{(chem)}$.
- **CCA with biological information (CCA bio):** The prediction is performed based only on $\mathbf{x}_{new}^{(bio)}$.
- **CCA with chemical and biological information (CCA chembio):** The prediction is performed based on the concatenated profile $\mathbf{x}^{(chembio)} = (\mathbf{x}^{(chem)\top}, \mathbf{x}^{(bio)\top})^\top$.

Note that "CCA chem" corresponds to the previous method [1].

15.3.2 Sparse canonical correlation analysis (SCCA)

The use of sparse canonical correlation analysis (SCCA) was proposed to predict drug side-effect profiles from chemical profiles [14] or from biological profiles [13]. We make a brief review of the SCCA-based method.

In ordinary CCA, the weight vectors α and β are not unique if p or q exceeds n. Most elements in the weight vectors α and β in the ordinary CCA are non-zeros, which makes it difficult to interpret the result. In practice, it is desirable to find weight vectors that have large correlation, but that are also sparse for easier interpretation.

To impose the sparsity on α and β, we consider adding the following L_1 penalty terms on the above maximization problem :

$$||\alpha||_1 \leq s_1\sqrt{p}, \quad ||\boldsymbol{\beta}||_1 \leq s_2\sqrt{q}, \tag{15.3}$$

where $||\cdot||_1$ is L_1 norm (the sum of all absolute values of the vector elements), and s_1 and s_2 are parameters to control the sparsity ($0 < s_1 \leq 1$ and $0 < s_2 \leq 1$). For simplicity, the same value is used for s_1 and s_2 in this study. The CCA with L_1 penalties is referred to as sparse canonical correlation analysis (SCCA) [23].

It should be pointed out that the statistical model of SCCA has high interpretability, because there are a limited number of non-zero elements in the weight vectors α_k and β_k ($k = 1, 2, \cdots, m$). Chemical substructures (resp. target proteins) and side-effects with non-zero weights in $\alpha^{(chem)}$ (resp. $\alpha^{(bio)}$) and β in the same canonical component can be considered to be associated with each other.

In practice, we consider three possible predictive models for a new drug candidate molecule with chemical profile $\mathbf{x}_{new}^{(chem)}$ and biological profile $\mathbf{x}_{new}^{(bio)}$, as follows:

- **SCCA with chemical information (SCCA chem):** The prediction is performed based only on $\mathbf{x}_{new}^{(chem)}$.
- **SCCA with biological information (SCCA bio):** The prediction is performed based only on $\mathbf{x}_{new}^{(bio)}$.
- **SCCA with chemical and biological information (SCCA chem-bio):** The prediction is performed based on the concatenated profile $\mathbf{x}^{(chembio)} = (\mathbf{x}^{(chem)\top}, \mathbf{x}^{(bio)\top})^\top$.

In each case, we use the same prediction score defined in ordinary CCA. Note that "SCCA chem" corresponds to the previous method [14] and "SCCA bio" corresponds to the previous method [13].

15.3.3 Kernel regression (KR)

The use of kernel regression (KR) was proposed to predict drug side-effect profiles from chemical profiles and biological profiles [27]. We make a brief review of the KR-based method.

We consider a regression model to predict q-dimensional feature vector $\mathbf{y}$ (response variables) from p-dimensional feature vector $\mathbf{x}$ (explanatory variables). Here we propose to apply a kernel regression (KR) model for multiple responses, formulated as follows:

$$\mathbf{y}_{new} = f(\mathbf{x}_{new}) = \sum_{i=1}^{n} k(\mathbf{x}_{new}, \mathbf{x}_i)\mathbf{w}_i + \epsilon, \qquad (15.4)$$

where $\mathbf{x}_{new}$ is the feature vector of a new object, $\mathbf{y}_{new}$ is the response vector of the new object, f is the projection $f : \mathcal{X} \to \mathbf{R}^q$, $\mathcal{X}$ is a set of objects, n is the number of objects in a training set, $\mathbf{x}_i$ is a feature vector of the i-th object in the training set, $\mathbf{w}_i \in \mathbf{R}^q$ is a weight vector, $k(\cdot, \cdot)$ is a positive definite kernel, that is, a symmetric function $k : \mathcal{X} \times \mathcal{X} \to \mathbf{R}$ satisfying $\sum_{i,j=1}^{n} a_i a_j k(\mathbf{x}_i, \mathbf{x}_j) \geq 0$ for any $a_i, a_j \in \mathbf{R}$, and ϵ is a noise vector. For example, if k is Gaussian RBF kernel, it is defined as $k(\mathbf{x}, \mathbf{x}') = \exp(-||\mathbf{x} - \mathbf{x}'||^2/2\sigma^2)$, where σ is the width parameter.

The fitting of the model can be done by finding $\mathbf{w}_i$ which minimizes the following penalized loss function:

$$L = ||Y - KW||_F^2 + \lambda ||W||_F^2, \tag{15.5}$$

where K is an $n \times n$ kernel similarity matrix $(K)_{ij} = k(\mathbf{x}_i, \mathbf{x}_j)$, $W = (\mathbf{w}_1, \cdots, \mathbf{w}_n)^T$, $Y = (\mathbf{y}_1, \cdots, \mathbf{y}_n)^T$, λ is a regularization parameter, and $|| \cdot ||_F$ is Frobenius norm. Note that we put the above penalty term to avoid over fitting. Once we have trained the KR model, which is computing W, we can apply the model to unseen objects (new drug candidate molecules in our case).

In practice, we consider four predictive models. The side-effect prediction for a new drug candidate molecule with chemical profile $\mathbf{x}_{new}^{(chem)}$ and biological profile $\mathbf{x}_{new}^{(bio)}$ is performed as follows:

- **KR with chemical information (KR chem):** The prediction is performed based only on $\mathbf{x}^{(chem)}$ as follows:

$$\mathbf{y}_{new} = \sum_{i=1}^{n} k(\mathbf{x}_{new}^{(chem)}, \mathbf{x}_i^{(chem)}) \mathbf{w}_i. \tag{15.6}$$

- **KR with biological information (KR bio):** The prediction is performed based only on $\mathbf{x}^{(bio)}$ as follows:

$$\mathbf{y}_{new} = \sum_{i=1}^{n} k(\mathbf{x}_{new}^{(bio)}, \mathbf{x}_i^{(bio)}) \mathbf{w}_i. \tag{15.7}$$

- **KR with chemical and biological information (KR chembio):** The prediction is performed based on the concatenated profile $\mathbf{x}^{(chembio)} = (\mathbf{x}^{(chem)\top}, \mathbf{x}^{(bio)\top})^\top$, as follows:

$$\mathbf{y}_{new} = \sum_{i=1}^{n} k(\mathbf{x}_{new}^{(chembio)}, \mathbf{x}_i^{(chembio)}) \mathbf{w}_i. \tag{15.8}$$

- **KR with the kernel integration of chemical and biological information (KR Kchem+Kbio):** The prediction is performed based on the sum of the two kernel functions as follows.

$$\mathbf{y}_{new} = \sum_{i=1}^{n} \{c_1 k_1(\mathbf{x}_{new}^{(chem)}, \mathbf{x}_i^{(chem)}) + c_2 k_2(\mathbf{x}_{new}^{(bio)}, \mathbf{x}_i^{(bio)})\}\mathbf{w}_i, \quad (15.9)$$

where k_1 and k_2 are kernel functions for chemical and biological profiles, respectively, and c_1 and c_2 are the weights for k_1 and k_2, respectively, in the data integration. For simplicity, the same weight is used in the kernel integration ($c_1 = c_2 = 0.5$) in this study.

In any predictive models if the j-th element in $\mathbf{y}_{new}$ has a high score, the new molecule is predicted to have the j-th side-effect ($j = 1, 2, \cdots, q$).

15.3.4 Multiple kernel regression (MKR)

An extension of kernel regression model (KR) for multiple data sources, which is called multiple kernel regression (MKR), was proposed to predict drug side-effect profiles from chemical profiles and biological profiles [27]. We make a brief review of the MKR-based method.

The objects are often represented by multiple different data sources. Suppose that each object is represented by m heterogeneous feature vectors as $\mathbf{x}^{(1)}, \cdots, \mathbf{x}^{(m)}$ ($m = 2$ in our case, since drugs are represented by chemical and biological profiles).

To deal with such a situation, we propose an extension of the kernel regression model, which we call multiple kernel regression (MKR), formulated as follows:

$$\mathbf{y}_{new} = f(\mathbf{x}^{(1)}, \cdots, \mathbf{x}^{(m)}) = \sum_{l=1}^{m} \sum_{i=1}^{n} k_l(\mathbf{x}_{new}^{(l)}, \mathbf{x}_i^{(l)})\mathbf{w}_i^{(l)} + \epsilon, \quad (15.10)$$

where $\mathbf{x}_{new}^{(l)}$ is the feature vector of a new object for the l-th data source, n is the number of objects in a training set, $\mathbf{x}_i^{(l)}$ is a feature vector of the i-th element for l-th data source in the training set, $\mathbf{w}_i^{(l)} \in \mathbf{R}^q$ is a weight vector, $k_l(\cdot, \cdot)$ is a kernel similarity function for $\mathbf{x}^{(l)}$, and ϵ is a noise vector. In this study we use Gaussian RBF kernel for each k_l.

The fitting of the model can be done by finding $\mathbf{w}_i^{(1)}, \cdots, \mathbf{w}_i^{(m)}$ which minimizes the following penalized loss function:

$$L = ||Y - \sum_{l=1}^{m} K_l W_l||_F^2 + \sum_{l=1}^{m} \lambda_l ||W_l||_F^2, \tag{15.11}$$

where K_l is an $n \times n$ kernel similarity matrix $(K_l)_{ij} = k_l(\mathbf{x}_i, \mathbf{x}_j)$, $W_l = (\mathbf{w}_1^{(l)}, \cdots, \mathbf{w}_n^{(l)})^T$, $Y = (\mathbf{y}_1, \cdots, \mathbf{y}_n)^T$, λ_l is a regularization parameter, and $||\cdot||_F$ is Frobenius norm. Once we have trained our model, which is computing $W_1, \cdots, W_m$, we can apply the model to unseen objects (new drug candidate molecules in practice).

In practice, we consider a predictive model for a new drug candidate molecule with chemical profile $\mathbf{x}_{new}^{(chem)}$ and biological profile $\mathbf{x}_{new}^{(bio)}$, as follows:

- **MKR based on chemical and biological information (MKR chem+bio):**

$$\mathbf{y}_{new} = \sum_{i=1}^{n} k_1(\mathbf{x}_{new}^{(chem)}, \mathbf{x}_i^{(chem)})\mathbf{w}_i^{(1)} + \sum_{i=1}^{n} k_2(\mathbf{x}_{new}^{(bio)}, \mathbf{x}_i^{(bio)})\mathbf{w}_i^{(2)}, \tag{15.12}$$

  where k_1 and k_2 are kernel functions for chemical and biological profiles, respectively.

If the j-th element in $\mathbf{y}_{new}$ has a high score, the new molecule is predicted to have the j-th side-effect ($j = 1, 2, \cdots, q$).

15.4 Results

15.4.1 Data

Side-effect terms were obtained from the SIDER database which contains information about side-effects of marked drugs [11]. In this review, we focused on drugs annotated as "small molecules" in the DrugBank database [22]. There are some side-effects which are associated with almost all the drugs (e.g., vomiting, nausea, dizziness), while there are some side-effects associated with very few drugs (e.g. agnosia, variant angina, aspergillosis). Therefore, we removed side-effects which lie at the top 10% in terms of frequency, and we also removed side-effects which are associated with only one drug (singletons), leading to a dataset consisting of 658 drugs, 969 side-effect terms, and 23,061 associations between drugs and side-effects. This dataset is used as gold standard data.

Chemical information about drugs was obtained from the PubChem database [4]. To encode the chemical structure of drugs, we used a fingerprint corresponding to the 881 chemical substructures defined in PubChem.

Biological information about target proteins of drugs was obtained from the DrugBank database [22] and the Matador database [7]. We focused on target proteins that are indicated as direct interactions, and we extracted 5074 drug-protein interactions involving 1368 unique target proteins.

15.4.2 Performance evaluation

We applied the statistical methods to predict drug side-effect profiles from chemical profiles and biological profiles. We tested 11 approaches: (1) “CCA chem”, (2) “CCA bio”, (3) “CCA chembio”, (4) “SCCA chem”, (5) “SCCA bio”, (6) “SCCA chembio”, (7) “KR chem”, (8) “KR bio”, (9) “KR chembio”, (10) “KR Kchem+Kbio” and (11) “MKR chem+bio” on their abilities to predict known side-effect profiles using 658 drugs in the gold standard data (see the Methods section for a description of each approach). We performed the following 5-fold cross-validation: Drugs in the gold standard set were split into five subsets of roughly equal size, each subset was then taken in turn as a test set, and we performed the training on the remaining 4 sets.

We evaluated the performance of each method by the Precision-Recall curve (PR curve), and summarized the performance by the area under the PR curve (AUPR). All parameters in each method (e.g., regularization parameters, number of components, kernel parameters) were optimized using grid search with the AUPR score as an objective function. To obtain a robust result we repeated the overall cross-validation procedure three times, and computed the average and standard deviation (S.D.) of the AUPR score.

The first column of Table 15.1 shows the AUPR score of each method on the 5-fold cross-validation experiment, where the prediction scores for all side-effects were merged and a global PR curve was drawn for each method. It is observed that SCCA works better than CCA for biological profiles, but SCCA works worse than CCA for chemical profiles. It is observed that KR-based methods outperform CCA-based methods and SCCA-based methods in using any data sources. The “KR Kchem+Kbio” and “MKR chem+bio” methods seem to work better than the “KR chem” and “KR bio” methods, which suggests that the integration of chemical and biological information is meaningful. However, the “KR chembio” method based on the vector concatenation procedure did not improve the performance, compared with

Table 15.1 Performance statistics based on 5-fold cross-validation. AUPR is the total area under the precision-recall curve. Top1 indicates the number of drugs having a known side-effect which is ranked highest in the prediction score. Top5 indicates the number of drugs having at least one known side-effect which is ranked among the top 5 high scoring side-effects. S.D. indicates the standard deviation over the repetition of the cross-validation. The best result in each column is shown in bold.

Method	AUPR ± S.D.	Top1 ± S.D.	Top5 ± S.D.
Random	0.0364 ± 0.0000	27 ± 2.6	99 ± 5.5
CCA chem	0.1493 ± 0.0015	203 ± 6.0	398 ± 8.5
CCA bio	0.1479 ± 0.0010	221 ± 9.8	418 ± 4.5
CCA chembio	0.1527 ± 0.0009	218 ± 5.6	412 ± 14.2
SCCA chem	0.1286 ± 0.0017	163 ± 6.0	348 ± 6.8
SCCA bio	0.1555 ± 0.0044	204 ± 12.8	392 ± 3.0
SCCA chembio	0.1263 ± 0.0006	170 ± 9.2	248 ± 8.0
KR chem	0.1817 ± 0.0025	250 ± 3.7	431 ± 1.5
KR bio	0.1933 ± 0.0011	259 ± 9.4	451 ± 10.9
KR chembio	0.1891 ± 0.0029	254 ± 1.5	441 ± 9.5
KR Kchem+Kbio	**0.2089** ± 0.0024	**280** ± 14.9	**461** ± 6.0
MKR chem+bio	0.2008 ± 0.0020	272 ± 5.2	**461** ± 8.1

the "KR chem" and "KR bio" methods, and a similar tendency was also observed in the vector concatenation procedure in CCA-based and SCCA-based methods as well. This result suggests that kernel-based data integration is more useful, compared with vector concatenation.

We investigated the prediction accuracy of predicted side-effects for each drug with a high level of confidence. We computed the number of drugs having a known side-effect which is ranked highest in the prediction score, and the number of drugs having at least one known side-effect which is ranked among the 5 highest ranking side-effects. Originally, these statistics were used in the previous study [1]. The second and third columns in Table 15.1 show the performance measures of each method. KR-based and MKR-based methods seem to outperform CCA-based and SCCA-based methods. For example, the "KR Kchem+Kbio" method ranked first one of the known side-effects of 41.7 % of the 658 reference drugs, and ranked a correct side effect among the top five scoring side-effects for 70.0 % of the 658 reference drugs. This implies that high scoring predicted side-effects are more reliable, and the integration of multiple data sources is meaningful in drug side-effect prediction. The "KR Kchem+Kbio" method worked slightly better than the "MKR chem+bio" method. One explanation for the better performance of the "KR Kchem+Kbio" method over the "MKR chem+bio" method is that MKR requires too many parameters to be estimated and the predictive model of the "MKR chem+bio" method may have the over-fitting problem.

15.5 Discussion and Conclusion

In this chapter, we reviewed several of recently proposed statistical methods for predicting potential side-effect profiles of drug candidate molecules based on their chemical structures and their target protein information. In this review we focused on canonical correlation analysis (CCA), sparse canonical correlation analysis (SCCA), kernel regression (KR) and an extension of KR (MKR). In the results we showed the usefulness of these statistical methods in the cross-validation experiments. In practical applications, these statistical methods may be useful in various ways and at various stages of the drug development process. At early stages, among several active drug candidates, the method could help to choose the molecules that should further continue the development process. It could also help to find novel drug indications of known drugs toward drug repositioning. Indeed, negative side-effects of drugs used in a given pathology can be viewed as a beneficial effect in another pathology.

These statistical methods depend highly on the pre-definition of chemical profiles, biological profiles and side-effect terms. Future development in CCA-based or SCCA-based methods would be to design more appropriate fingerprints for drugs and target proteins. The algorithm of KR-based and MKR-based methods belong to a class of kernel methods [17], so the performance could be improved by using more sophisticated kernel similarity functions designed for drug structures and target proteins. In this review, we did not take into account the difference of side-effects between individuals. In reality, all patients do not always suffer from the same side-effects; some side-effects occur often, and others rarely occur. Currently, only a limited number of drugs have publicly available information on the likelihood of the occurrence of certain side-effects. We hope that such information becomes available for more drugs to conduct more detailed analysis in near future.

A serious limitation of the biological target-based approach and the integrative approach is that the target protein information is not always available or not complete for all drug candidate compounds of interest. Recently, a variety of *in silico* chemogenomic approaches have been developed to predict drug-target interactions or compound-protein interactions, and the usefulness of the chemogenomic approach has been shown in many previous works [25, 5, 9, 2]. The chegemonomic approach is based on the fact that similar compounds tend to interact with similar proteins, and the prediction is performed based on compound chemical structures, protein sequences and all possible known compound-protein interactions. It would be interesting to apply the chemogenomic approach for estimating potential target proteins of drugs and use the results for the drug side-effect prediction.

Acknowledgements

We thank Jean-Philippe Vert, Véronique Stoven, and Edouard Pauwels for useful discussions. This work is supported by The Japan Society for the Promotion of Science; JSPS Kakenhi (25700029). This work was also supported by the Program to Disseminate Tenure Tracking System, MEXT, Japan, and Kyushu University Interdisciplinary Programs in Education and Projects in Research Development.

References

[1] Nir Atias and Roded Sharan. An algorithmic framework for predicting side-effects of drugs. J Comput Biol., 18(3):207–218, 2011.

[2] K. Bleakley and Y. Yamanishi. Supervised prediction of drug-target interactions using bipartite local models. Bioinformatics, 25: 2397–2403, 2009.

[3] M. Campillos, M. Kuhn, A.C. Gavin, L.J. Jensen, and P. Bork. Drug target identification using side-effect similarity. Science, 321(5886): 263–6, 2008.

[4] B. Chen, D. Wild, and R. Guha. PubChem as a source of polypharmacology. J Chem Inf Model., 2009.

[5] J.L. Faulon, M. Misra, S. Martin, K. Sale, and R Sapra. Genome scale enzyme-metabolite and drug-target interaction predictions using the signature molecular descriptor. Bioinformatics, 24: 225–233, 2008.

[6] K. M. Giacomini, R. M. Krauss, D. M. Roden, M. Eichelbaum, M. R. Hayden, and Y. Nakamura. When good drugs go bad. Nature, 446 (7139): 975–977, 2007.

[7] S. Gunther, M. Kuhn, M. Dunkel, M. Campillos, C. Senger, E. Petsalaki, J. Ahmed, GE. Urdiales, A. Gewiess, LJ. Jensen, R. Schneider, R. Skoblo, RB. Russell, PE. Bourne, P. Bork, and R. Preissner. Supertarget and matador: resources for exploring drug-target relationships. Nucleic Acids Res, 36: D919–D922, 2008.

[8] H. Hotelling. Relations between two sets of variates. Biometrika, 28: 321–377, 1936.

[9] L. Jacob and J.-P. Vert. Protein-ligand interaction prediction: an improved chemoge-nomics approach. Bioinformatics, 24: 2149–2156, 2008.

[10] M. Kanehisa, S. Goto, M. Hattori, K.F. Aoki-Kinoshita, M. Itoh, S. Kawashima, T. Katayama, M. Araki, and M. Hirakawa. From genomics to chemical genomics: new developments in kegg. Nucleic Acids Res., 34(Database issue):D354–357, Jan 2006.
[11] M. Kuhn, M. Campillos, I. Letunic, L.J. Jensen, and P. Bork. A side effect resource to capture phenotypic effects of drugs. Mol Syst Biol, 6: 343, 2010.
[12] P. Mahe´, N. Ueda, T. Akutsu, J.L. Perret, and J.P. Vert. Graph kernels for molecular structure-activity relationship analysis with support vector machines. J Chem Inf Model., 45(4): 939–951, 2011.
[13] S. Mizutani, E. Pauwels, V. Stoven, S. Goto, and Y. Yamanishi. Relating drug-protein interaction network with drug side-effects. Bioinformatics, 28:i522–i528, 2012.
[14] Edouard Pauwels, Veronique Stoven, and Yoshihiro Yamanishi. Predicting drug side-effect profiles: a chemical fragment-based approach. BMC Bioinformatics, 12(1): 169, 2011.
[15] J. Scheiber, B. Chen, M. Milik, S.C. Sukuru, A. Bender, D. Mikhailov, S. Whitebread, J. Hamon, K. Azzaoui, L. Urban, M. Glick, J.W. Davies, and J.L. Jenkins. Gaining insight into off-target mediated effects of drug candidates with a comprehensive systems chemical biology analysis. J Chem Inf Model, 49(2): 308–17, 2009.
[16] J. Scheiber, J.L. Jenkins, S.C. Sukuru, A. Bender, D. Mikhailov, M. Milik, K. Azzaoui, S. Whitebread, J. Hamon, L. Urban, M. Glick, and J.W. Davies. Mapping adverse drug reactions in chemical space. J Med Chem, 52(9): 3103–7, 2009.
[17] B. Scho¨lkopf, K. Tsuda, and J.P. Vert. Kernel Methods in Computational Biology. MIT Press, MA, 2004.
[18] N.P. Tatonetti, T. Liu, and R.B. Altman. Predicting drug side-effects by chemical systems biology. Genome Biol, 10(9): 238, 2009.
[19] A. Varnek. Fragment descriptors in structure-property modeling and virtual screening. Methods Mol Biol., 672: 213–243, 2011.
[20] A. Varnek, N. Kireeva, I.V. Tetko, I.I. Baskin, and V.P. Solov'ev. Exhaustive qspr studies of a large diverse set of ionic liquids: how accurately can we predict melting points? J Chem Inf Model., 47(3): 1111–1122, 2007.
[21] S. Whitebread, J. Hamon, D. Bojanic, and L. Urban. Keynote review: invitro safety phar-macology profiling: an essential tool for successful drug development. Drug Discovery Today, 10 (21): 1421–1433, 2005.

[22] D.S. Wishart, C. Knox, A.C. Guo, D. Cheng, S. Shrivastava, D. Tzur, B. Gautam, and M. Hassanali. Drugbank: a knowledgebase for drugs, drug actions and drug targets. Nucleic Acids Res., 36: D901–D906, 2008.

[23] D.M. Witten, R. Tibshirani, and T. Hastie. A penalized matrix decomposition, with appli-cations to sparse principal components and canonical correlation analysis. Biostatistics, 2009.

[24] L. Xie, J. Li, L. Xie, and P.E. Bourne. Drug discovery using chemical systems biology: identification of the protein-ligand binding network to explain the side effects of CETP inhibitors. PLoS Comput Biol, 5: e1000387, 2009.

[25] Y. Yamanishi, M. Araki, A. Gutteridge, W. Honda, and M. Kanehisa. Prediction of drug-target interaction networks from the integration of chemical and genomic spaces. Bioinformatics, 24:i232–i240, 2008.

[26] Y. Yamanishi, M. Kotera, M. Kanehisa, and S. Goto. Drug-target interaction prediction from chemical, genomic and pharmacological data in an integrated framework. Bioinformatics, 26:i246–i254, 2010.

[27] Y. Yamanishi, E. Pauwels, and M. Kotera. Drug side-effect prediction based on the integration of chemical and biological spaces. J Chem Inf Model., 52(12): 3284–3292, 2012.

About the Editors

Kishore R Sakharkar is an adjunct Professor at the Department of Biotechnology, Pune, India. He is also a research Director at OmicsVista, Singapore.

Meena K Sakharkar is an Associate Professor at the Department of Pharmacy and Nutrition, University of Saskatchewan, Canada. She was a Professor at the Graduate School of Life and Environmental Sciences, University of Tsukuba, Japan from 2010.03–2014.02.

Ramesh Chandra was the former Vice-Chancellor of Bundelkhand University, Jhansi, India. He is the Founder Director Dr.B.R.Ambedkar Center for Biomedical Research and Professor of Chemistry, University of Delhi.

About the Contributors

Martin Blythe is Senior Bioinformatics Fellow at Deep seq: The next-generation sequencing facility at The University of Nottingham.

Vladimir Brusic is the Director of Bioinformatics at the Cancer Vaccine Centre, Dana-Farber Cancer Institute and a Professor of Computer Science at Metropolitan College, Boston University. His education includes BEng (mechanical), MEng (biomedical), MAppSci (information technology), MBA (executive), and PhD (bioinformatics). His work focuses on computational modeling of biological systems, data mining, and pattern recognition with applications in immunology, vaccinology and cancer. He published more than 200 scientific articles and serves as a member of several journal editorial boards.

Ramesh Chandra was the former Vice-Chancellor of Bundelkhand University, Jhansi, India. He is the Founder Director Dr.B.R.Ambedkar Center for Biomedical Research and Professor of Chemistry, University of Delhi.

Matthew N Davies is a EuroBATS Research Fellow in Bioinformatics at the Institute for Psychiatry, Kings College London.

Irini A Doytchinova is a Professor in the Faculty of Pharmacy, Medical University of Sofia, Bulgaria.

Darren Flower is Reader in Pharmacy at the University of Aston.

Junaid Gamieldien, PhD is a genomicist and has been involved in bioinformatics since 1997. After completing his PhD in 2001, he joined Electric Genetics as chief bioinformaticist and has been the head of a research group at the South African National Bioinformatics Institute/MRC Capacity Development Unit, based at the University of the Western Cape, since mid-2010. His primary research focus is on the application of next generation sequencing and semantic technology in the elucidation of the genetic causes of diseases and in biomarker discovery.

Yuka Hara, PhD is a post-doctoral research fellow at Universiti Kebangsaan Malaysia, Malaysia. Her research focuses mainly on vaccine development and understanding the host immune responses against Burkholderia pseudomallei, a causative agent of melioidosis.

PremKumar Jayaraman has a PhD in Biotechnology and a Masters in Biomedical Engineering from the from the Nanyang Technological University (NTU), Singapore. He is currently a post-doctoral student at the NTU.

Anil G. Jegga is an Associate Professor of Biomedical Informatics and Pediatrics, Cincinnati Children's Hospital Medical Center, and University of Cincinnati. His research interests include biomedical data integration and mining focusing on systems biology of disease, drug networks and drug repositioning.

Varun Khanna is an Assistant Professor at the Institute of Life Sciences, Ahmedabad University, Gujarat, India.

Nilay Mahida is an independent Pharmacist.

Milsee Mol is a graduate student under the guidance of Dr. Shailza Singh.

Monica Moschioni, PhD was born in Udine, Italy, on August 17, 1973. She received a degree (cum laude) in Chemistry and pharmaceutical technology and a Doctor in Molecular and cellular biology degree from the Universities of Padua and Bologna, respectively. In 2001 she joined Novartis Vaccines, formerly Chiron Vaccines, Siena, where she started working as a scientist in the Molecular biology department of the Research center. Her principal areas of specialization and interest include molecular epidemiology, molecular pathogenesis and interaction with Public health agencies. She has published over 30 papers in peer reviewed journals and book chapters.

Sheila Nathan, DPhil is a Professor at the School of Biosciences and Biotechnology, Faculty of Science & Technology, Universiti Kebangsaan Malaysia. Her research interests focus on the tropical pathogen Burkholderia pseudomallei where she and her group have been investigating the mode of pathogenesis as well as the host-bacteria relationship.

Surendra Nimesh, PhD. is an internationally recognized expert of nanotechnology for biological applications with specialization in drug and gene delivery. He received his M.S. in Biomedical Science from Dr. B.R. Ambedkar Center for Biomedical Science Research (ACBR), University of Delhi, Delhi, India in 2001. He completed his PhD. in Nanotechnology at ACBR and Institute of Genomics and Integrative Biology (CSIR), Delhi, India in 2007. After completing his postdoctoral studies at Ecole Polyetchnique of Montreal, Montreal in 2009, he joined Clinical Research Institute of Montreal, Montreal, Canada as Postdoctoral Fellow. He worked for a short duration at McGill University, Montreal and thereafter joined Health Canada, Canada as NSERC visiting fellow in 2012. He joined Central University of Rajasthan, India as UGC-Assistant Professor at School of life Sciences in 2013. He has authored more than 14 research papers, 7 review articles in international peer reviewed journal, 7 book chapters and 3 books. His research interests include nanoparticles mediated gene, siRNA and drug delivery for therapeutics.

Lars Rønn Olsen is a postdoctoral fellow at the Bioinformatics Centre at Copenhagen University, where he works on computational modelling of the immune response against cancer, and cancer characterization using high-throughput genomics and proteomics methods. He holds a PhD degree in bioinformatics from University of Copenhagen and an engineering degree in Systems Biology from the Technical University of Denmark.

Janmejay Pandey, PhD is an expert of Applied Microbiology & Biotechnology with specialization in Microbial Metabolism, Microbial Physiology, Molecular Microbial Ecology and Microbial Genetics- Genomics. He received his M.Sc. in Biotechnology from Kurukshetra University, Kurukshetra, Haryana India in 2002. He completed his PhD. in Microbial Biotechnology from Institute of Microbial Technology – (IMTECH CSIR), Chandigarh, India in 2009. During his Ph.D. studies, he visited Ecole Polytechnique Federal de Lausanne (EPFL), Lausanne, Switzerland and Commonwealth Scientific and Industrial Research Organisation (CSIRO), Canberra, Australia as visiting scholar. After completing his Ph.D. studies, he did Postdoctoral research at Gordon Centre for Integrative Sciences- Institute of Biophysical Dynamics, University of Chicago, Chicago- IL, USA and Nanomedicine Centre for Nucleoproteins, Georgia Health Sciences University, Augusta- GA, USA.

He joined Central University of Rajasthan, Ajmer- Rajasthan, India as an Assistant Professor in Department of Microbiology, School of life Sciences in July 2012. Subsequently, he joined Department of Biotechnology at Central University of Rajasthan in March, 2013. He has authored 19 research papers, 2 review articles in international peer reviewed journal and 5 book chapters. His current research interest includes studies on extremophilic actinobacteria for discovery of novel bioactives and therapeutics.

Deepak Perumal acquired his PhD degree from the Nanyang Technological University. He is currently a post-doctoral candidate at the Moffit cancer centre, USA.

Shoba Ranganathan is Professor of Bioinformatics at Macquarie University, Sydney, Australia. She is also currently the President of the Asia Pacific Bioinformatics Network.

Rino Rappuoli, PhD is Global Head of R&D at Novartis Vaccines and is based in Siena, Italy. He earned his PhD in Biological Sciences at the University of Siena and was visiting scientist at Rockefeller University and Harvard Medical School. Prior to the present position he was head R&D of Sclavo and then head of vaccine research and Chief Scientific Officer of Chiron Corporation.

Roberto Rosini, PhD was born in Italy, on May 14, 1977. He received a Biology degree (cum laude) from the University of Siena in 2003. He joined the Research Center of Chiron Vaccines in 2004 (Siena, Italy), acquired by Novartis Vaccines in 2006. During this period he worked on the Streptococcus agalatiae vaccine development program using the Reverse Vaccinology approach. He contributed to the discovery of novel Streptococcus agalatiae protective antigens known as pili. Based on this work he received a Doctorate degree from the University of Siena in 2006. Afterwards and up to now, he joined the Molecular Epidemiology unit of the Novartis Vaccine Research Center (Siena) working on the molecular characterization of pathogenic E. coli and Streptococcus agalactiae bacteria.

Kishore R Sakharkar is an adjunct Professor at the Department of Biotechnology, Pune, India. He is also a research Director at OmicsVista, Singapore.

Meena K Sakharkar is an Associate Professor at the Department of Pharmacy and Nutrition, University of Saskatchewan, Canada. She was a Professor at the

Graduate School of Life and Environmental Sciences, University of Tsukuba, Japan from 2010.03–2014.02.

Mohammad Imran Siddiqi, PhD. is a senior scientist at the Central Drug Research Institute, Lucknow, India and is the group leader for the Computational Biology & Bioinformatics Group.

Christian Simon is a PhD Student at the Novo Nordisk Foundation Center for Protein Research at Copenhagen University. His work focuses on integrating, combining, and modelling electronic patient records with biochemical and genomic data in a temporal context. He holds an engineering degree in System Biology from the Technical University of Denmark.

Dr. Shailza Singh has completed her Ph.D from IIT Delhi and is currently Incharge of Bioinformatics and High performance Computing Facility at National Centre for Cell Science, Pune. Her research involves from Molecular Simulation to Biochemical Network Perturbation in infectious disease dealing with the stability and stochasticity in the synthetic circuit.

Sinosh Skariyachan is an Assistant Professor in the Department of Biotechnology Engg, Dayananda Sagar College of Engineering, Bangalore, India. After a B. Sc in Microbiology he persuaded two Master degrees, M. Sc in Microbiology and M. Sc in Bioinformatics. He has also completed Ph.D from Visveswarya Technological University, Belgaum, India. He has a decade of experience in teaching and research. His key research areas are Bioinformatics, Molecular Modeling and Simulations, Computer aided drug designing, Genomics & Proteomics, Microbial and Environmental Biotechnology and Food Biotechnology. He is member of several scientific societies such as American Society for Microbiology, Association of Microbiologist of India, Society of Biological Chemist, Environmental Mutagen Society of India, IAENG Society of Bioinformatics and International Association of Computer Science and Information Technology. He has published 24 International papers in reputed journals and presented more than 45 scientific papers in various National and International conferences and symposium. He has authored a text book "Introduction to Food Biotechnology" and several book chapters for reputed publishers. He has reviewed many research papers in reputed journals. He has guided more than 50 students for their dissertation work for BE/B.Tech in Biotechnology and 05 post graduate students (M. Sc and M. Tech). He is a member of board examiner in many universities. He is content reviewer of several academic programme. He has aplenary speaker for several National conferences and symposium on emerging trends in Bioinformatics and allied disciplines.

Jing Sun is a research fellow at the Cancer Vaccine Center, Dana-Farber Cancer Institute. Her Ph.D. thesis is analysis of B-cell epitope and antibody. Her current projects focuses on computational application on immunology.

John L Telford, PhD obtained a PhD in Molecular Biology from the University of Zurich and worked as a staff scientist at The European Molecular Biology Laboratory in Heidelberg, Germany before joining the Sclavo Research Center (now Novartis Vaccines Research Center) in Siena, Italy in 1985. He is currently Head of Microbial Molecular Biology, Italy in the Novartis Vaccines and Diagnostics research organization. His major scientific interests over the last 20 years have been in research aimed at understanding bacterial pathogenesis and in particular the identification of vaccine candidates for diseases caused by Helicobacter pylori, Group B Streptococcus and Group A Streptococcus. Recent research has centered around the discovery of complex pilus-like structures in the streptococci and their role in host cell interaction and pathogenesis.

Nicki Tiffin, PhD is a Principle Investigator at the South African National Bioinformatics Institute/Medical Research Council of South Africa Bioinformatics Capacity Unit, at the University of the Western Cape in South Africa. She currently researches bioinformatics approaches to understanding genetic factors contributing to disease in African populations, as well as ethical and legal considerations for genomics in Africa.

Chow T.K Vincent is an Associate Professor in the Department of Microbiology at the National University of Singapore.

Chao Wu is a Ph.D. candidate at the Department of Computer Science, University of Cincinnati and his research focuses on data mining and developing algorithms to integrate and analyze large heterogeneous biological networks.

Yoshihiro Yamanishi received his Ph.D. from Kyoto University in 2005. He was post-doctoral fellow at Ecole des Mines de Paris from 2005 to 2006. He was assistant professor at Kyoto University from 2006 to 2007. He was permanent researcher at Mines ParisTech and Curie Institute from 2008 to

2012. Since 2012, he has been associate professor at the Medical Institute of Bioregulation, Kyushu University and Institute for Advanced Study, Kyushu University. He is working on machine learning in bioinformatics, chemoinformatics, and genomic drug discovery.

Guang Lan Zhang is an Assistant Professor at Department of Computer Science, Metropolitan College, Boston University. She is also an adjunct member of the Cancer Vaccine Centre, Dana-Farber Cancer Institute. Her work focuses on health informatics, computational modeling of biological systems, data mining, and pattern recognition with applications in immunology, vaccinology and cancer.

Cheng Zhu received his Ph.D. in computer sciences from University of Cincinnati. His research is focused on leveraging data mining, machine learning, statistics and network based approaches for knowledge-mining from "omics" data. He is currently a research informatics scientist at Genzyme Corporation, Boston, USA.

Index

Lightning Source UK Ltd.
Milton Keynes UK
UKOW06n1147270117
293027UK00002B/42/P

9 788793 102842